European Trend Atlas of Extreme Temperature and Precipitation Records

Deliang Chen • Alexander Walther
Anders Moberg • Phil Jones • Jucundus Jacobeit
David Lister

European Trend Atlas of Extreme Temperature and Precipitation Records

Deliang Chen
Department of Earth Sciences
University of Gothenburg
Gothenburg
Sweden

Alexander Walther
Department of Earth Sciences
University of Gothenburg
Gothenburg
Sweden

Anders Moberg
Department of Physical Geography and
Quaternary Geology
Stockholm University
Stockholm
Sweden

Phil Jones
University of East Anglia
Norwich
United Kingdom

Jucundus Jacobeit
Department of Geography
University of Augsburg
Augsburg
Germany

David Lister
University of East Anglia
Norwich
United Kingdom

ISBN 978-94-017-9311-7 ISBN 978-94-017-9312-4 (eBook)
DOI 10.1007/978-94-017-9312-4

Library of Congress Control Number: 2014953837

Springer Dordrecht Heidelberg New York London

Printed on acid-free paper

Springer is part of Springer Science+Business Media (www.springer.com)

Preface

Climate change not only includes changes in mean conditions, but also covers changes in extremes. For impact on society, extreme climatic conditions are often much more important than mean climate. Thus adaptation to climate change needs to take historical changes in the extreme climatic conditions into account. Europe is one of a few places in the world where the longest instrumental meteorological records exist, which provides an important opportunity to reveal long term changes in the extremes. Over the past decade, several research projects at the European level have dealt with changes in extreme climatic conditions in Europe.

One of these projects is EMULATE (European and North Atlantic daily to MULtidecadal climATE variability) that was supported by the European Commission under the Fifth Framework Programme. The project contributed to the implementation of the Key Action "Global change, climate and biodiversity" within the Environment, Energy and Sustainable Development. It was coordinated by Prof. Phil Jones, Climatic Research Unit at the University of East Anglia, UK and had eight participating institutions across Europe. One of the outcomes of the EMULATE project was a systematic mapping of the observed trends of 64 temperature and precipitation indices based on daily instrumental records in Europe. The majority of the indices describe extreme climatic conditions.

In 2006, this systematic mapping was published as an internal report at the University of Gothenburg, which was one of the participating organizations of EMULATE. However, given the nature of the report, the accessibility is limited. Over recent years, needs for information about past changes in extreme climatic conditions have significantly increased. This is particularly true for really extreme conditions and for extended time perspectives. This in 2012, the authors of the report decided to publish an atlas of those indices that represent the most extreme conditions of climate in Europe. This idea resulted in an atlas attempting to show a subset of all the EMULATE indices to a much wider audience than the internal report does.

This atlas presents information in the form of maps, time series and tables for a selection of 27 indices. Four of them represent mean climate conditions while the remaining indices represent climate extremes. All indices were derived from daily temperature and precipitation data at European meteorological stations with records starting before 1901. Since the updating of the daily records for the stations after 2000 has not been finished, this atlas was prepared by using the trends only until 2000 for all the stations. Seasonal trends of the indices during three periods (1801–2000, 1851–2000, and 1901–2000) were shown and their significance was tested. The stations for 1901–2000 were also grouped into three regions (Northern, Central, and Southern Europe) and regional means were calculated. The trend which this atlas provides is an easy way to show spatial patterns for a given time period, region, season, and index. There is strong evidence that climate in Europe has changed during the three periods analyzed, such that the occurrence and intensity of warm temperature extremes have increased. Precipitation extremes have also changed, but with a less clear pattern compared to the temperature extremes.

The atlas should be interesting to and useful for researchers who are from or interested in Earth and Environmental Sciences and practitioners from many sectors in society who are concerned with climate changes in the mean and extreme conditions in Europe.

On behalf of the writing team, I would like to express our thanks to the European Commission for their financial support to EMULATE with the contract EVK2-CT-2002-00161. The Swedish Research Council and the Swedish Rescue Services Agency are thanked for their support and grants. Dr. Elodie J. Tronche and Mariëlle Klijn from Springer are acknowledged for their support and understanding throughout the preparation of the atlas.

Gothenburg
October 2013

Deliang Chen

Contents

1 Introduction

The impact of climate change on society is of fundamental importance for future planning and management. Statistics of extreme events is considered as a part of the climate, and their changes often have much higher impact on society than changes in the mean climate. As the mean climate changes, the characteristics of extremes may also change (e.g. Trenberth 1999; Easterling et al. 2000; Beniston and Stephenson 2004). For many impact applications and decision support systems, extreme events are much more important than the mean climate (e.g. Mearns et al. 1984; WISE 1999). Changes in extremes may be due to changes in the mean (Wigley 1985), changes in the variance (Katz and Brown 1992), or a combination of both factors (e.g. Brown and Katz 1995; Beniston 2004).

Extreme climate events can be defined as events that occur with extraordinary low frequency during a certain period of time (rarity), events with high magnitude (intensity) or duration, and events causing sizeable impacts such as direct damages to assets, cultural heritage, ecosystem service and loss of human lives. According to The Intergovernmental Panel on Climate Change (IPCC) (IPCC 2001, p. 790), "An extreme weather event is an event that is rare within its statistical reference distribution at a particular place. Definitions of 'rare' vary, but an extreme weather event would normally be as rare as or rarer than the 10th or 90th percentile. By definition, the characteristics of what is called extreme weather may vary from place to place. An extreme climate event is an average of a number of weather events over a certain period of time, an average which itself is extreme (e.g. rainfall over a season)." This definition has also been used in the fourth IPCC report published in 2007 (IPCC 2007).

Indices have been developed to measure the degree of exceedance of specific thresholds (which define the rarity of such an event) for maxima/minima during specific time-periods (Jones et al. 1999). Examples are the number of very warm and very cold days for the time of year, the number of heavy rainfall days, and number of frost days (Frich et al. 2002). Some extremes are defined by natural thresholds, while the majority of extremes are determined by the data's own distribution. The majority of indices relate to counts of individual daily extremes, but a few are determined by spells of exceptionally warm/cold temperatures or wet/dry periods or the first/last occurrence of an event during a season (like spring/autumn frosts, beginning/end of the summer dry season etc.). With respect to temperature and rainfall, spells of extreme weather generally have large societal and economic impacts. Examples of short-lived extremes that may cause extensive damage are windstorms, hailstorms and extensive and heavy snowfall.

Recently, there has been an international effort towards developing a suite of standardized indices so that researchers around the world can calculate the indices in exactly the same manner. This is important for detecting and monitoring changes in the extreme climates and allows for comparison of observations and model simulations at the global scale. These analyses can then be combined into the regional and global perspectives (Karl and Easterling 1999; Peterson et al. 2001; Frich et al. 2002). However, the definition of several of the suggested indices is somewhat cumbersome and some of the indices still exist in various forms. Consequently, different software may produce slightly different results. Several EU research projects either use the indices in the European Climate Assessment (Klein-Tank et al. 2002a; Klein-Tank and Können 2003) or a program developed within the STARDEX project (STAtistical and Regional dynamical Downscaling of EXtremes for European regions; Haylock and Goodess (2004)). In this atlas we use the extremes indices software developed within EMULATE (European and North Atlantic daily to MULtidecadal climATE variability; Moberg et al. (2006)).

While some of the climatic extremes are well described by meteorological variables/indices, others may not be easily defined with data for a single meteorological variable only. This is true when the combined impact is involved (Pellikka and Järvenpää 2003). For example, freezing rain is a special combination of low temperature and rain that produces major damages through ice loading on wires and structures. Other examples are snow damage on forests (Solantie 1994) and

D. Chen et al., *European Trend Atlas of Extreme Temperature and Precipitation Records*, DOI 10.1007/978-94-017-9312-4_1

the occurrence of landslides that are known to be caused by heavy rains (e.g. Schuster 1996). However, geological and geomorphologic settings may also play an important role here (Brunsden 1999). Yet another type of climate extreme is that some weather conditions, which may seem more or less normal from a purely anthropocentric perspective, nevertheless could induce a strong impact on some other species. Thus, such ecological climate extremes may not be easy to identify as climate extremes. The dependence on factors other than meteorological data makes it difficult to disentangle the specific contribution of weather/climate in producing the impacts and to describe the combination of extremes. A possible solution to this problem is interactions with affected and interested groups and individuals, such as the insurance industry and design engineers.

Damage reports from (re)insurance companies may also be useful. Nevertheless, since there is no long-term homogeneous data yet in this regard, the use of this approach in climate change studies is not feasible at this point. Therefore, this atlas will not discuss these kinds of extremes. Rather, the focus will emphasize a set of indices well defined by meteorological data that are available for relatively long time periods, i.e. temperature and precipitation.

While an increasing number of climate model studies indicate that rising contents of greenhouse gases in the atmosphere will probably lead to more severe weather conditions in the future, it is of great importance to increase our understanding regarding the occurrence of climate extremes in the recent and more remote past. During recent years, a large number of studies have been carried out focusing on various aspects of climate extremes (mainly temperature and precipitation) in different regions of the world (e.g. Trenberth 1999; Easterling et al. 2000; Beniston and Stephenson 2004). Depending on research tasks, different measures were used to quantify extremes, but which do not always allow direct comparison of results. A general conclusion of many of these studies is, however, that changes in extreme temperature and precipitation have occurred world-wide during the past century along with the ongoing climate change in terms of the mean temperature (Donat et al. 2013). Yet, it is still hard to draw a firm conclusion from these studies concerning the extent to which these changes are due to natural variability or caused by anthropogenic activity (IPCC 2007; Field et al. 2012), although more recent studies indicate that increase in extreme events is indeed linked to man-made global warming (Hansen et al. 2012).

In an attempt to summarize the changes around the globe, Groisman et al. (1999) and Groisman et al. (2005) studied the probability distribution of daily precipitation in eight countries located on different continents and concluded that increased mean precipitation is associated with an increase in heavy rainfalls. In their near-global analysis, Frich et al. (2002) found regions with both negative and positive changes in extremes, with parts of Europe having more robust positive changes. Using extremes indices similar to some of those produced by the EMULATE project, Moberg and Jones (2005) investigated trends in daily temperature and precipitation extremes across Europe over the past century and found that both mean and extreme precipitation have increased mainly during winter. Also Klein-Tank and Können (2003) found an increase in the annual number of moderate and very wet days between 1946 and 1999.

Several studies have also been carried out at the national level. Fowler and Kilsby (2003) studied multi-day rainfall events in the UK since 1961 and found significant but regionally varying changes in the 5- and 10-day events, which they consider as having important implications for the design and planning of flood control measures. In Central Europe, Schmidli and Frei (2005) found significant increasing trends in winter and autumn rainfall in Switzerland and Hundecha and Bárdossy (2005) found increasing precipitation extremes across western Germany since 1958. The increase in Central European daily precipitation beyond the 98th percentile has occurred during all seasons except summer (Jacobeit et al. 2009). In northern Europe, Achberger and Chen (2006) studied the spatial patterns and long-term trends of precipitation indices in Sweden and Norway on an annual and seasonal basis for the years 1961 to 2004. These indices are based on daily data from 471 stations. Analysis of the trends of the various indices for the period 1961–2004 shows that the magnitude and sign of the trends varies depending on index, region and season. A clear majority of stations show increasing trends, though the fraction having statistically significant trends is small. In Norway, positive trends are most common during winter, while at Swedish stations, positive trends are most frequent in spring and summer. Autumn has the highest number of stations in both countries with negative trends. The findings are generally in line with results from other studies concluding that regions at middle and higher latitudes are becoming wetter and extreme temperatures and precipitation are becoming more frequent and more intense.

The importance of monitoring and analyzing climate extremes has been highlighted by the past assessment reports of the IPCC, and during a special IPCC meeting on climate extremes held in Beijing in June 2002 (Houghton et al. 2002). Since then, there has been increased research around the world, particularly in the US and European countries, that aims at a better understanding of the observed extremes (e.g. Arndt et al. 2010), detection and attribution of extremes (e.g. Christidis et al. 2005; Morak et al. 2011; Zwiers et al. 2011), factors that influence extremes (e.g. Haylock and Goodess 2004; Vautard and Yiou 2009; Zhang et al. 2010), and interpretation of observed extremes in a climate context (e.g. Peterson et al. 2012). In 2012 IPCC published a Special Report on Extremes (Field et al. 2012) which summarized and assessed studies with regard to changes in climate

Extremes and their Impacts on the natural physical environment and human systems and ecosystems, as well as managing the risks from climate extremes at the Local Level round the world.

Instrumental data sets enabling studies of changes in extreme climate events in Europe have been collected by ECA&D (European Climate Assessment and Dataset) (Klein-Tank et al. 2002b), and by the STARDEX project (Haylock and Goodess 2004). This atlas represents an effort within the EMULATE project. The EMULATE project partly built on the ECA&D data set, but also on European station data from other projects and data obtained through direct contacts with several national weather services and additionally presented a set of newly digitized and collected daily data at many Spanish stations (Moberg et al. 2006). While the analysis in Moberg et al. (2006) focused on a rather small subset of all the indices (19 out of 64) during only the summer and winter seasons within the period 1901–2000, Chen et al. (2006) presented results for all the 64 indices for all seasons and three different time periods. The focus of this new atlas is on a systematic trend analysis of a subset of 27 out of the 64 indices that together represent the most extreme climatic conditions for all seasons and also climatic mean conditions for comparison. The trends at individual stations are presented for the periods 1801–2000, 1851–2000, and 1901–2000. Results for the latter period are also aggregated into three sub-regions of Europe (Northern, Central, and Southern Europe).

This study is motivated by the need for easy access to information about changes in the extreme climatic conditions at the European meteorological stations with the longest daily records. This need is becoming more and more important as adaptation work in response to climate change has being implemented in Europe. The report is organized as follows. Chapter 2 describes the data and method used. Chapter 3 presents the results of the trend analysis for all the selected indices, stations and regions in maps and figures. Some statistics for the stations and regions are developed in Chapter 4 to summarize these results in tabular form. Finally, Chapter 5 provides the conclusions that summarize the information shown by the tables and figures. The appendix lists all estimated seasonal trends for all the indices at all stations as a quick overview.

2 Data & Methods

The methods used in this study follow those used by Moberg et al. (2006) and Chen et al. (2006). A brief description is given below, while detailed information can be found in these references, except for the regional division used here. For the regional analysis, we followed the regional divisions used by the recent IPCC special report (IPCC 2012), rather than the approach used by Moberg et al. (2006) and Chen et al. (2006).

EMULATE collected a dataset containing daily temperature and precipitation observations over European locations having data starting before 1901. This dataset is the basis for the analyses in this atlas. Up to four daily climate variables are available: minimum and maximum temperature (Tmin/Tmax), mean temperature (Tmean), and precipitation (Prec). We focus on three periods: 1801–2000, 1851–2000, and 1901–2000. Since a trend analysis is extremely sensitive to the starting and ending of the time series, particular attention is paid to the missing data in these beginning and ending periods. A data completeness criterion has been applied to filter out stations with insufficient data for the proposed analyses. We divided each analysis period in question into three sub periods: one 20-year period at the beginning, one 20-year period at the end of the series and the entire period in between (160, 110, or 60 years according to the period analyzed). A station passed the filter if it did not have more than 4 % of missing values in the two 20 year blocks at the beginning and end and at most 6 % missing values in the longer block in between. All the stations that passed the criterion and are used in this work are listed in Table 2.1 and shown in Figs. 2.1, 2.2 and 2.3.

For the 200-year period (1801–2000) we can use only three stations with Tmin/Tmax, and seven stations with Tmean measurements. No precipitation observations are available for this period. Looking at the 150-year period (1851–2000) the number of observations is increased to nine for Tmin/Tmax, thirteen for Tmean, and nine for precipitation. For these two periods the analyses were carried out for each station with sufficient data. The number of stations for these two periods is too few to undertake any averaging or regionalization approaches. For the 100-year period (1901–2000) significantly more stations are available: we have 57 observation sites for Tmin/Tmax, 54 for Tmean, and 100 for Precipitation. The stations are fairly unevenly distributed over the study area, although we find the highest station density in Central Europe. The relatively high number of observation sites for this period provides the possibility to group the stations into regions, which enables a regional analysis and inter comparison between regions.

Using regional divisions for the regional analysis facilitates direct comparison of the regional trends with other estimates and model simulations. As indicated in Figs. 2.1, 2.2 and 2.3, the following three regions were created: NEU (Northern Europe), CEU (Central Europe), and SEU (Southern Europe).

The climate indices described in the next section were computed for all stations belonging to a particular region. Afterwards the time series have been averaged arithmetically. The resulting averaged time series was taken as the regional mean and was subject to a linear trend analysis over the whole 100-year period.

The quality of the station series selected for this study varies. All series have been corrected for obvious errors whereas efforts towards homogenizing the database could not be undertaken. Relatively few series have undergone intensive testing and correction for inhomogeneities, among them the very long records for some Swedish stations (Moberg and Bergström 1997). Some of the series have been used more frequently than others in the literature and are in this context also more quality controlled. We have to keep in mind that inhomogeneous data can cause errors in estimated trend values as demonstrated by Venema et al. (2012). In particular with respect to the method of linear regression, which is more sensitive to values at the beginning and end of the respective time series analyzed for linear trend. During the recent past substantial efforts have been undertaken to improve the availability and quality of long-term climate observations. For example, 557 monthly long-term observation series for the Greater Alpine Region where collected,

D. Chen et al., *European Trend Atlas of Extreme Temperature and Precipitation Records*, DOI 10.1007/978-94-017-9312-4_2,

Table 2.1 Data availability for the three groups of observation at the meteorological stations used

ID	Lat	Lon	Station	Country	Tmin/Tmax	Tmean	Prec
1	48.08	15.45	Graz university	Austria	x	x	x
2	47.27	11.40	Innsbruck university	Austria	x	x	x
3	48.05	14.13	Kremsmünster	Austria	x	x	x
4	47.80	13.00	Salzburg	Austria	x	x	x
5	47.05	12.95	Sonnblick	Austria	x	x	x
6	48.23	16.35	Wien	Austria	x	x	x
7	50.90	4.53	Brussels-Uccle	Belgium	x	x	x
8	43.85	18.38	Sarajevo	Bosnia	x	x	x
9	45.17	14.70	Crikvenica	Croatia		x	
10	45.82	15.98	Zagreb	Croatia	x	x	x
11	50.08	14.42	Prag	Czech	x	x	x
12	55.28	14.78	Hammer odde fyr	Denmark	x	x	x
13	55.45	8.40	Nordby	Denmark	x	x	x
14	55.85	10.60	Tranebjerg	Denmark	x	x	x
15	56.77	8.32	Vestervig	Denmark	x	x	x
16	60.32	24.97	Helsinki	Finland	x	x	x
17	43.31	5.40	Marseille	France	x	x	x
18	48.82	2.34	Paris parc montsouris	France	x	x	x
19	49.88	10.88	Bamberg	Germany	x	x	x
20	52.45	13.30	Berlin	Germany	x	x	x
21	53.03	8.78	Bremen	Germany	x	x	x
22	48.73	9.72	Göppingen	Germany			x
23	49.28	9.17	Gundelsheim	Germany			x
24	53.48	10.25	Hamburg-Bergedorf	Germany	x	x	x
25	53.55	9.97	Hamburg-Fuhlsbüttel	Germany	x	x	x
26	53.15	11.03	Hitzacker	Germany			x
27	47.80	11.00	Hohenpeissenberg	Germany	x	x	x
28	48.42	8.67	Horb-Betra	Germany			x
29	47.68	10.05	Isny	Germany			x
30	50.93	11.58	Jena	Germany	x	x	x
31	49.03	8.35	Karlsruhe	Germany	x	x	x
32	49.20	8.10	Landau/Pfalz	Germany			x
33	48.85	12.92	Metten	Germany			x
34	48.17	11.50	Muenchen	Germany	x	x	x
35	49.22	9.52	Oehringen	Germany			x
36	52.38	13.07	Potsdam	Germany	x	x	x
37	48.72	9.22	Stuttgart	Germany	x	x	x
38	48.02	8.82	Tuttlingen	Germany			x
39	47.87	11.78	Valley-Mühltal	Germany			x
40	53.78	7.90	Wangerooge	Germany			x
41	52.90	8.43	Wildeshausen	Germany			x
42	47.42	10.98	Zugspitze	Germany	x	x	x
43	37.97	23.72	Athens	Greece	x	x	x
44	65.08	−22.73	Stykkisholmur	Iceland		x	x
45	53.37	−6.35	Dublin	Ireland			x
46	44.48	11.50	Bologna	Italy	x	x	x
47	44.82	11.50	Ferrara	Italy			x
48	45.15	10.75	Mantova	Italy			x
49	45.50	9.20	Milano	Italy	x	x	x
50	38.11	13.36	Palermo	Italy			x
51	45.38	10.87	Verona-Villafranca	Italy			x

Table 2.1 (continued)

ID	Lat	Lon	Station	Country	Tmin/Tmax	Tmean	Prec
52	52.10	5.18	De Bilt	Netherlands	*x*	*x*	
53	52.97	4.76	De Kooy	Netherlands			*x*
54	53.13	6.58	Eelde	Netherlands			*x*
55	53.18	6.60	Groningen	Netherlands			*x*
56	52.40	6.05	Heerde	Netherlands			*x*
57	52.31	4.70	Hoofddorp	Netherlands			*x*
58	52.64	5.07	Hoorn	Netherlands			*x*
59	51.68	3.86	Kerkwerve	Netherlands			*x*
60	51.57	4.53	Oudenbosch	Netherlands			*x*
61	51.18	5.97	Roermond	Netherlands			*x*
62	52.88	7.06	Terapel	Netherlands			*x*
63	53.37	5.22	West-Terschelling	Netherlands			*x*
64	51.98	6.70	Winterswijk	Netherlands			*x*
65	60.65	6.22	Bulken	Norway			*x*
66	59.12	11.38	Halden	Norway			*x*
67	63.22	11.12	Lien I Selbu	Norway			*x*
68	38.72	−9.15	Lisboa Geofisica	Portugal	*x*	*x*	
69	44.42	26.10	Bucuresti	Romania			*x*
73	51.65	36.18	Kursk	Russia		*x*	*x*
77	59.97	30.30	St Petersburg	Russia	*x*	*x*	*x*
79	38.37	−0.50	Alicante	Spain	*x*	*x*	*x*
80	38.88	−6.83	Badajoz	Spain	*x*	*x*	*x*
81	42.36	−3.72	Burgos	Spain			*x*
82	36.50	−6.23	Cadiz	Spain	*x*	*x*	
83	37.14	−3.63	Granada	Spain	*x*	*x*	*x*
84	40.41	−3.68	Madrid	Spain	*x*	*x*	
85	37.98	−1.12	Murcia	Spain	*x*	*x*	*x*
86	37.42	−5.90	Sevilla	Spain			*x*
87	41.44	−2.48	Soria	Spain	*x*	*x*	*x*
88	41.64	−4.77	Valladolid	Spain	*x*	*x*	*x*
89	60.62	15.67	Falun	Sweden	*x*	*x*	*x*
90	65.07	17.15	Stensele	Sweden	*x*		*x*
91	59.35	18.05	Stockholm	Sweden	*x*	*x*	
92	59.87	17.63	Uppsala	Sweden	*x*	*x*	*x*
93	56.87	14.80	Växjö	Sweden	*x*		*x*
94	47.55	7.58	Basel	Switzerland	*x*	*x*	*x*
95	46.93	7.42	Bern	Switzerland	*x*	*x*	*x*
96	47.05	6.99	Chaumont	Switzerland			*x*
97	46.25	6.13	Geneva	Switzerland	*x*	*x*	*x*
98	46.00	8.97	Lugano	Switzerland	*x*	*x*	*x*
99	47.25	9.35	Säntis	Switzerland	*x*	*x*	*x*
100	46.23	7.37	Sion	Switzerland		*x*	*x*
101	47.38	8.57	Zurich	Switzerland	*x*	*x*	*x*
102	54.35	−6.65	Armagh	UK	*x*		*x*
103	52.20	0.13	Cambridge	UK			*x*
104	52.42	−1.83	CET	UK	*x*	*x*	
105	51.77	−1.27	Oxford	UK	*x*		*x*
106	58.33	−6.32	Stornoway	UK	*x*		
107	45.03	35.38	Feodosija	Ukraine	*x*		*x*
108	50.40	30.45	Kiev	Ukraine	*x*	*x*	*x*
109	49.60	34.55	Poltava	Ukraine	*x*	*x*	

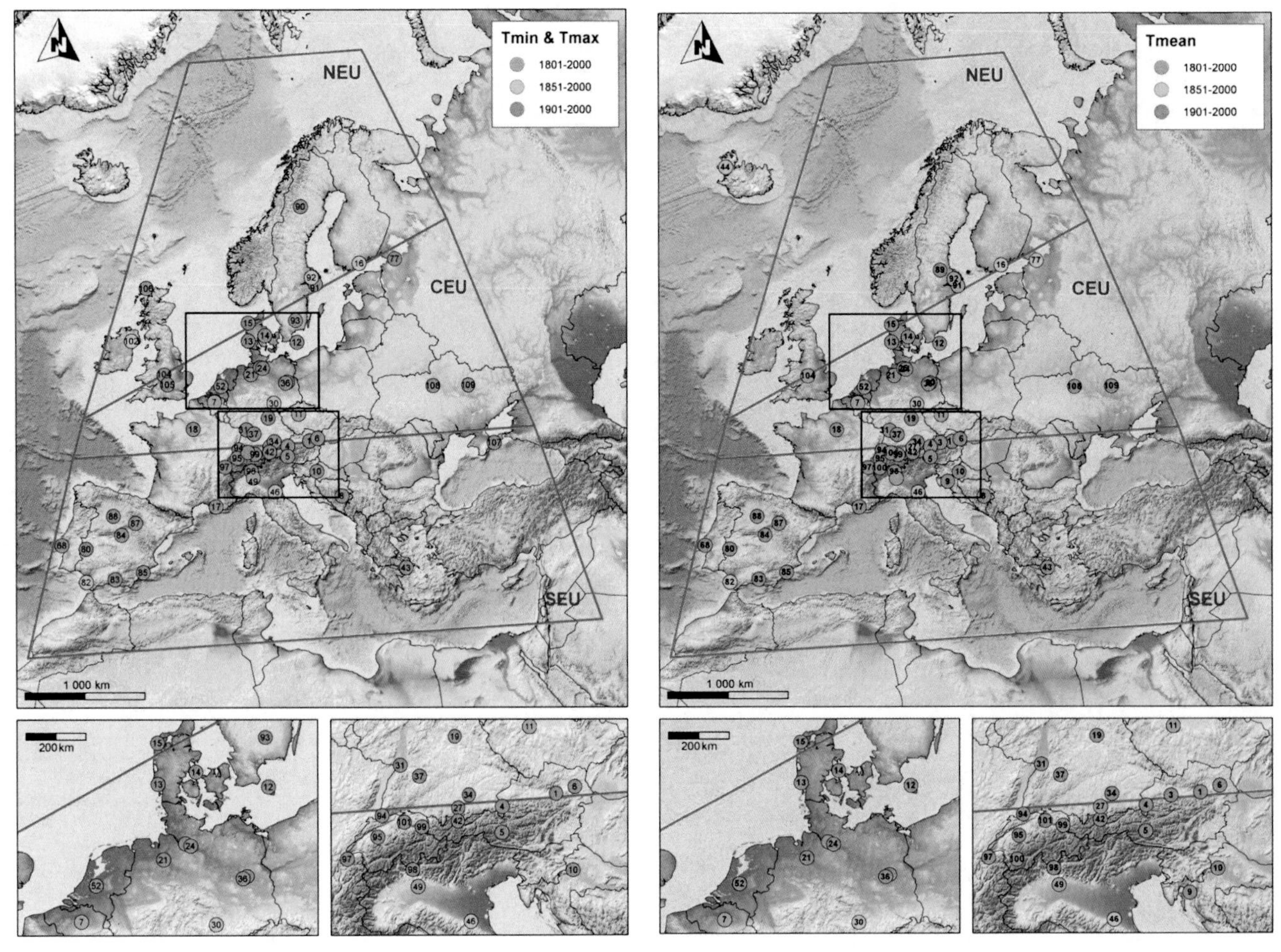

Fig. 2.1 Overview map showing the meteorological Tmin and Tmax observations used for each period and the regional division used for the period 1901–2000

Fig. 2.2 Overview map showing the meteorological Tmean observations used for each period and the regional division used for the period 1901–2000

quality controlled and homogenized within the HISTALP project (Auer et al. 2007). There is an obvious need for increased resources and efforts going into further developing homogenization methods in order to improve data quality especially for observations with temporal resolution higher than monthly, last but not least with respect to climatology beyond mean conditions. Brunet and Jones (2011) estimate that about 80 % of past climate data is still unavailable to the research community hampering a more robust assessment of climate parameters.

2.1 Climate Indices Calculated

All the selected indices have been computed for each of the three periods described above. Typically there was one value per season per index comprising the index timeseries (100, 150 or 200 years) which was then used for the trend estimation. The indices are listed and explained in Table 2.2. Most indices are either percentiles or percentile-based. With percentiles the most extreme values in both tails of

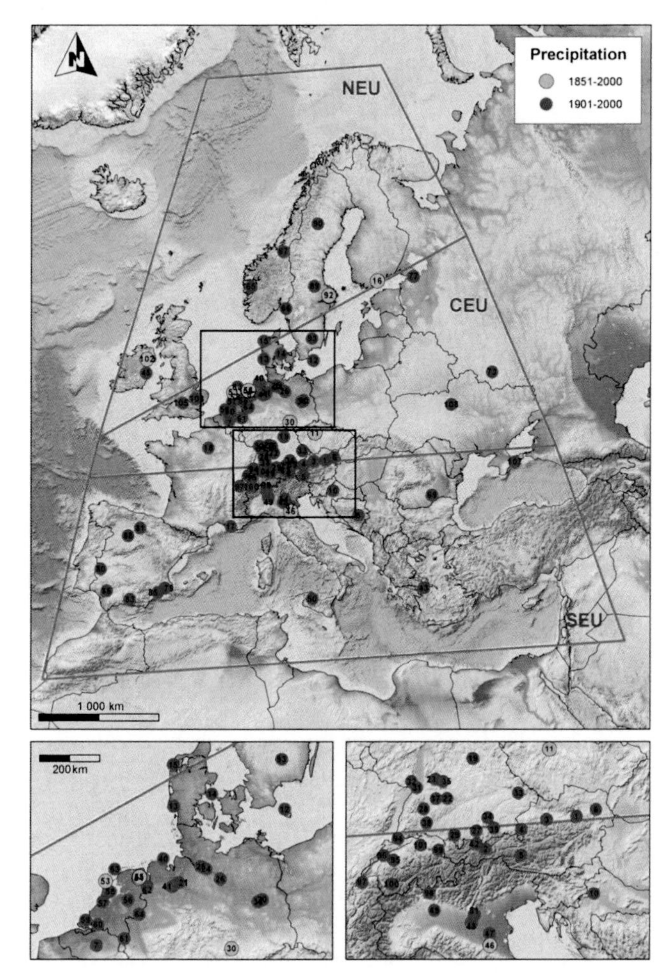

Fig. 2.3 Overview map showing the meteorological precipitation observations used for each period and the regional division used for the period 1901–2000

the frequency distribution of a variable can be captured. For example, the 2 % highest values exceed the 98th percentile. Some indices use thresholds calculated for a reference period to be exceeded or fallen below. For those 1961–1990 was used as reference. For the temperature variables both the 2nd and 98th percentile were chosen in order to capture extremely cold and warm conditions, whereas for precipitation only the 98th percentile was utilized. In the literature the usage of 1st/99th percentile (instead of 2nd and 98th) is more common. However, using somewhat lower thresholds can yield more robust results. The index time series to deal with are relatively short, e.g. one index value per season yields 150 values for the 150-year period. In this way three values exceed the threshold (98th) instead of one (99th). Considering the characteristics of extremes connected to issues related to data quality this can be a useful measure.

Not all the indices are real extremes indices. Some of them represent mean conditions to make it possible to illustrate similarities and differences between changes in extremes and in mean conditions. In addition to the mean indices, the following indices are not based on percentiles. The Heat Wave Duration Index (HWDI), daily precipitation intensity indices (SDII98 and SDII), highest 5-day total rainfall (R5d), highest daily rainfall (R1d) and the highest number of consecutive dry days (CDD).

All indices were computed separately for each of the 3-month seasons (MAM = March–May, JJA = June–August, SON = September–November, and DJF = December–February). Further details on the indices used can be found in Moberg et al. (2006) and Beniston et al. (2007).

2.1.1 Trend Analysis

The ordinary least squares method was used to estimate linear trends of the temperature and precipitation indices. The trend was computed for each particular period in question: 200 years, 150 years, and 100 years. For easy comparison among different periods, all trends were given in *unit/100years*. The significance of a particular trend estimated was determined by a t-test on the estimated slope of the regression as done in Moberg et al. (2006). The lag-1 autocorrelation in each series has been taken into account to adjust the degrees of freedom accordingly. Trends that are determined to be significant at the 5 and 1 % level were flagged in diagrams and tables that present the results.

Table 2.2 Extremes indices calculated

#	Variable	Identifier	Parameter	Unit
1	*Tmin/Tmax*	MEANTN	Mean Tmin	°C
2		MEANTX	Mean Tmax	°C
3		TN2P	Tmin 2nd percentile	°C
4		TN98P	Tmin 98th percentile	°C
5		TN2N	[a]# of days below the reference 2nd Tmin percentile	[days]
6		TN98N	[a]# of days exceeding the reference 98th Tmin percentile	[days]
7		CSDI10	[b]Cold spell duration index	[days]
8		TX2P	Tmax 2nd percentile	°C
9		TX98P	Tmax 98th percentile	°C
10		TX2N	[a]# of days below the reference 2nd Tmax percentile	[days]
11		TX98N	[a]# of days exceeding reference 98th Tmax percentile	[days]
12		WSDI90	[c]Warm spell duration index	[days]
13		HWDI	[d]Heat wave duration index	[days]
14	*Tmean*	MEANTG	Mean Tmean	°C
15		TG2P	Tmean 2nd percentile	°C
16		TG98P	Tmean 98th percentile	°C
17		TG2N	[a]# of days below reference 2nd Tmean percentile	[days]
18		TG98N	[a]# of days exceeding reference 98th Tmean percentile	[days]
19	*Precipitation*	PRECTOT	Precipitation total	[mm]
20		PREC98P	Precipitation 98th percentile	[mm]
21		R98N	[a]# of days exceeding reference 98th precipitation percentile	[days]
22		R98T	Fraction of precipitation above the reference 98th precipitation percentile	[%]
23		SDII98P	Daily rainfall intensity of rainfall events above 98th reference precipitation percentile	[mm]
24		SDII	Simple daily rainfall intensity index	[mm]
25		R5d	Greatest 5-day total rainfall	[mm]
26		R1d	Greatest daily total rainfall	[mm]
27		CDD	max # of consecutive dry days (< 1 mm)	[days]

[a] Reference percentile based on 1961–1990

[b] CSDI10: Cold Spell Duration Index. Counted are the total number of at least 6 consecutive days with Tmin below long-term 10th percentile (1961–1990)

[c] WSDI90: Warm Spell Duration Index. Counted are the total number of at least 6 consecutive days with Tmax exceeding long-term 90th percentile (1961–1990)

[d] HWDI: Heat Wave Duration Index ($Tx_{ij} > Tx_{inorm} + 5$). Let Txij be the daily maximum temperature at day *i* of period *j*. Let Tx_{inorm} be the calendar day mean calculated for a 5 day window centered on each calendar day during the base period (1961-1990).Then counted are the total number of spells of at least 6 consecutive days exceeding Tx_{inorm}+5 °C

3 Atlas of the Trend Analysis

In this chapter, the results of the trend analysis for each index, station or region are presented in the form of maps and time-series plots. The maps, however, are only used for the most recent period (1901–2000) due to the limited number of stations in earlier periods. For each period, seasonal indices starting with spring (MAM) and ending with winter (DJF) are presented. The order of presentation is Tmin/Tmax, followed by Tmean and Precipitation (see Table 2.2). To make it easier to find an index in a given season and for a given period, a header is put on each page to indicate the period, season, and index group. The circles in the maps indicate station locations for the two longer periods where results for individual stations are presented. All the maps contain a colour bar symmetric around zero and a title (Fig. 3.1). The range is determined by the highest absolute value appearing in the map. An index dependent general colour scheme is used throughout the whole atlas. For temperature indices, the colour scale ranges blue-green-red where red colours indicate warming and blue colours cooling conditions. For example, an increase in Tmean would be shown in red as well as a decreased number of days below the second percentile. Trends in precipitation indices are visualized using a brown-yellow-green colour scale where brownish colour indicate drier and greenish colours wetter conditions. Trend significance is indicated for each symbol marking a station (circles) or regional averages (squares) using significance levels of $p<0.05$ and $p<0.01$. Analysis period, season, index and the unit of the trend are provided in every figure title.

For the time-series plots, each individual station's data are separately shown for the two longer periods (1801–2000 and 1851–2000), while the regional means are displayed for the period 1901–2000 on background plots containing all contributing individual station time series (in light grey lines). Each figure shows annual values of an index for a season, its low-frequency variation and long-term linear trend (Fig. 3.2). The smoothed curve emphasizing low-frequency variation is a 10-year Gaussian filter applied to the original time-series suppressing variations on time-scales less than 10 years. The straight solid line shows the linear trend estimated by linear regression for the period in question. The statistical trend significance is indicated as follows: '*' significant at $p<0.05$, '**' significant at $p<0.01$, and '()' not significant. In each figure analysis period, the names of stations or region, season, index name and trend unit are indicated. To save space, two (and sometimes three) indices or two trend lines for two periods are plotted in the same figure whenever appropriate and feasible (Figs. 3.3, 3.4, 3.5, 3.6, 3.7, 3.8, 3.9, 3.10, 3.11, 3.12, 3.13, 3.14, 3.15, 3.16, 3.17, 3.18, 3.19, 3.20, 3.21, 3.22, 3.23, 3.24, 3.25, 3.26, 3.27, 3.28, 3.29, 3.30, 3.31, 3.32, 3.33, 3.34, 3.35, 3.36, 3.37, 3.38, 3.39, 3.40, 3.41, 3.42, 3.43, 3.44, 3.45, 3.46, 3.47, 3.48, 3.49, 3.50, 3.51, 3.52, 3.53, 3.54, 3.55, 3.56, 3.57, 3.58, 3.59, 3.60, 3.61, 3.62, 3.63, 3.64, 3.65, 3.66, 3.67, 3.68, 3.69, 3.70, 3.71, 3.72, 3.73, 3.74, 3.75, 3.76, 3.77, 3.78, 3.79, 3.80, 3.81, 3.82, 3.83, 3.84, 3.85, 3.86, 3.87, 3.88, 3.89, 3.90, 3.91, 3.92, 3.93, 3.94, 3.95, 3.96, 3.97, 3.98, 3.99, 3.100, 3.101, 3.102, 3.103, 3.104, 3.105, 3.106, 3.107, 3.108, 3.109, 3.110, 3.111, 3.112, 3.113, 3.114, 3.115, 3.116, 3.117, 3.118, 3.119, 3.120, 3.121, 3.122, 3.123, 3.124, 3.125, 3.126, 3.127, 3.128, 3.129, 3.130, 3.131, 3.132, 3.133, 3.134, 3.135, 3.136, 3.137, 3.138, 3.139, 3.140, 3.141, 3.142, 3.143, 3.144, 3.145, 3.146, 3.147, 3.148, 3.149, 3.150, 3.151, 3.152, 3.153, 3.154, 3.155, 3.156).

D. Chen et al., *European Trend Atlas of Extreme Temperature and Precipitation Records*, DOI 10.1007/978-94-017-9312-4_3

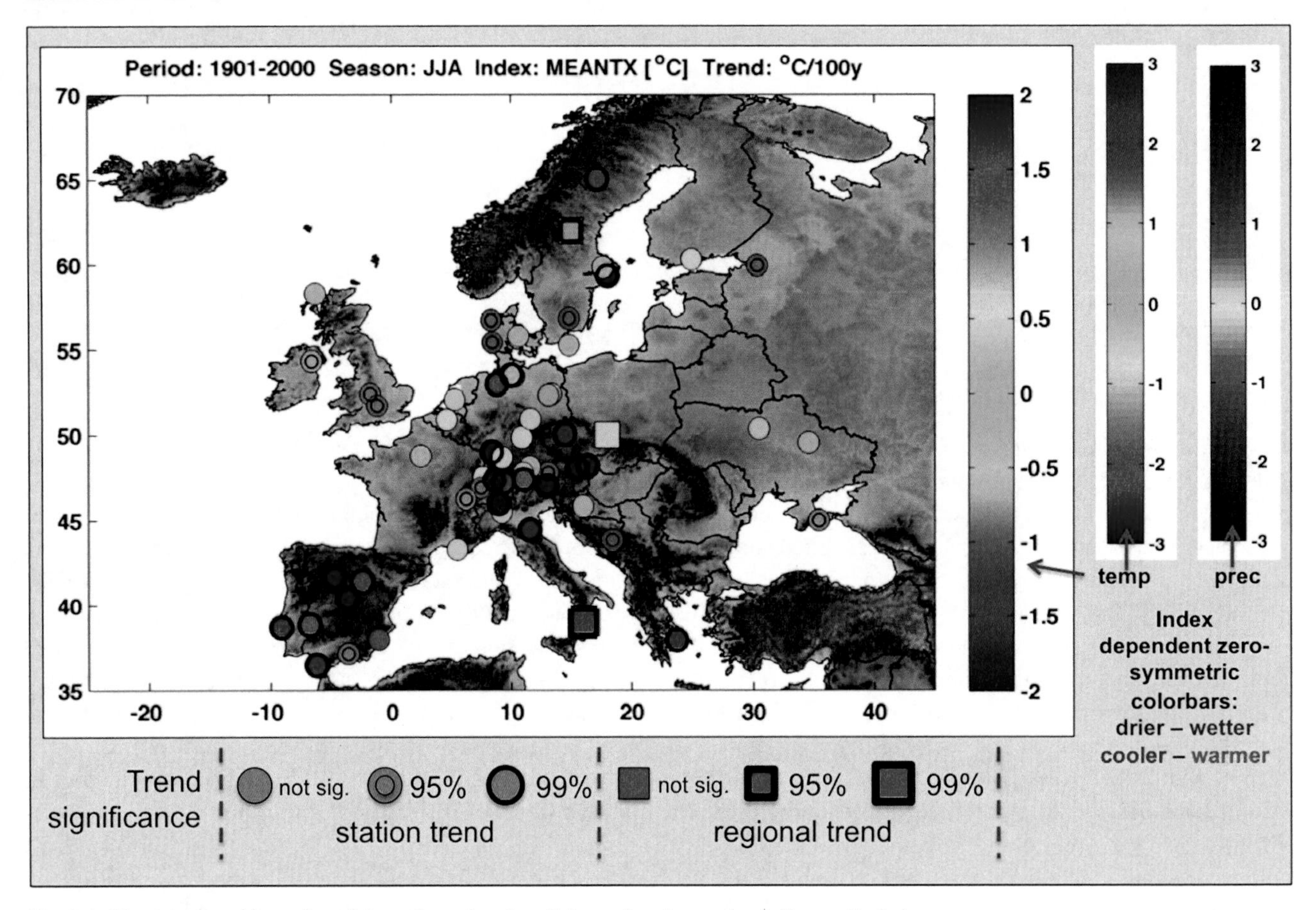

Fig. 3.1 Map layout and legend used throughout the atlas. Color scales change depending on the index

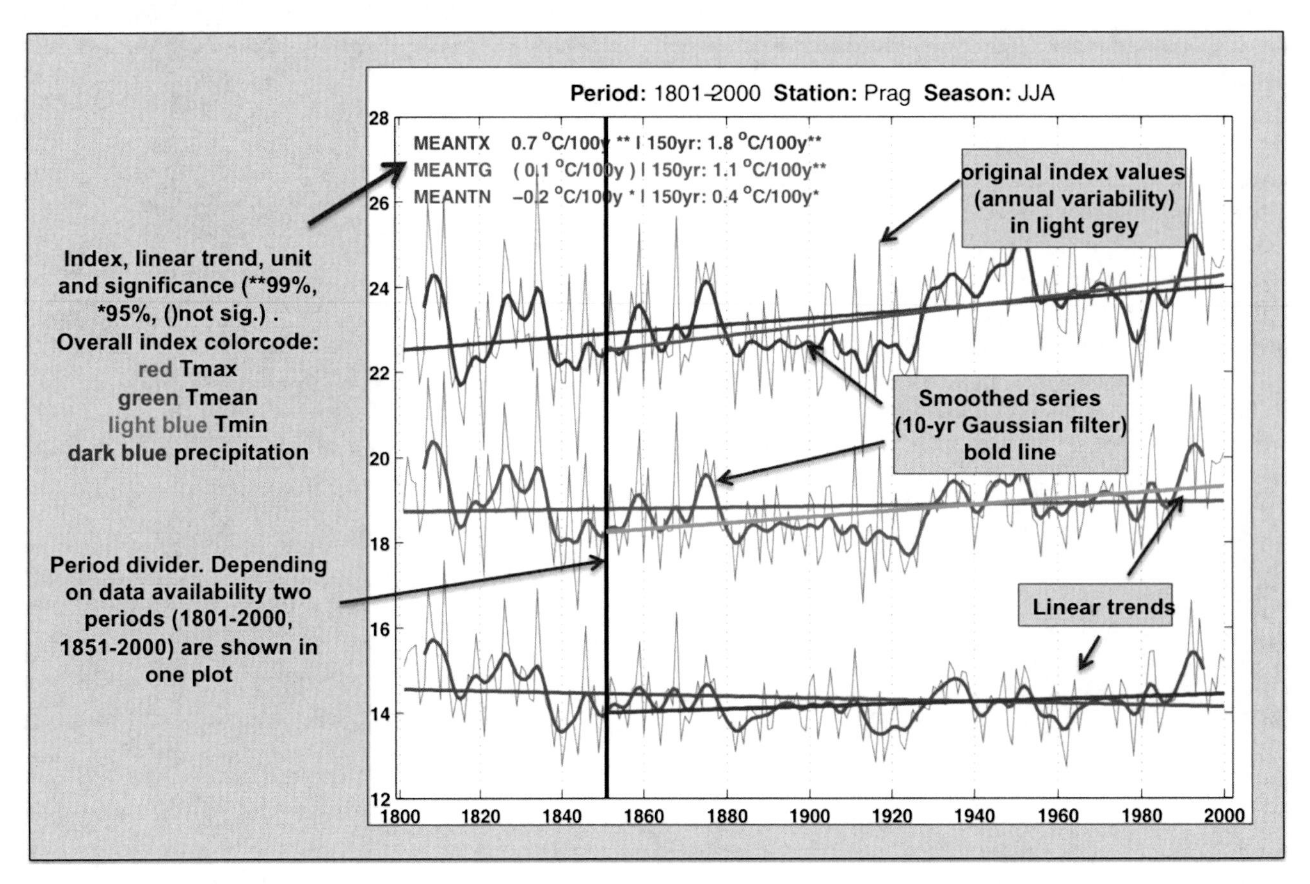

Fig. 3.2 Layout and legend for the time-series plots

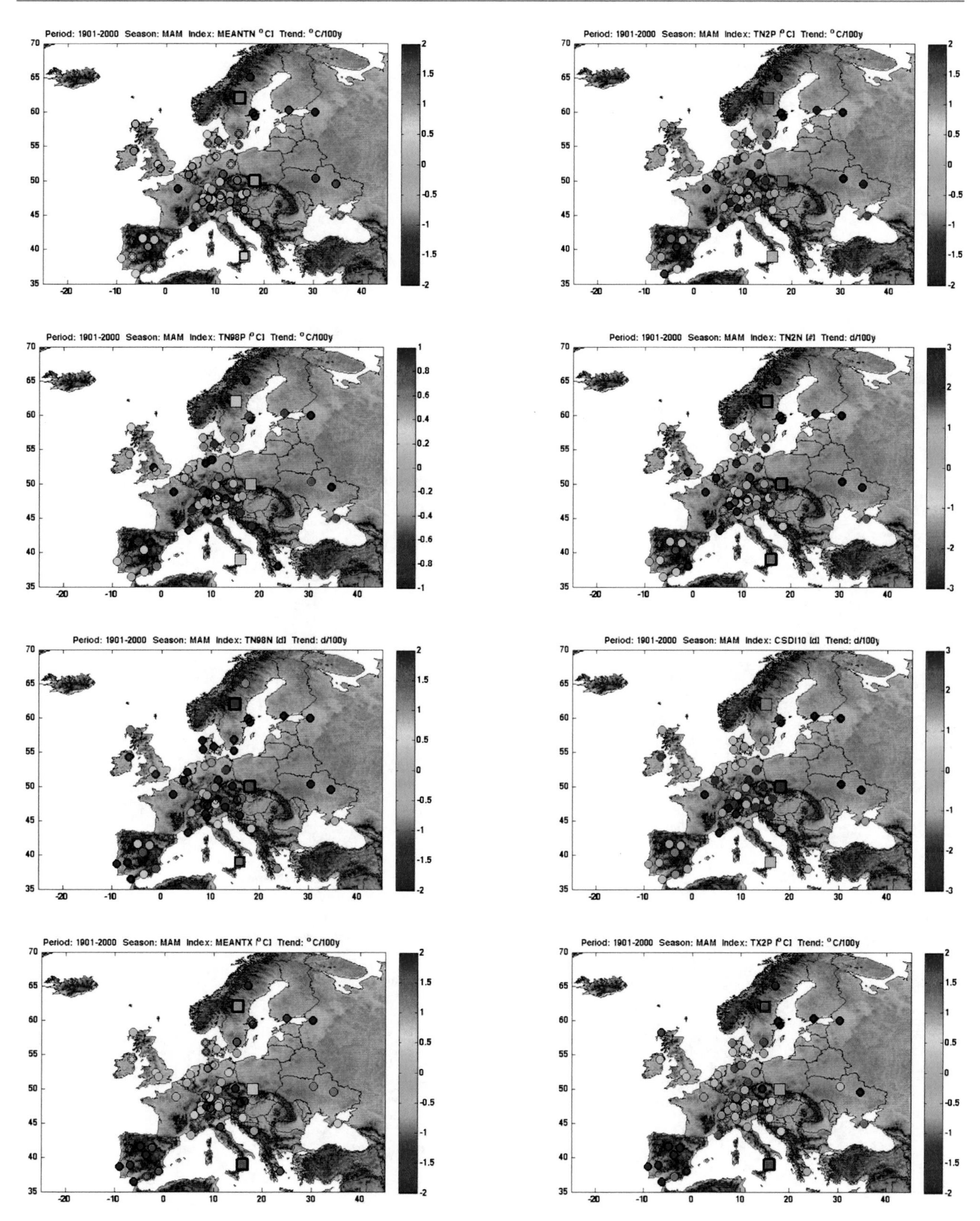

Fig. 3.3 1901–2000 MAM Tmin

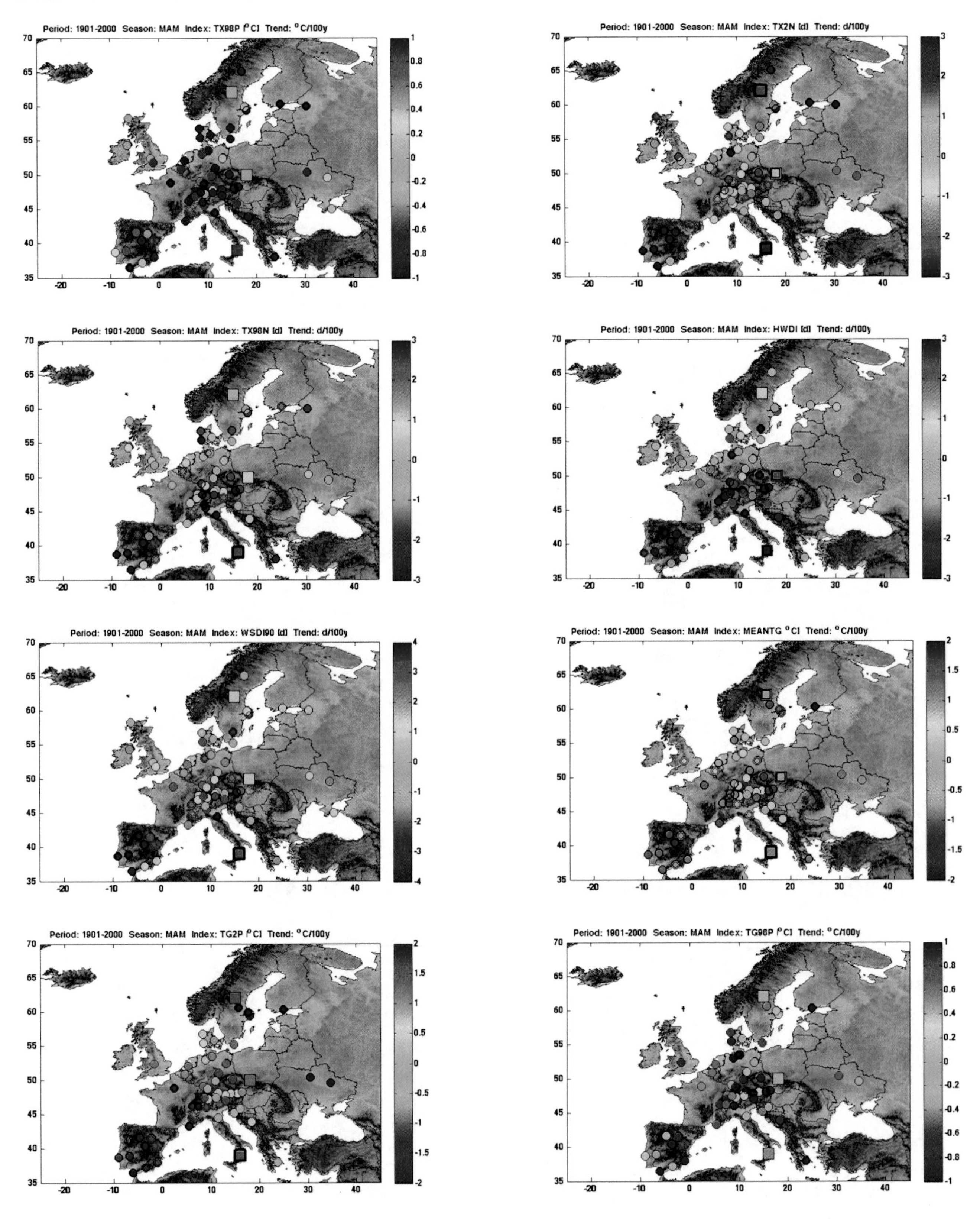

Fig. 3.4 1901–2000 MAM Tmax

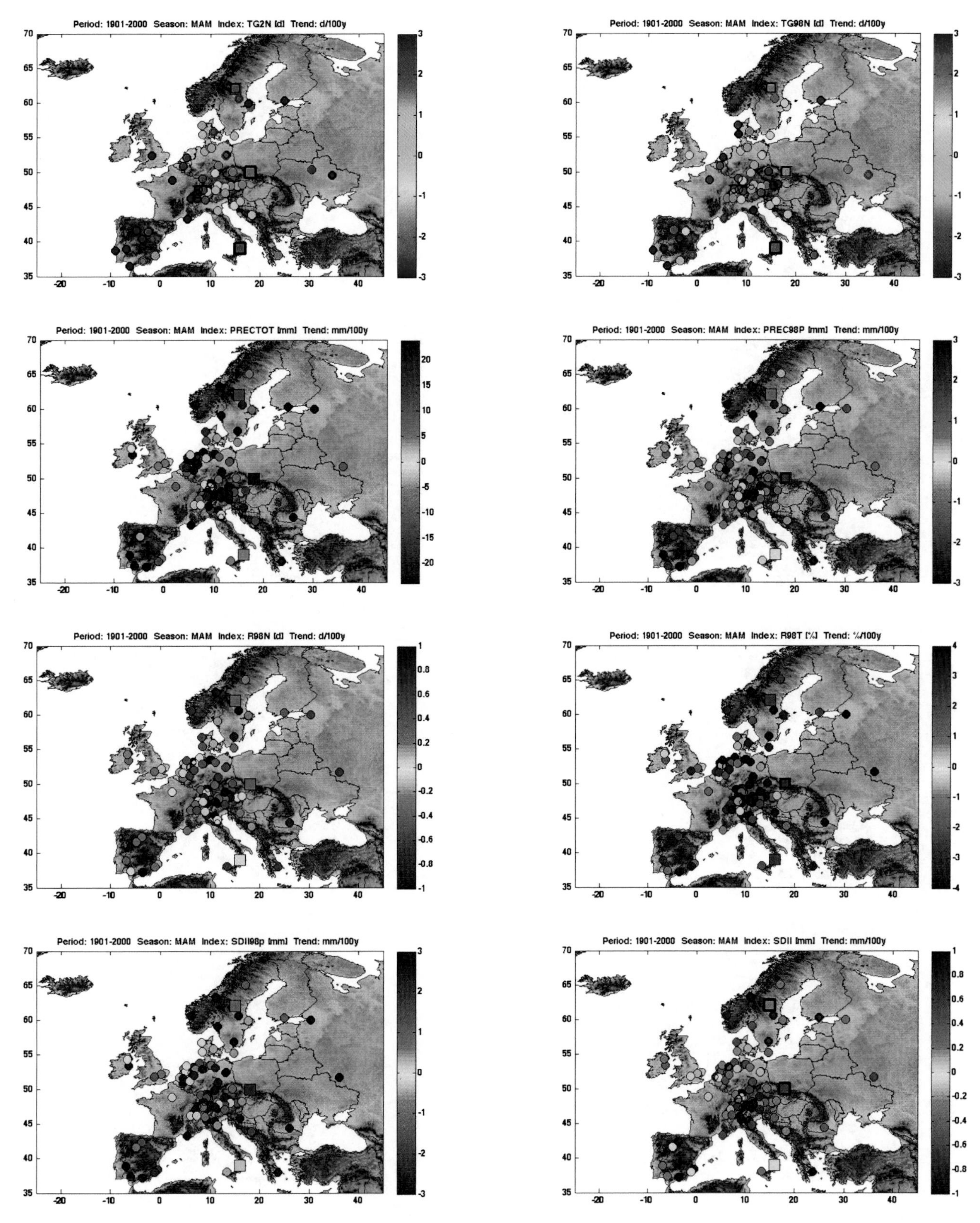

Fig. 3.5 1901–2000 MAM Tmean

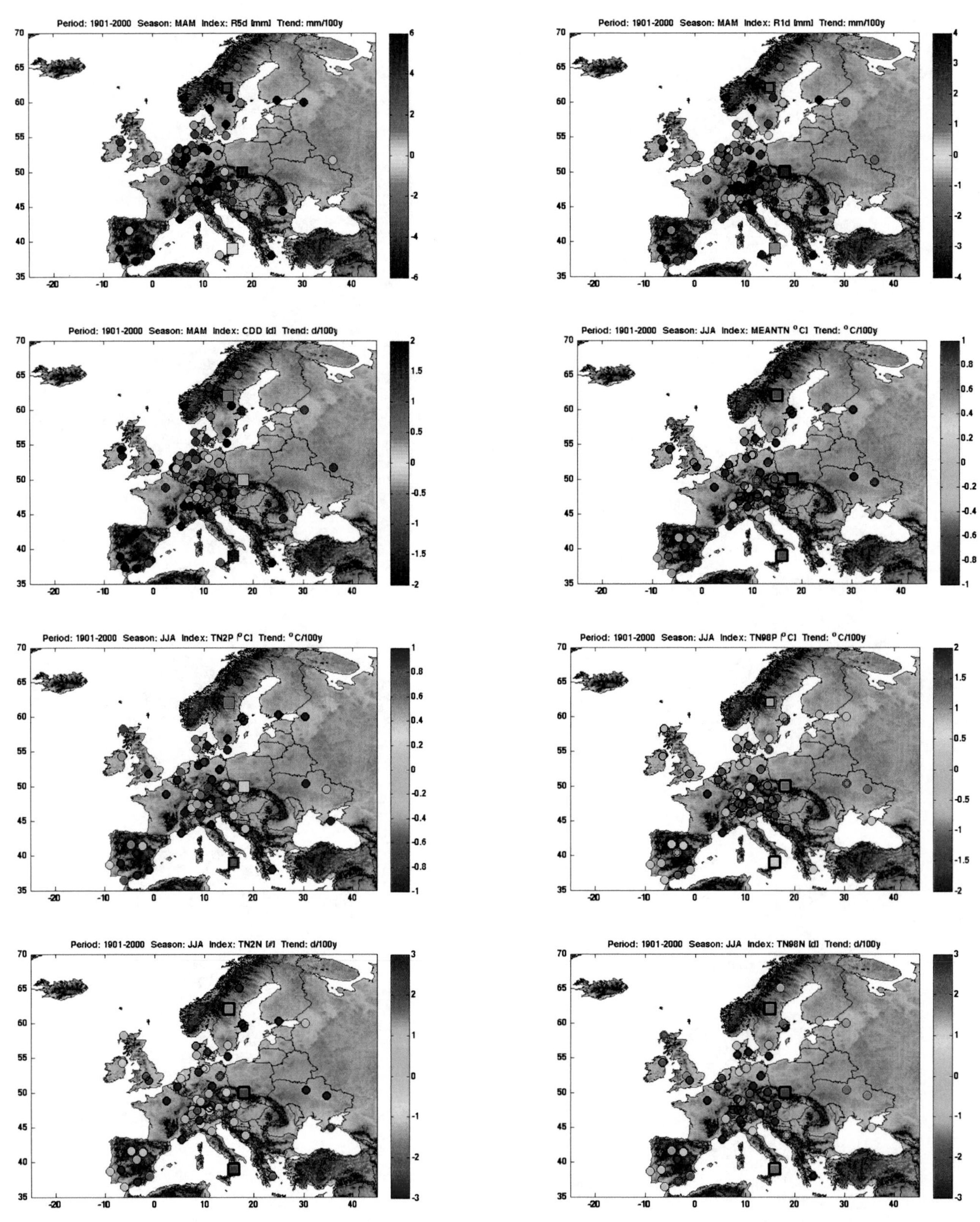

Fig. 3.6 1901–2000 MAM Prec

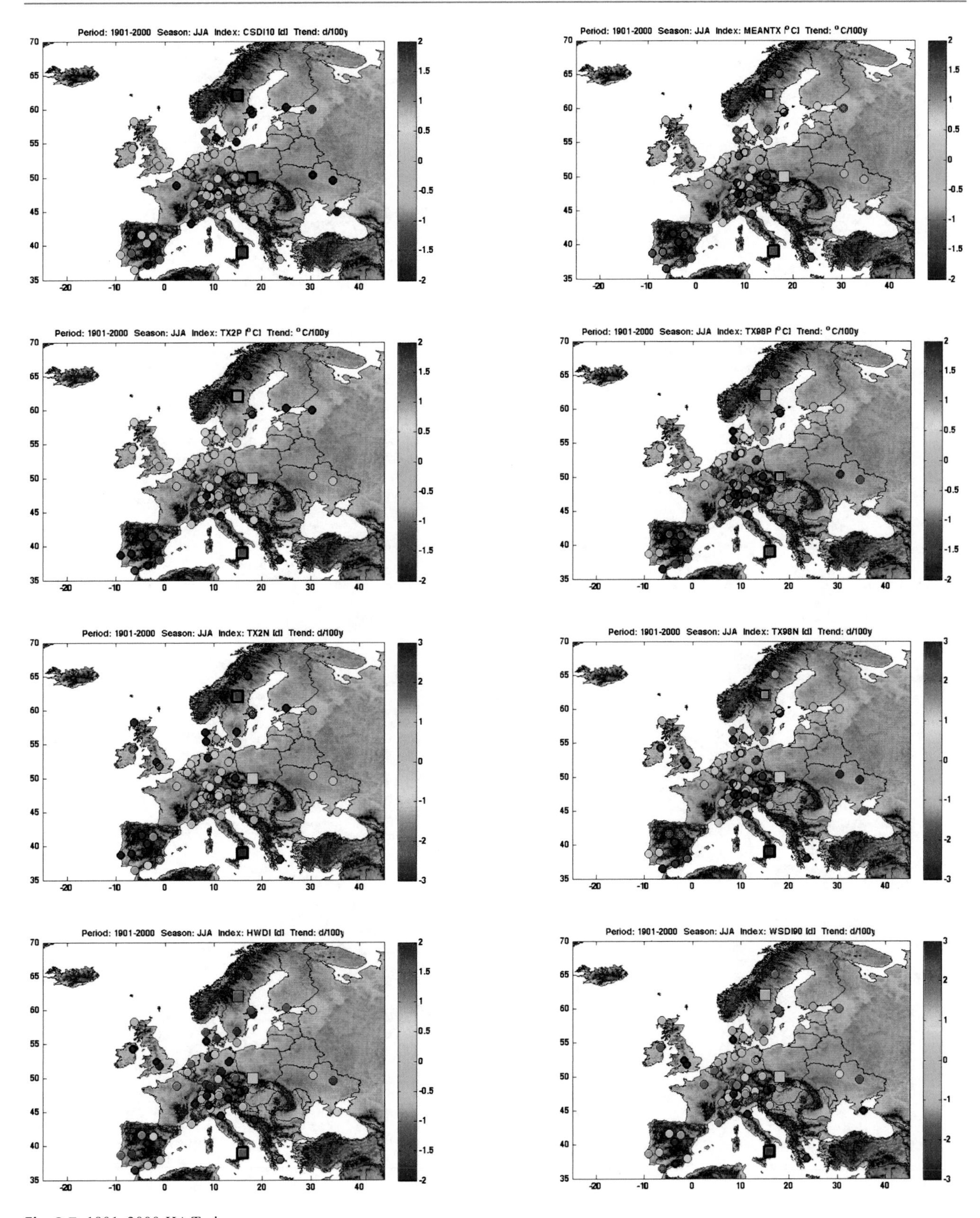

Fig. 3.7 1901–2000 JJA Tmin

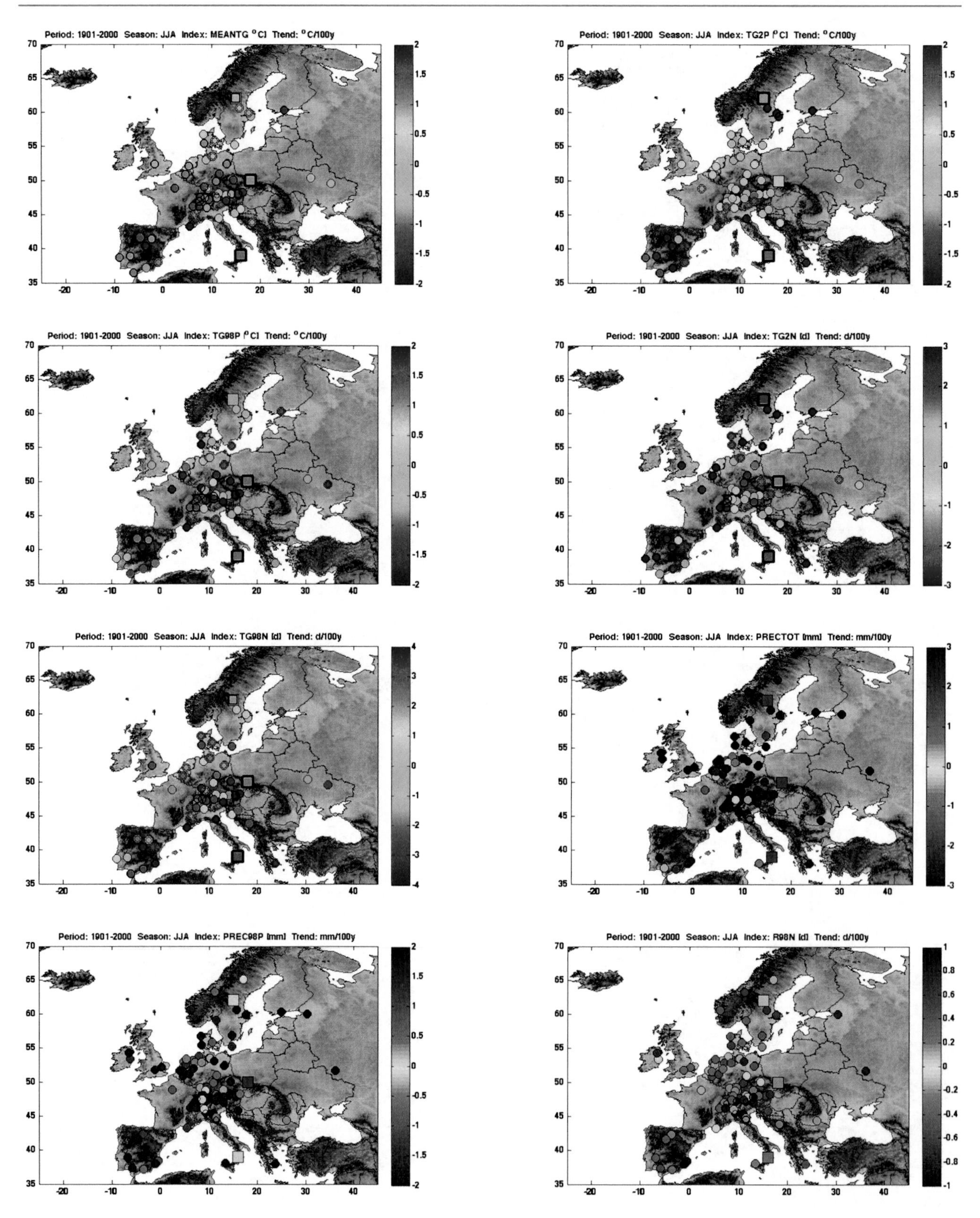

Fig. 3.8 1901–2000 JJA Tmean

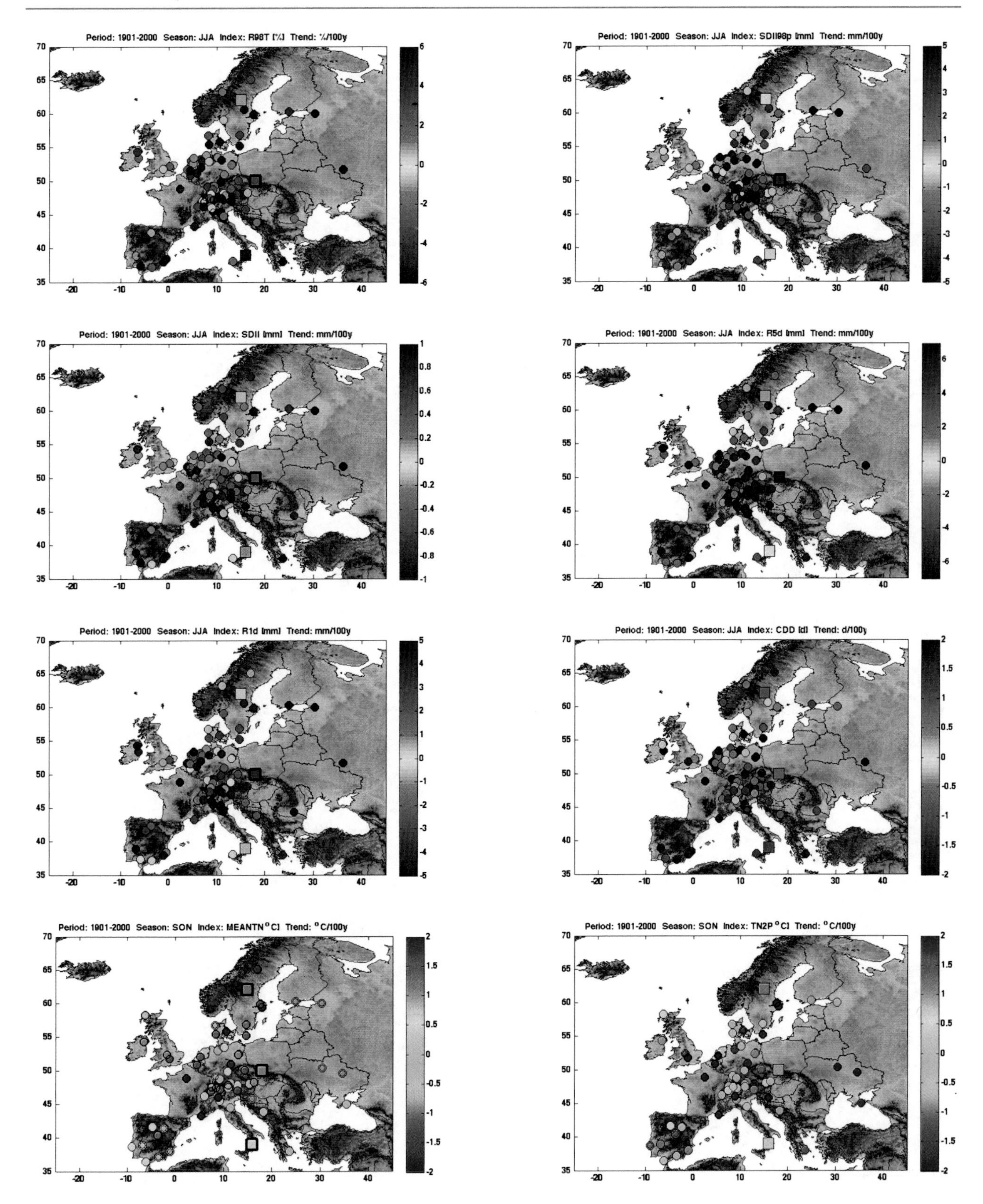

Fig. 3.9 1901–2000 JJA Prec

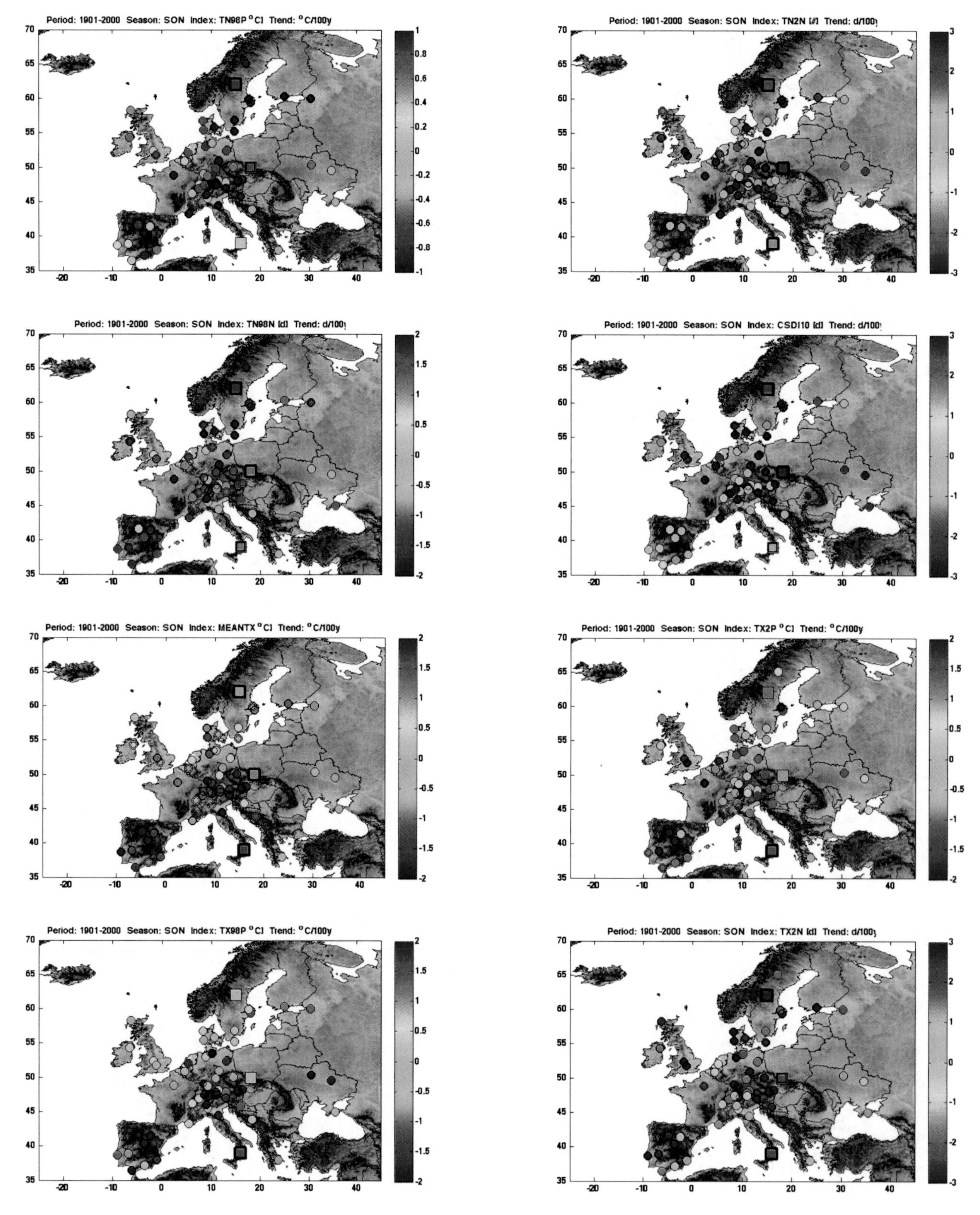

Fig. 3.10 1901–2000 SON Tmin

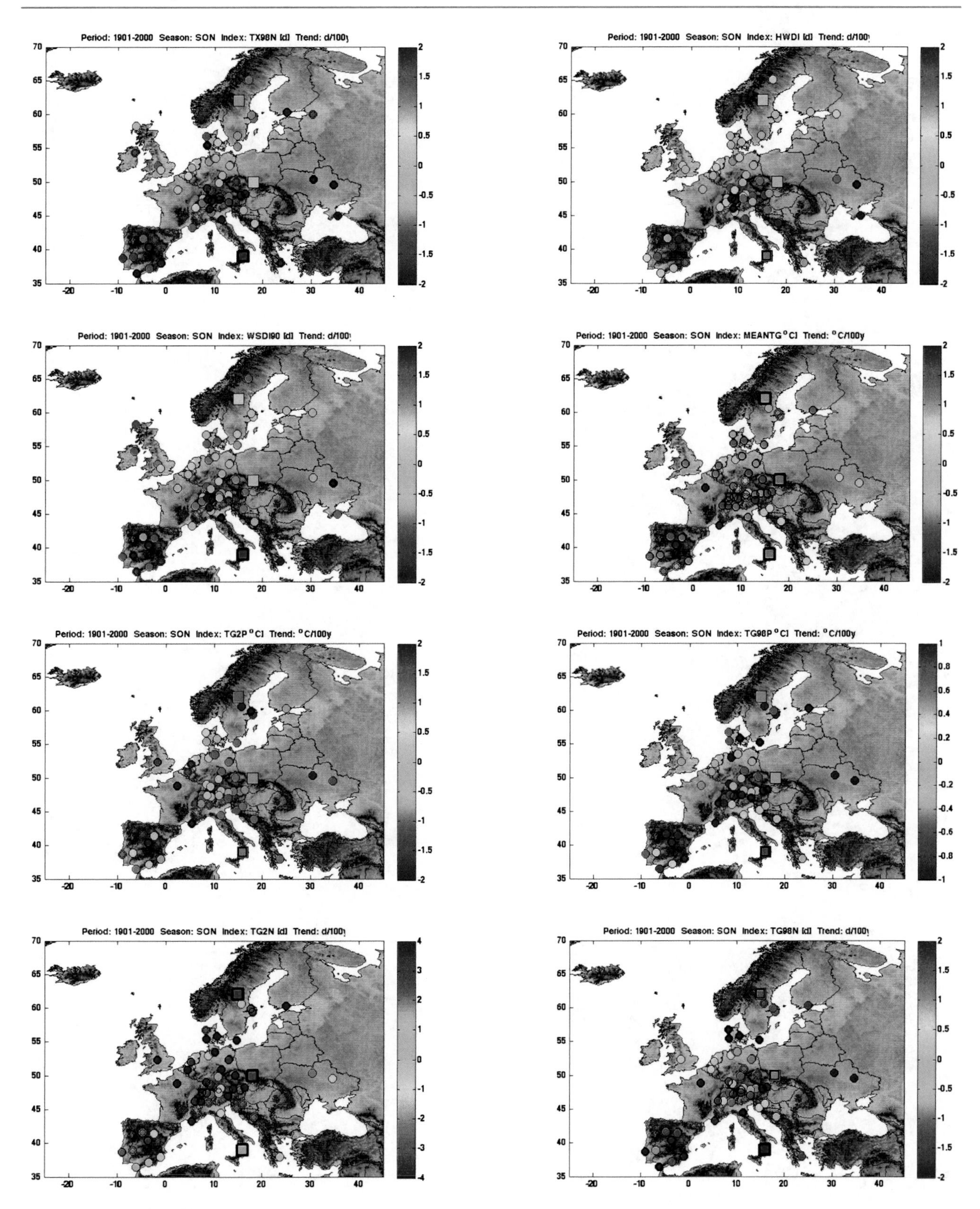

Fig. 3.11 1901–2000 SON Tmax

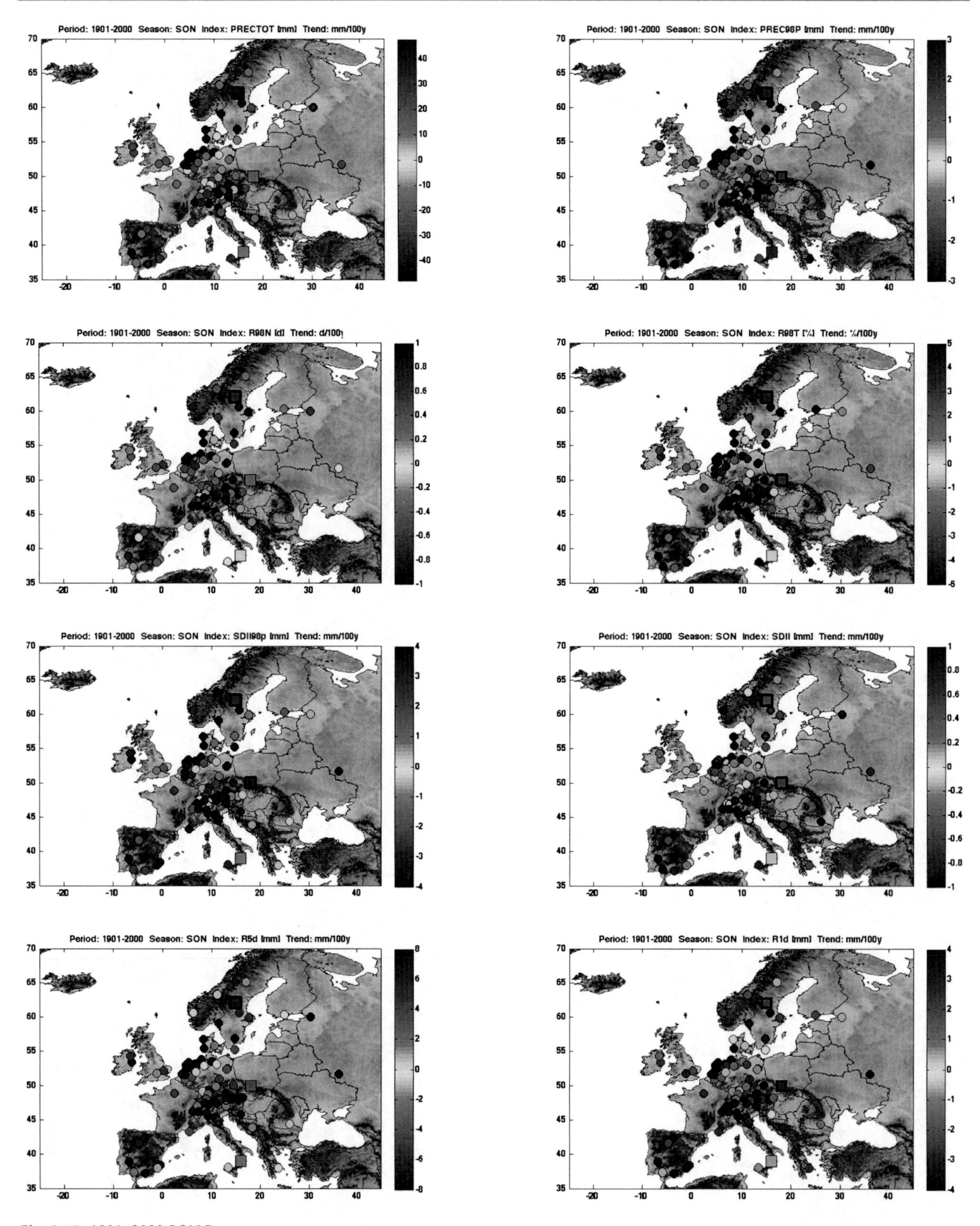

Fig. 3.12 1901–2000 SON Prec

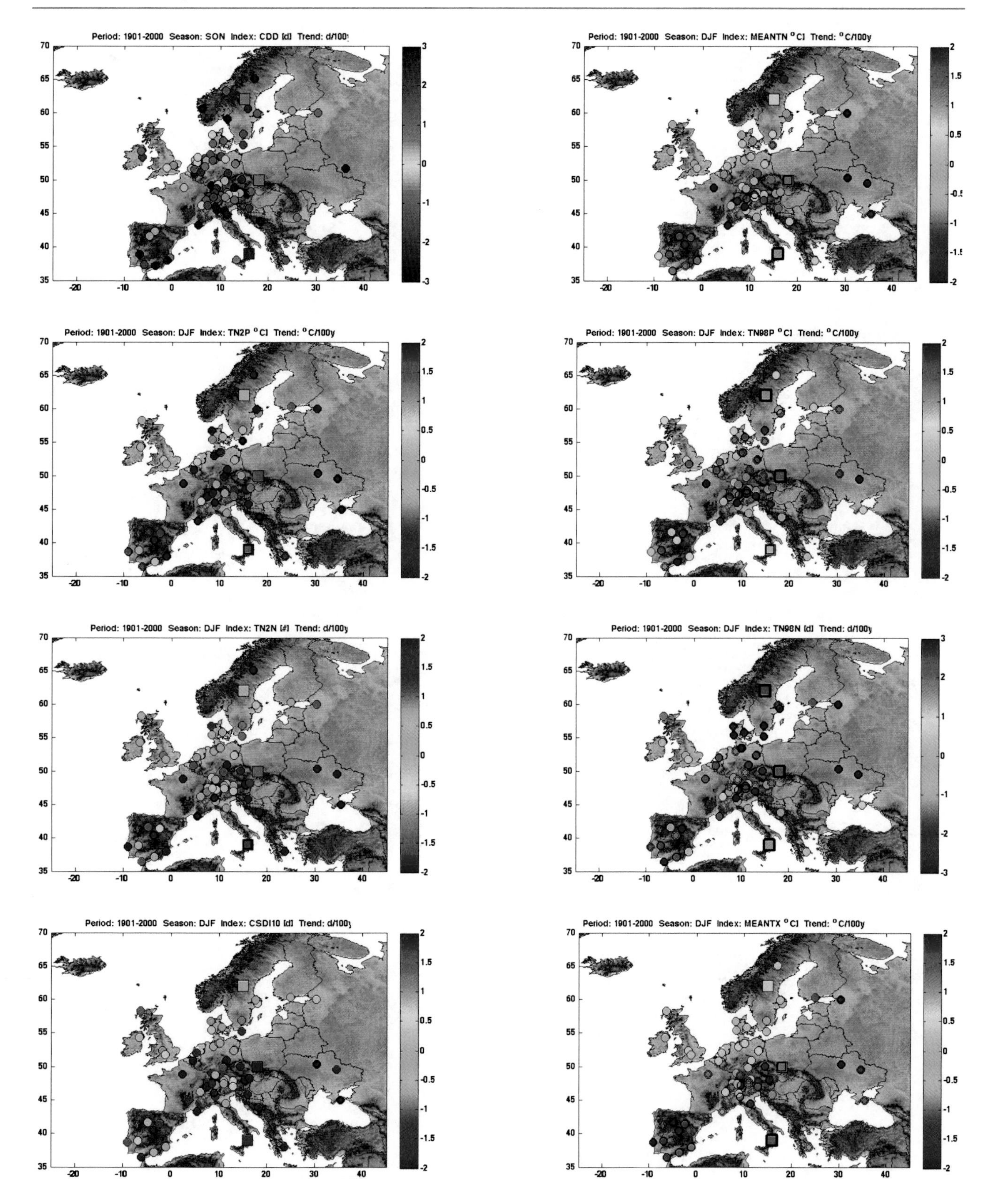

Fig. 3.13 1901–2000 SON Prec

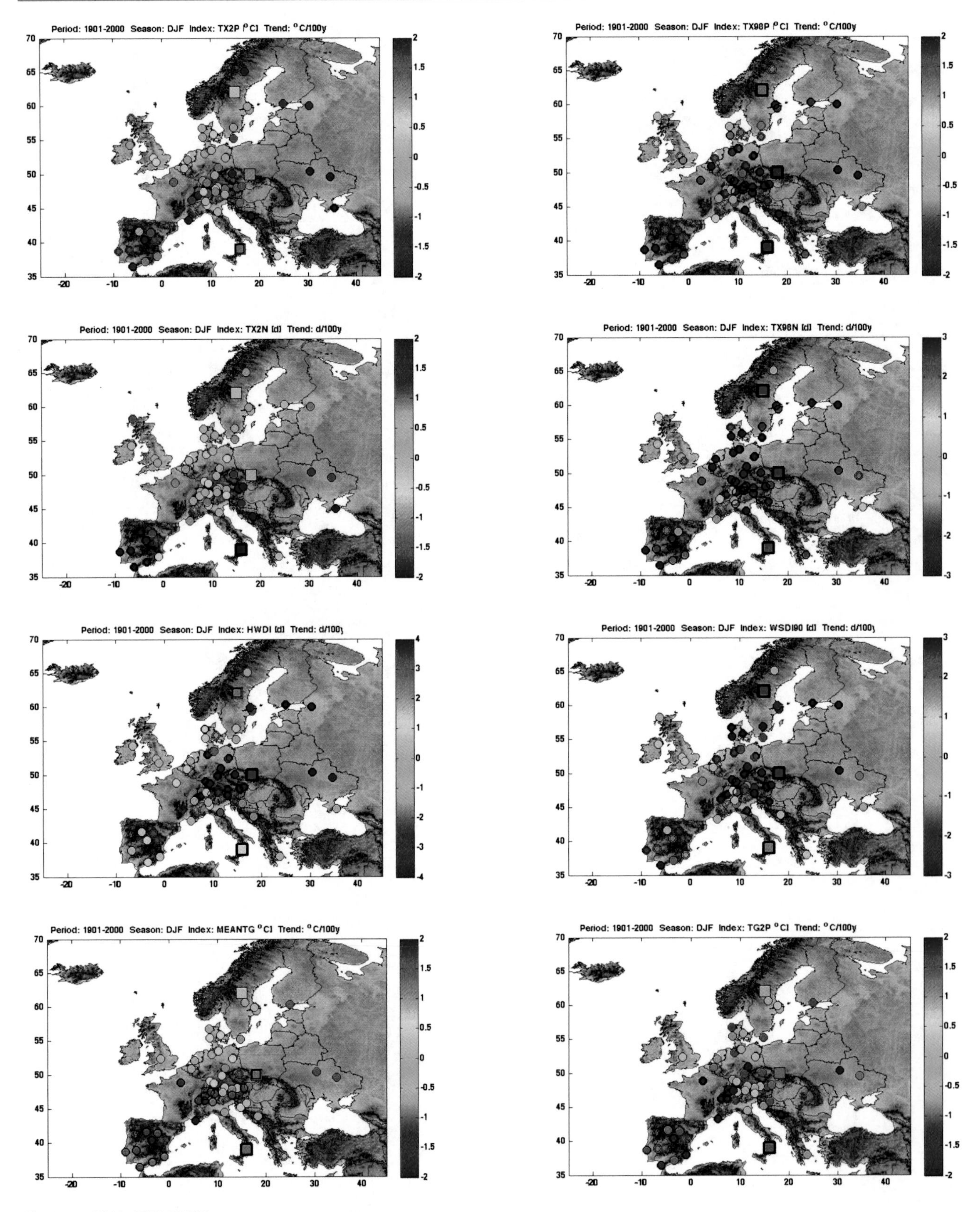

Fig. 3.14 1901–2000 DJF Tmax

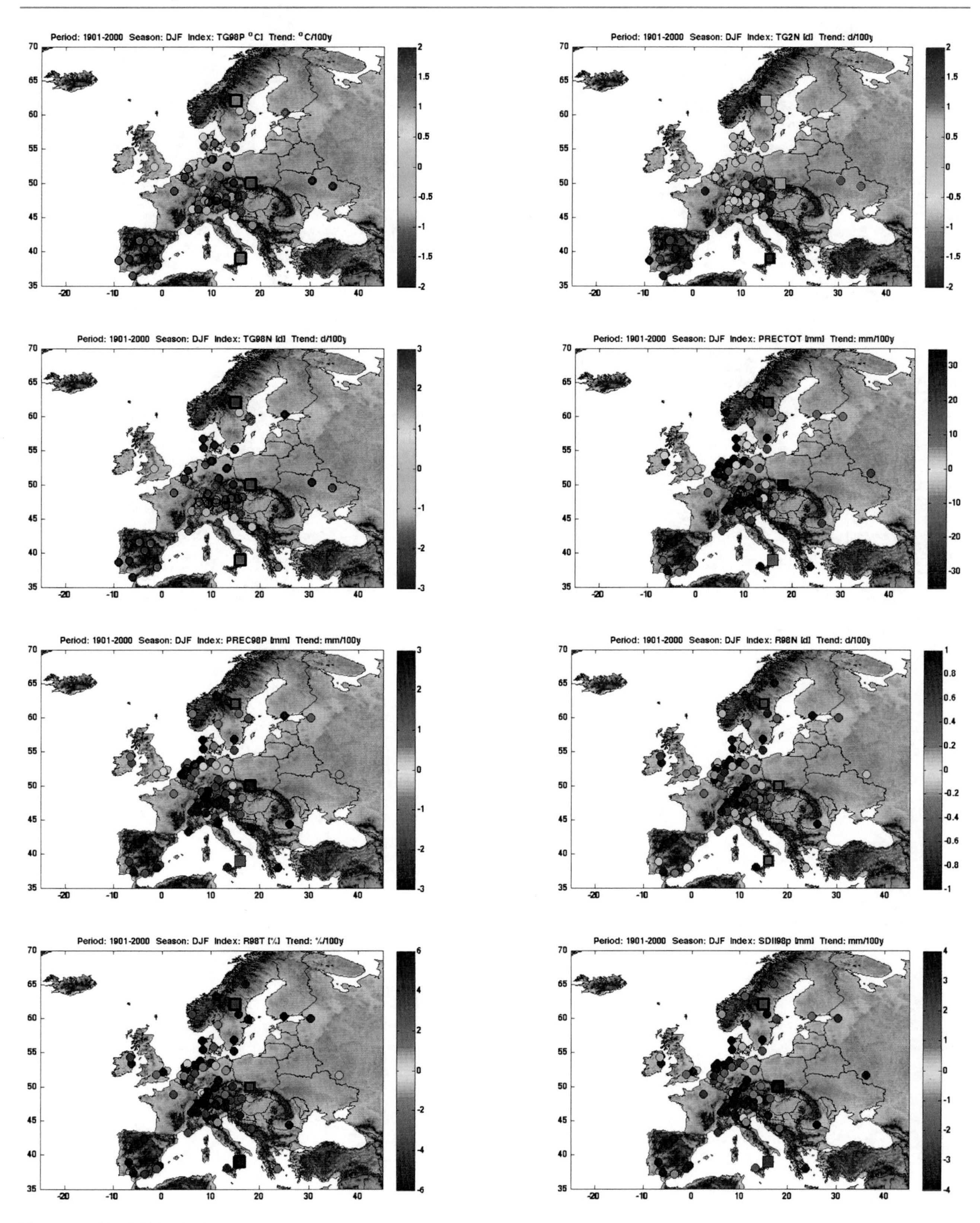

Fig. 3.15 1901–2000 DJF Tmean

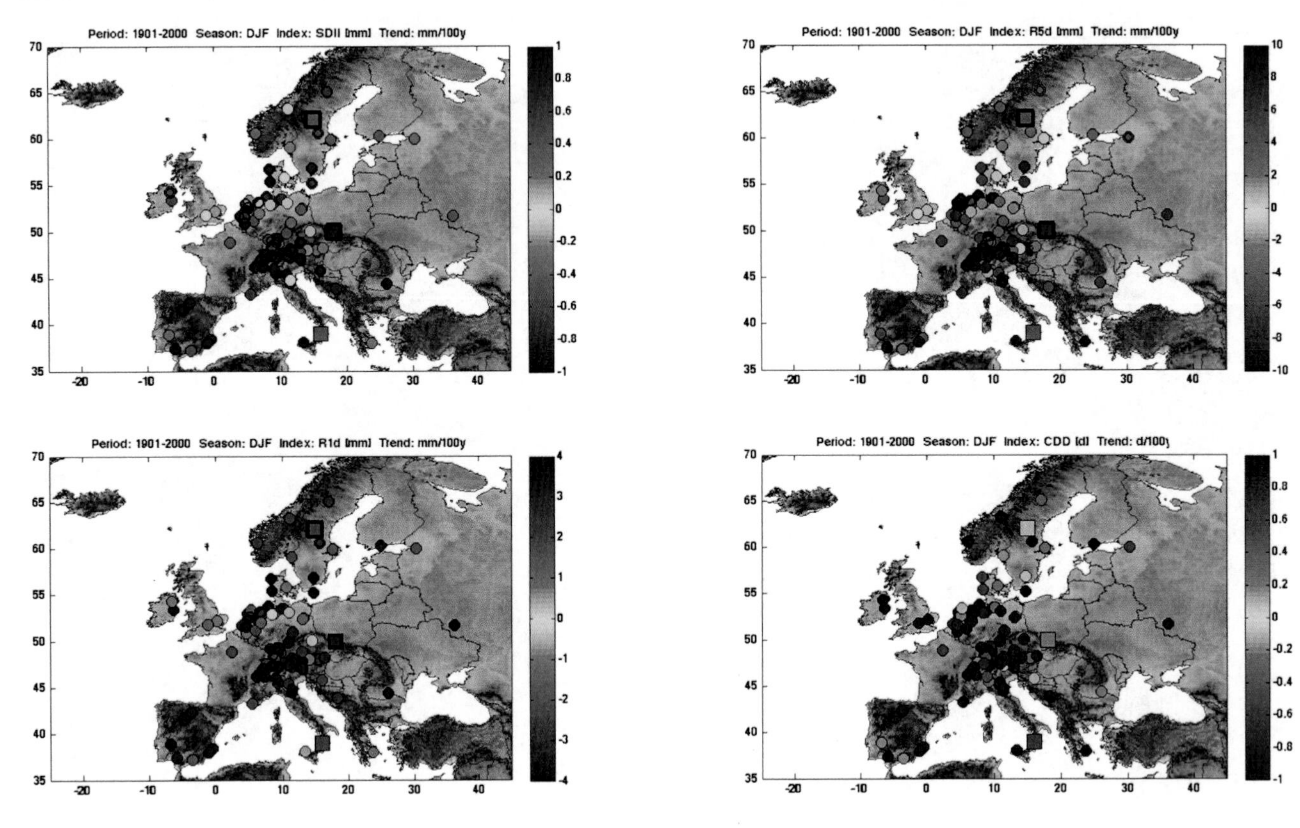

Fig. 3.16 1901–2000 DJF Prec

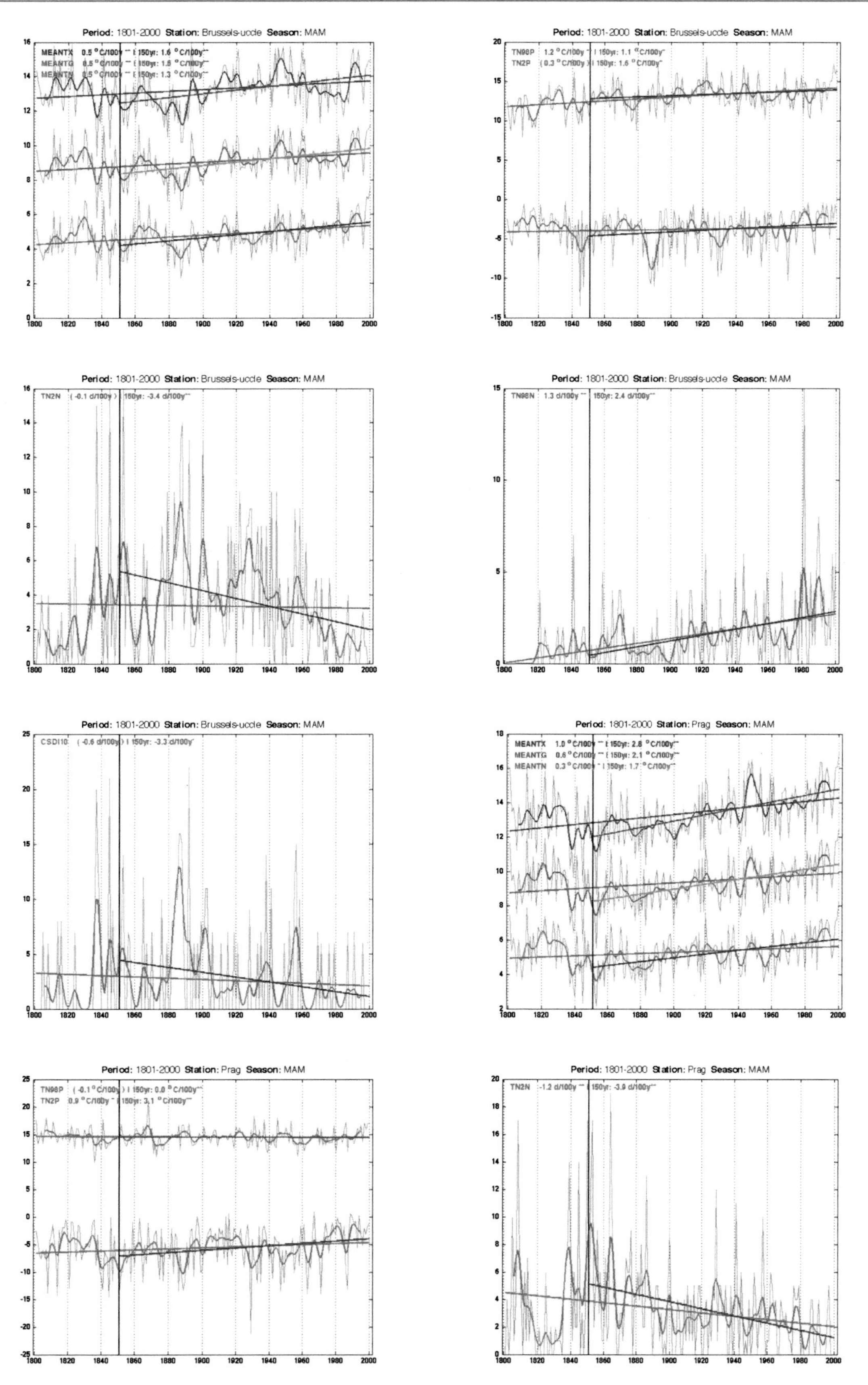

Fig. 3.17 1851–2000 MAM Tmin Brussels-uccle

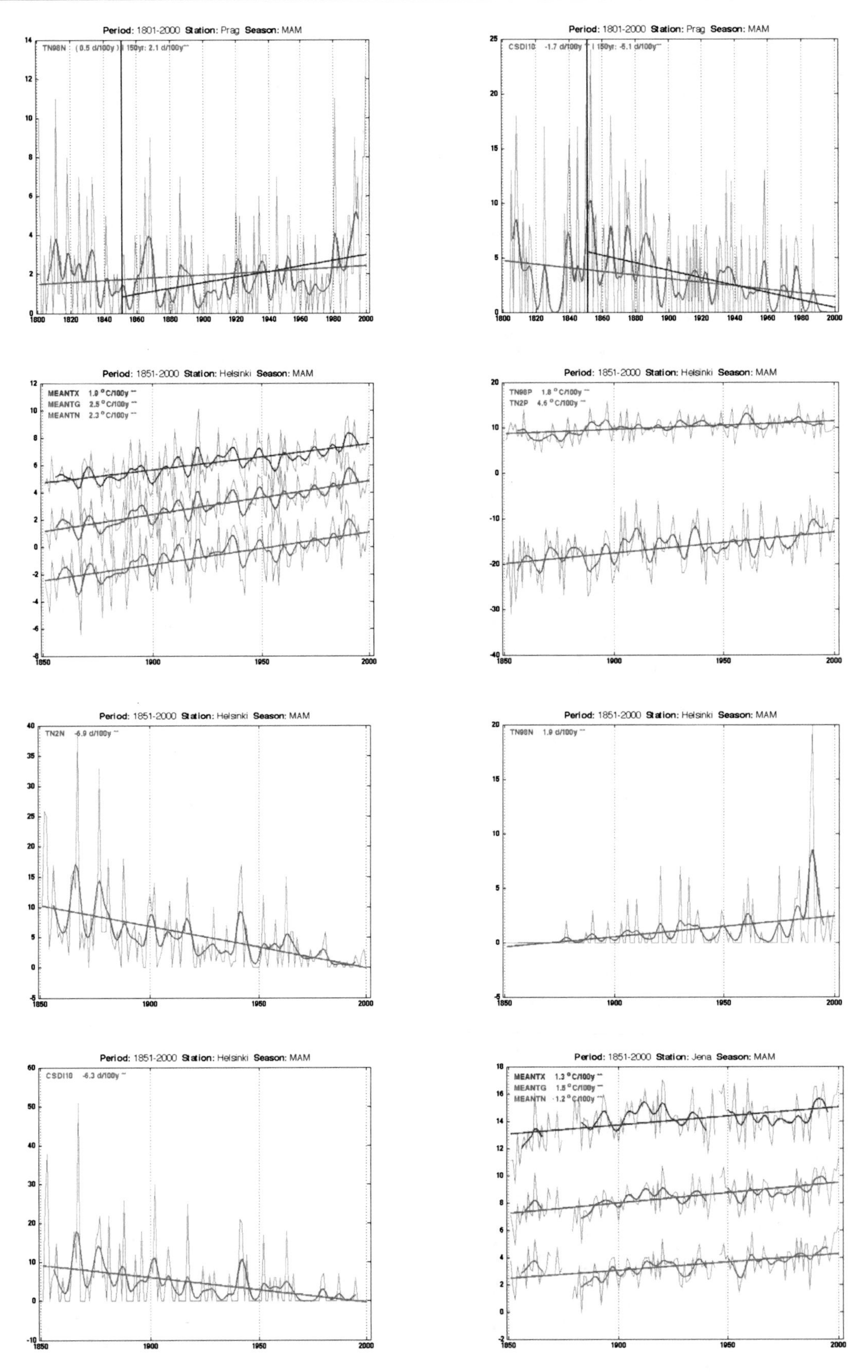

Fig. 3.18 1851–2000 MAM Tmin Prag

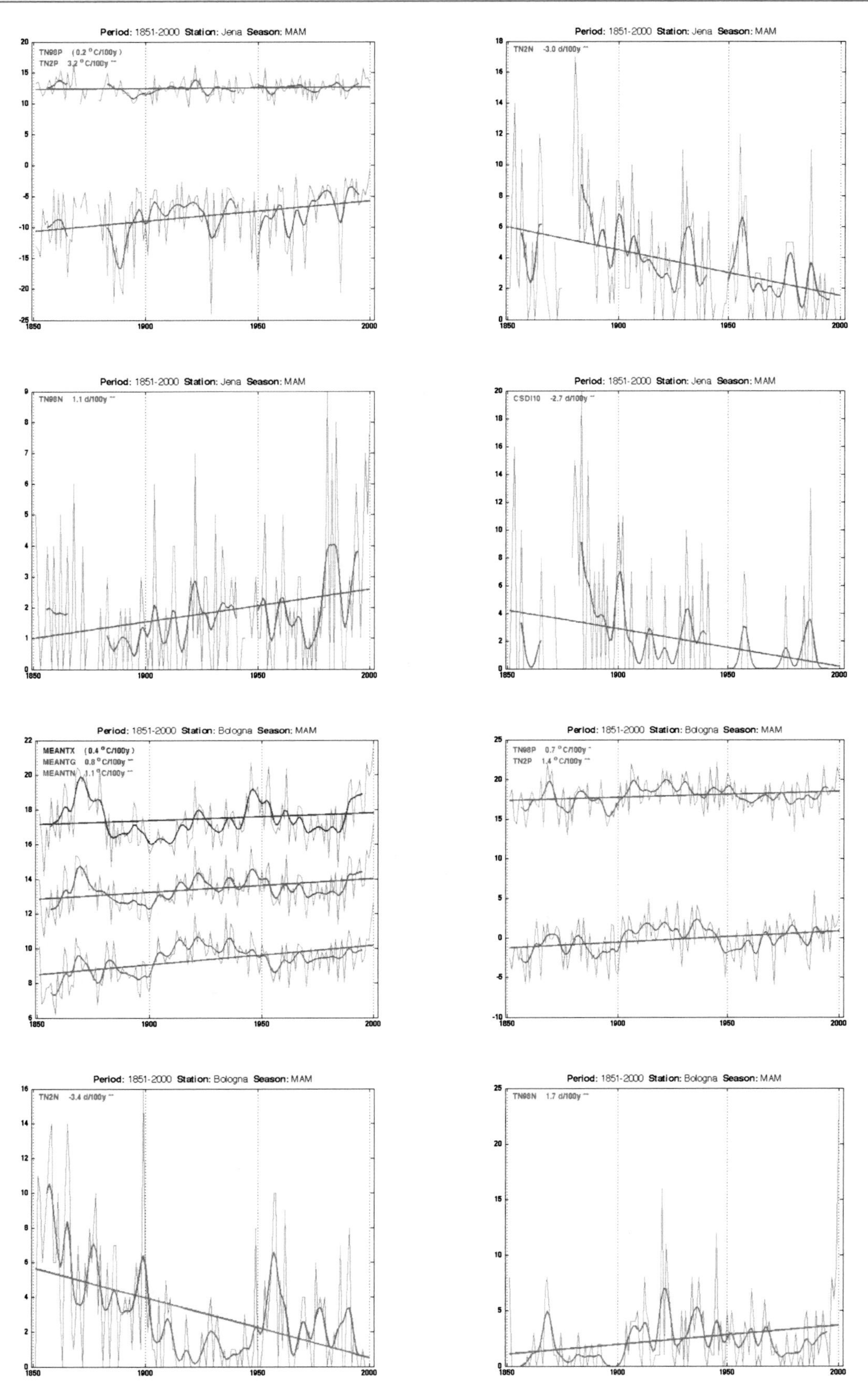

Fig. 3.19 1851–2000 MAM Tmin Jena

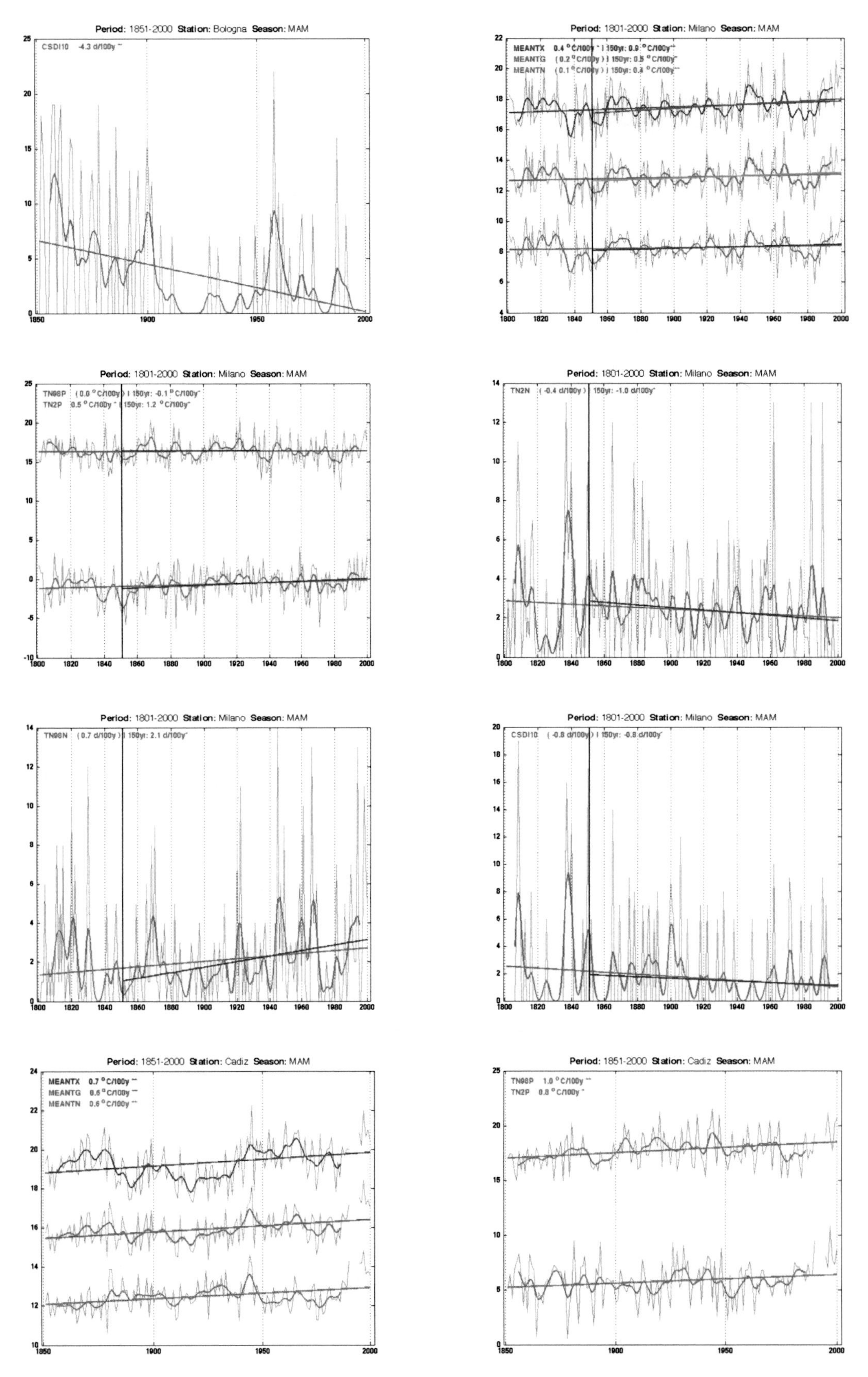

Fig. 3.20 1851–2000 MAM Tmin Bologna

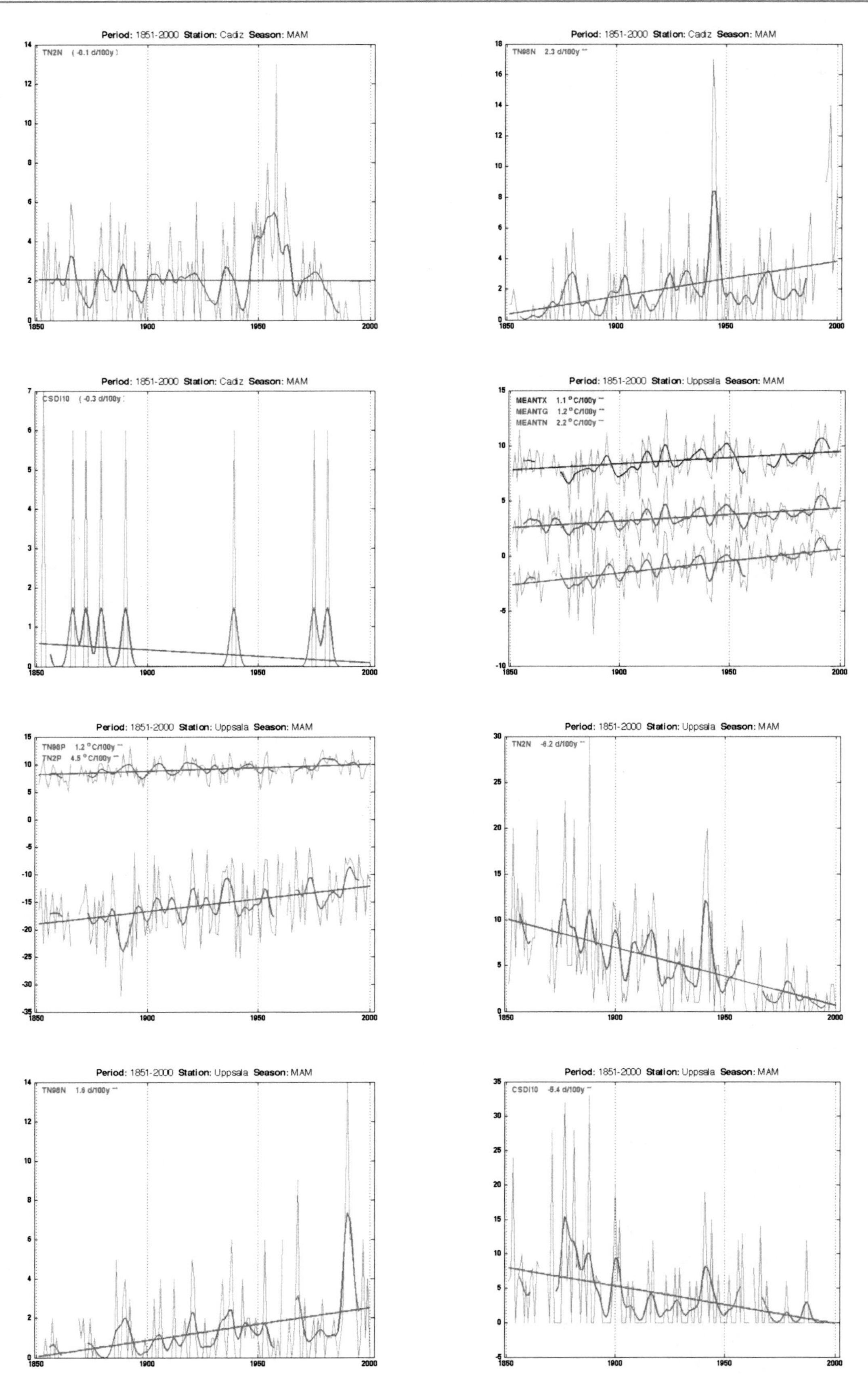

Fig. 3.21 1851–2000 MAM Tmin Cadiz

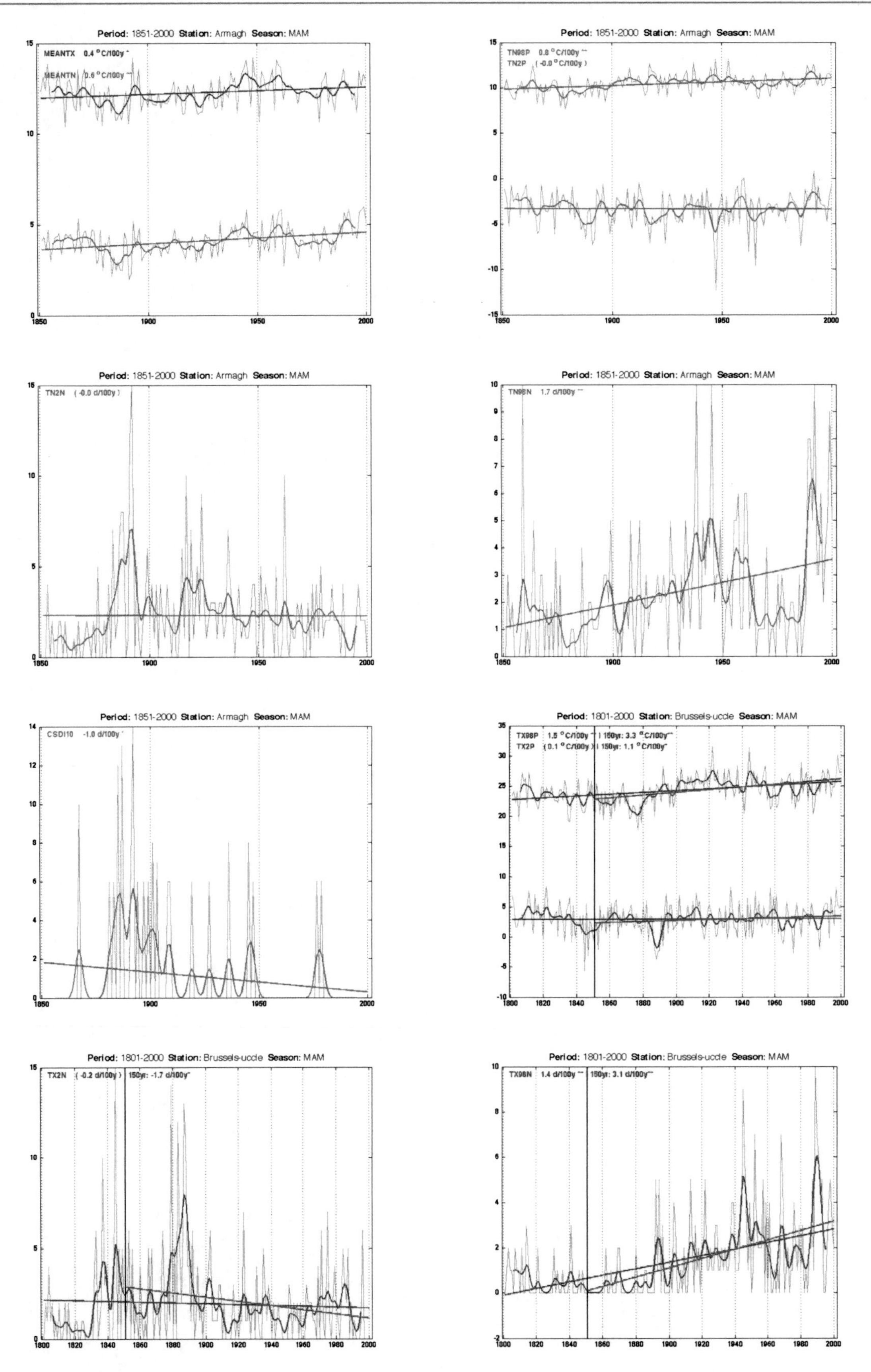

Fig. 3.22 1851–2000 MAM Tmin Armagh

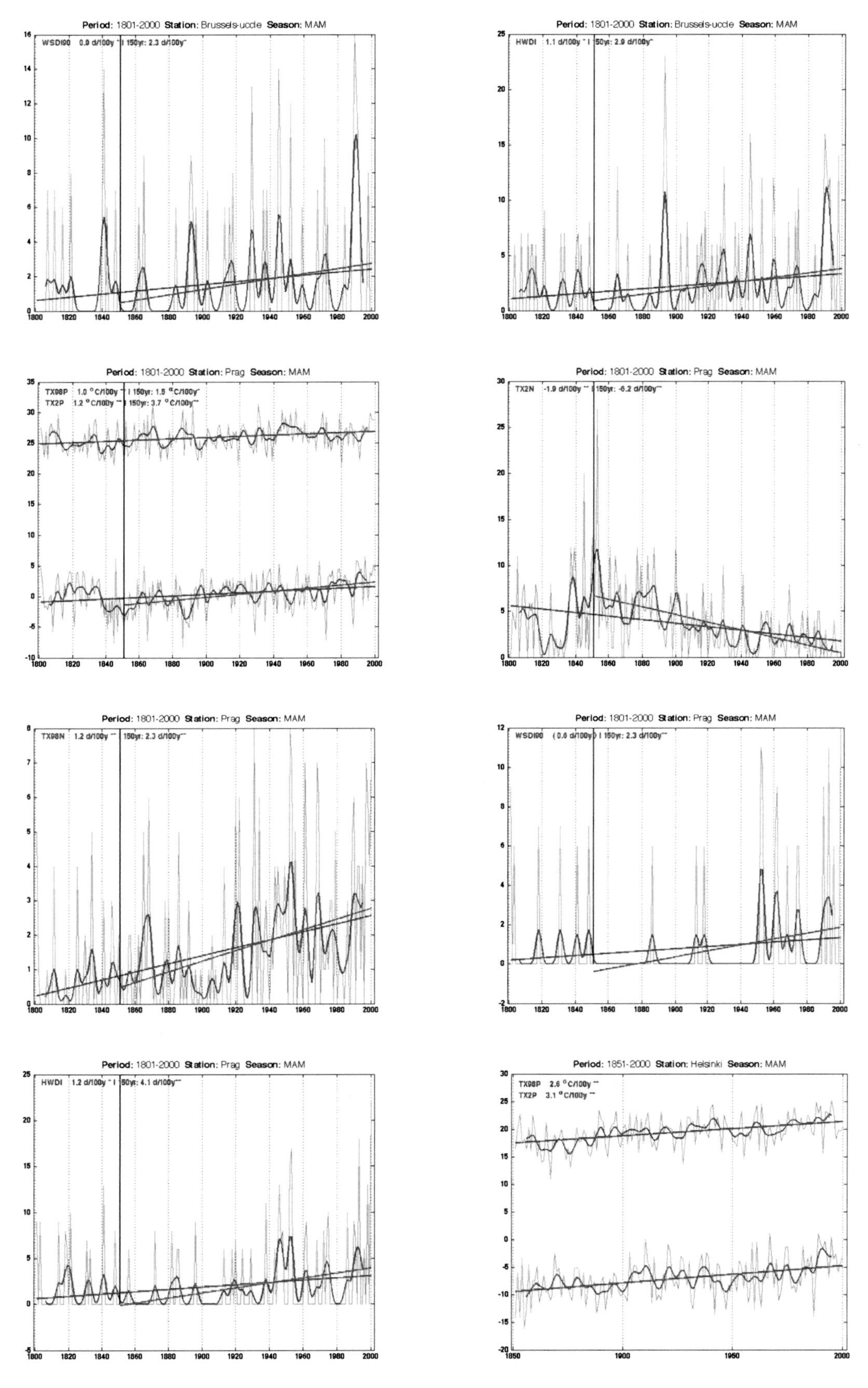

Fig. 3.23 1851–2000 MAM Tmax Brussels-uccle

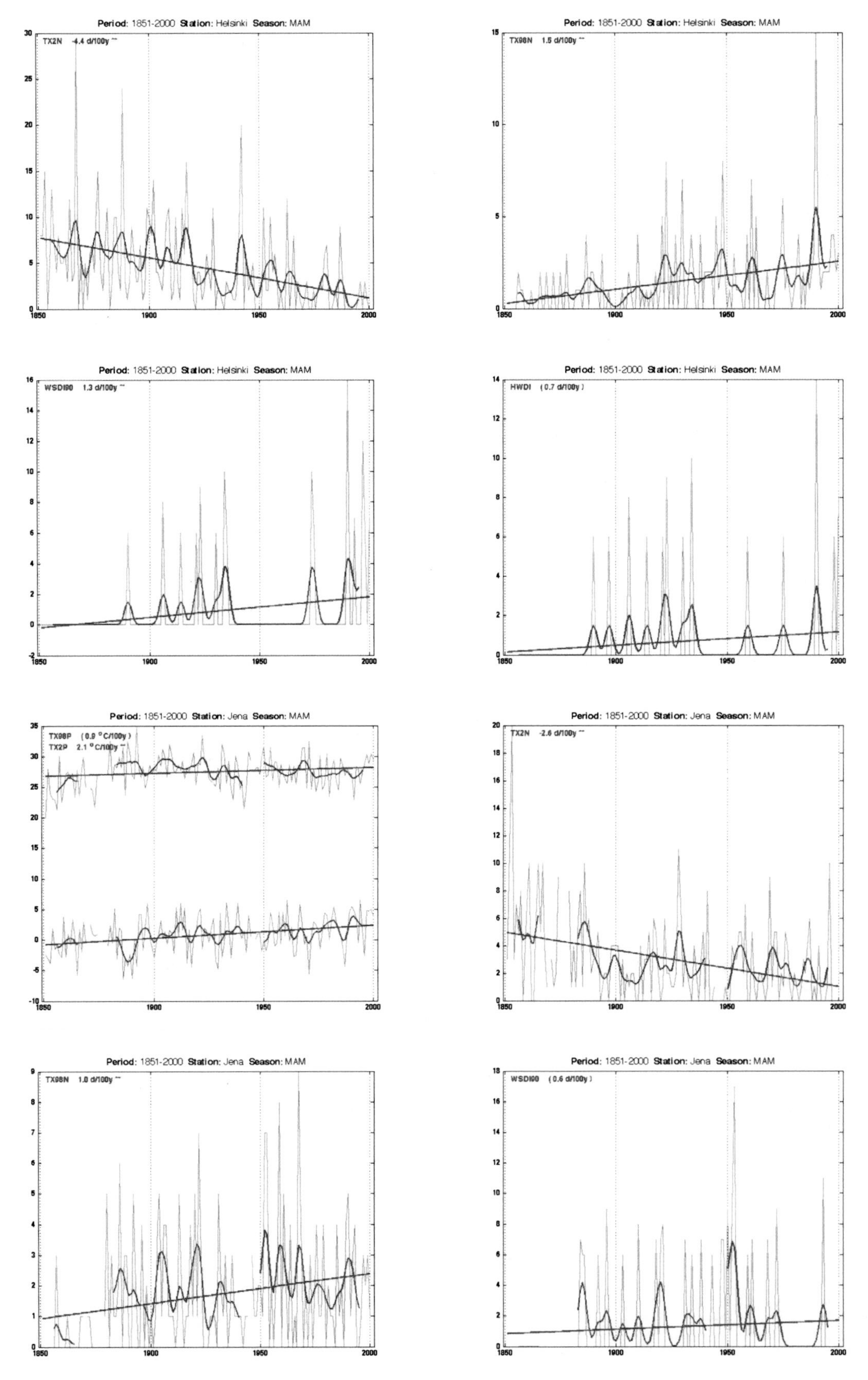

Fig. 3.24 1851–2000 MAM Tmax Helsinki

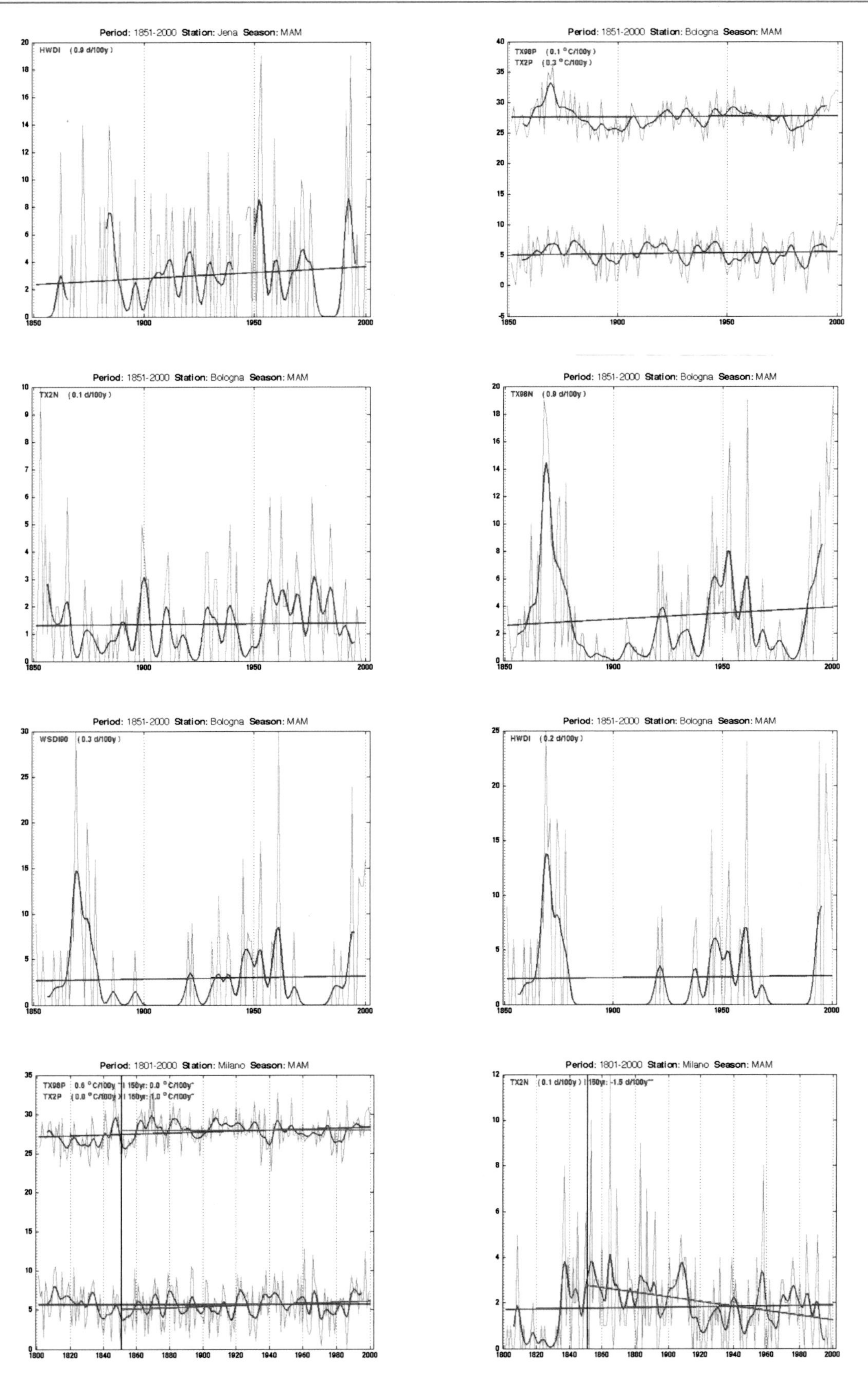

Fig. 3.25 1851–2000 MAM Tmax Jena

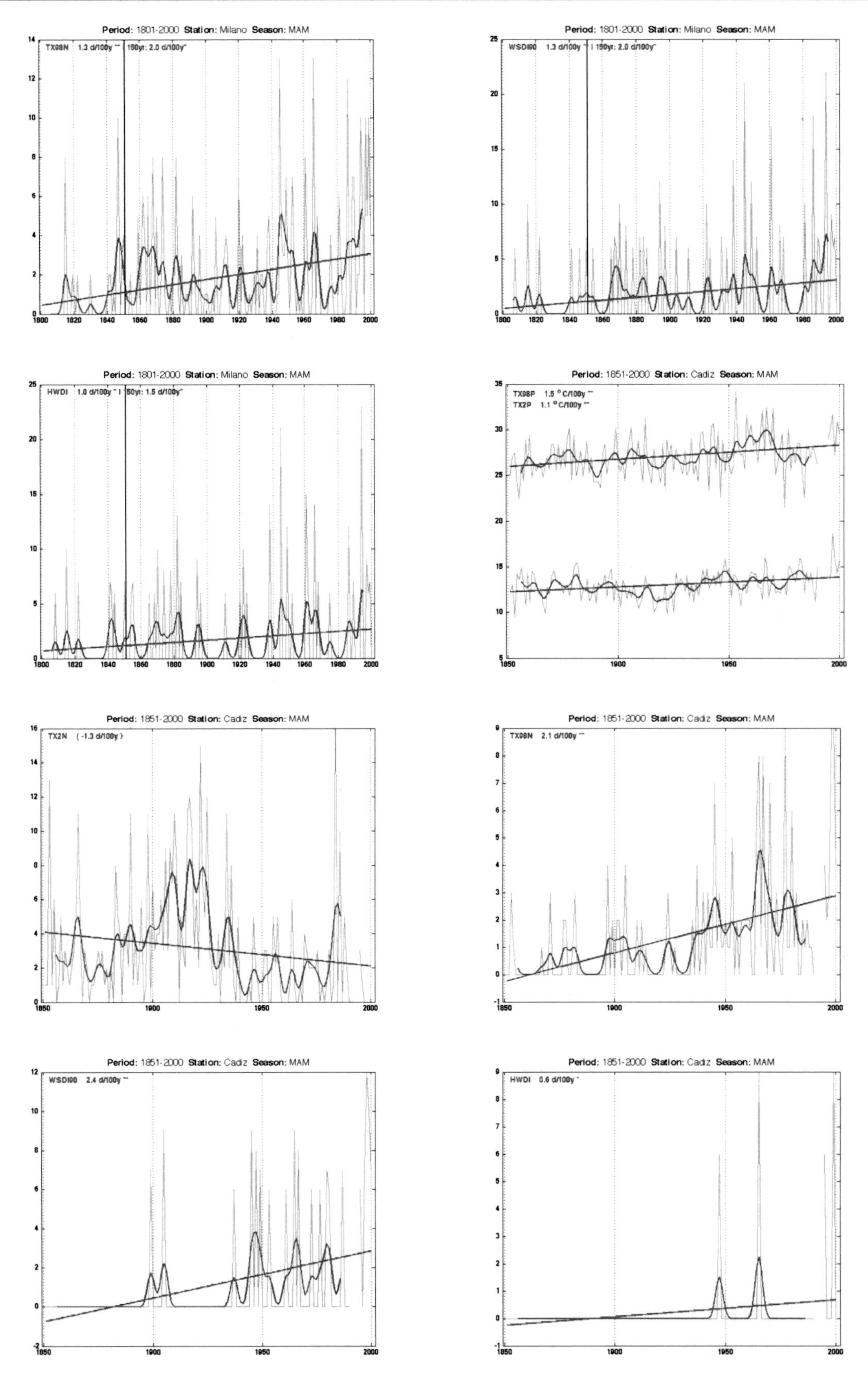

Fig. 3.26 1851–2000 MAM Tmax Milano

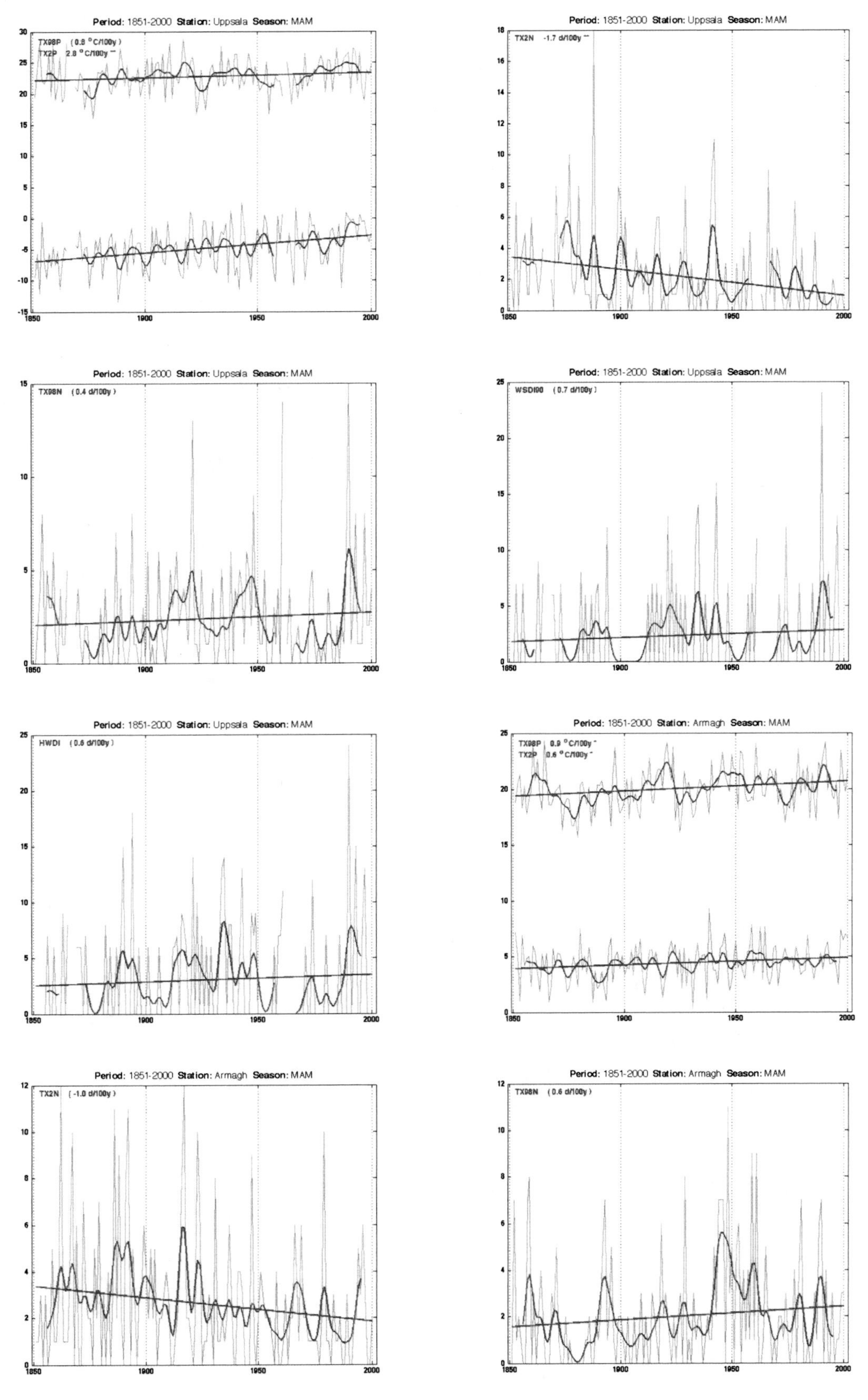

Fig. 3.27 1851–2000 MAM Tmax Uppsala

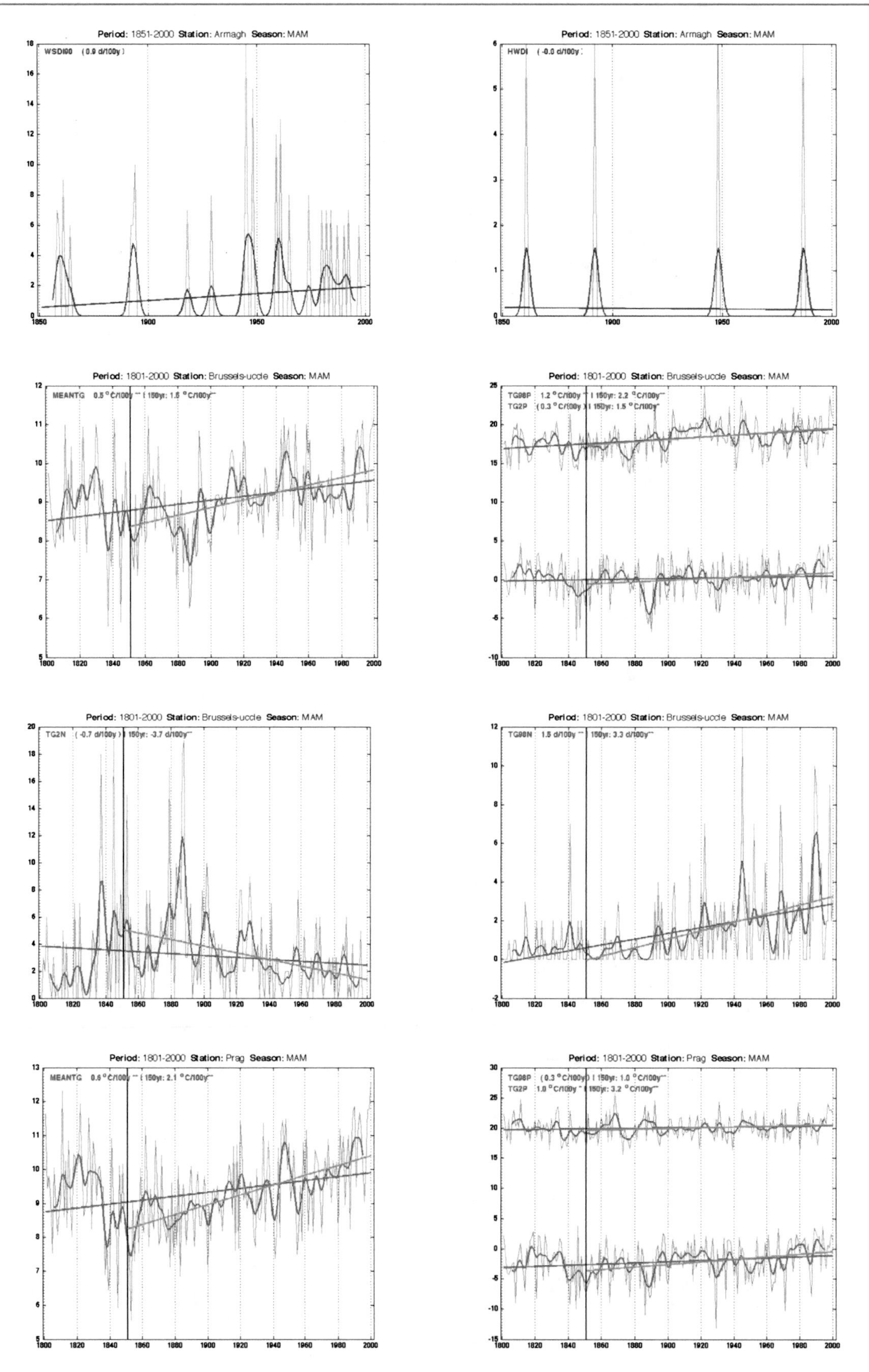

Fig. 3.28 1851–2000 MAM Tmax Armagh

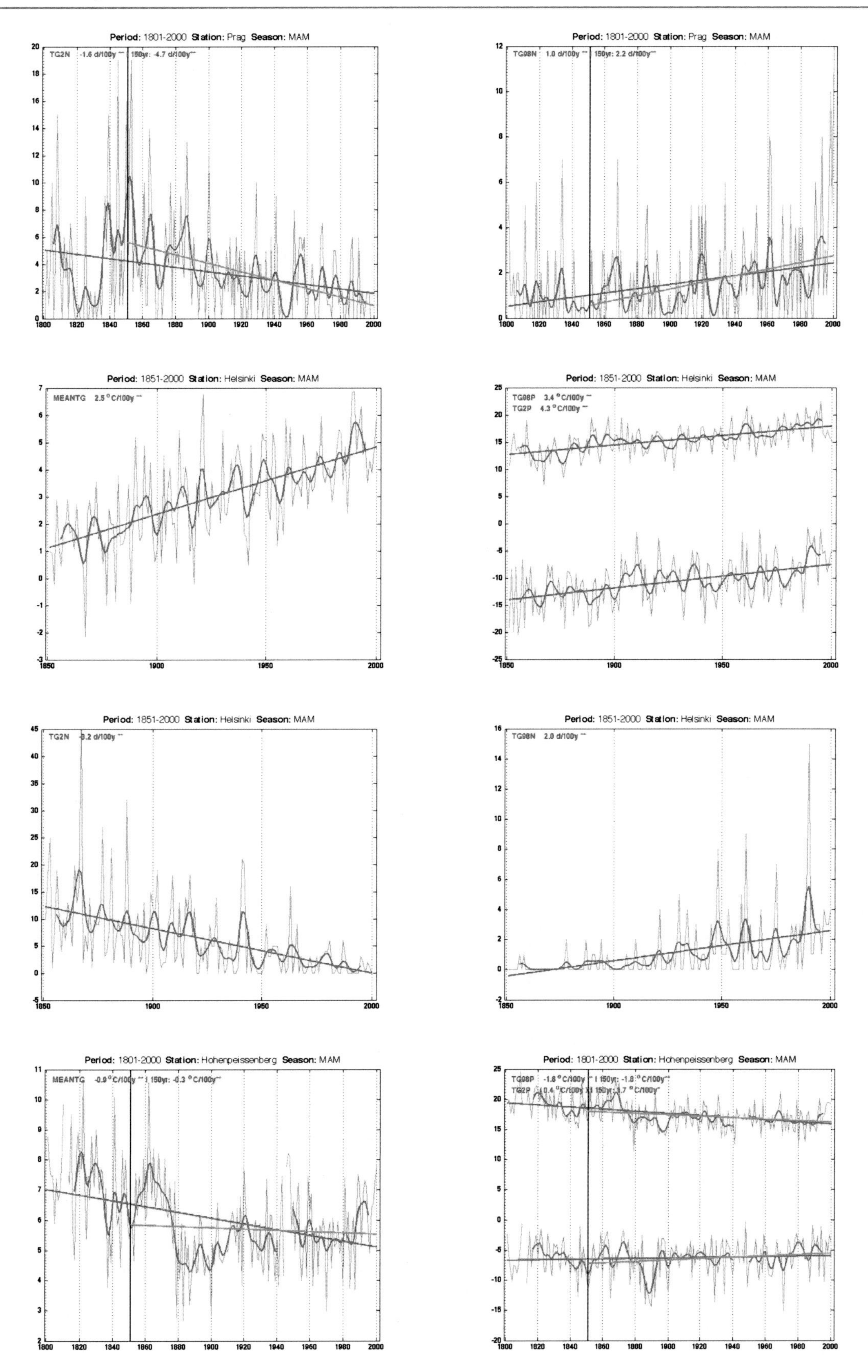

Fig. 3.29 1851–2000 MAM Tmean Prag

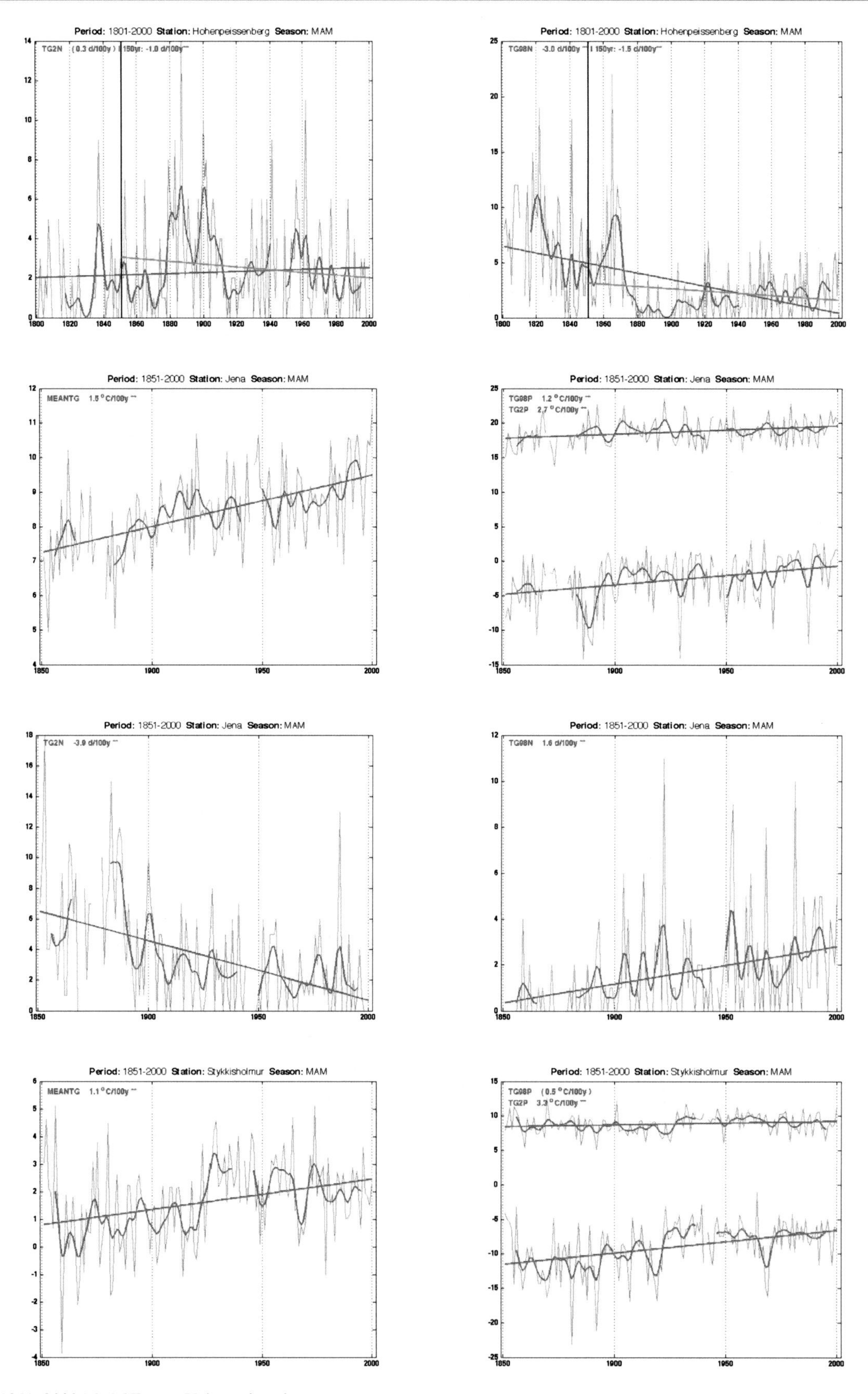

Fig. 3.30 1851–2000 MAM Tmean Hohenpeissenberg

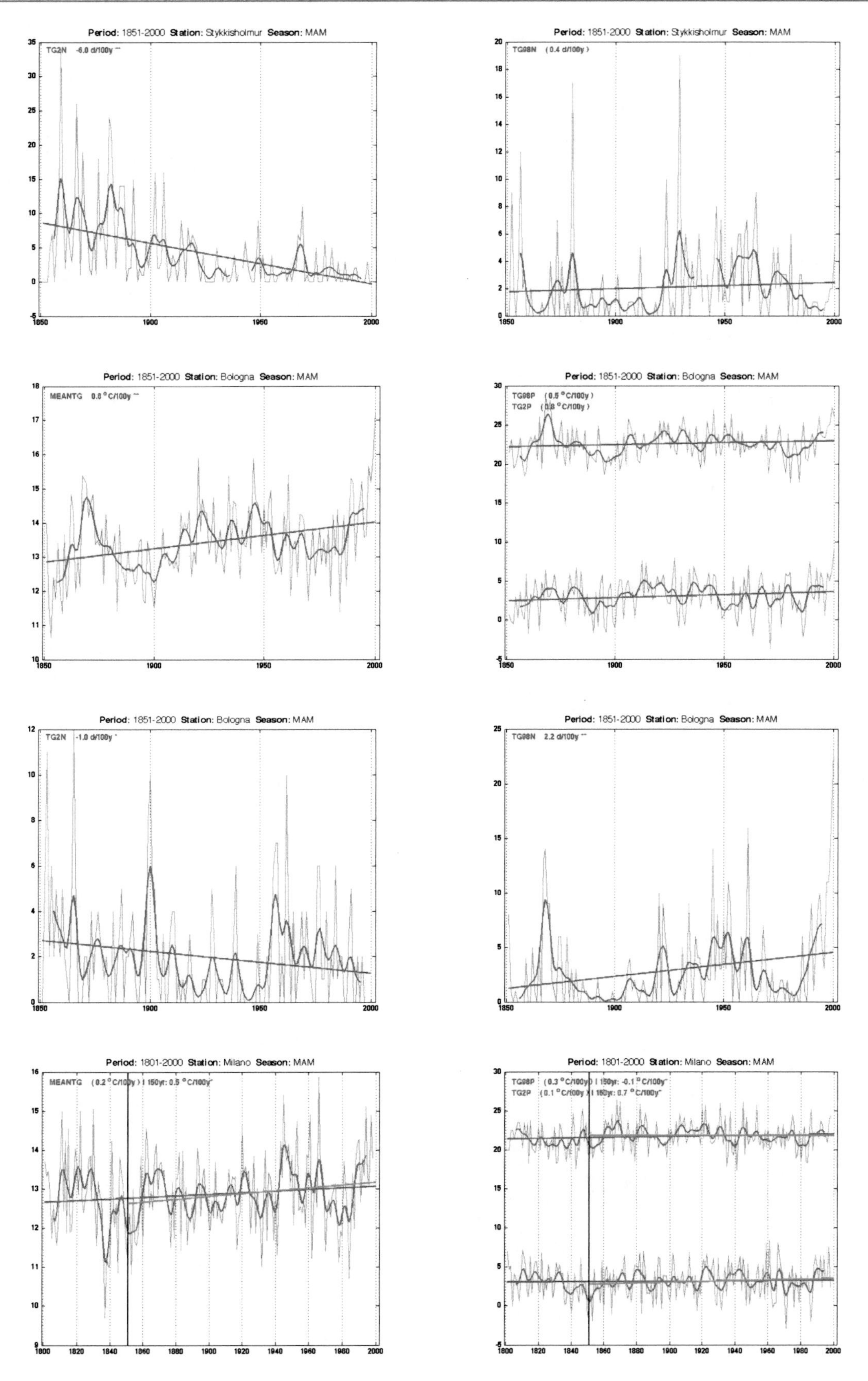

Fig. 3.31 1851–2000 MAM Tmean Stykkisholmur

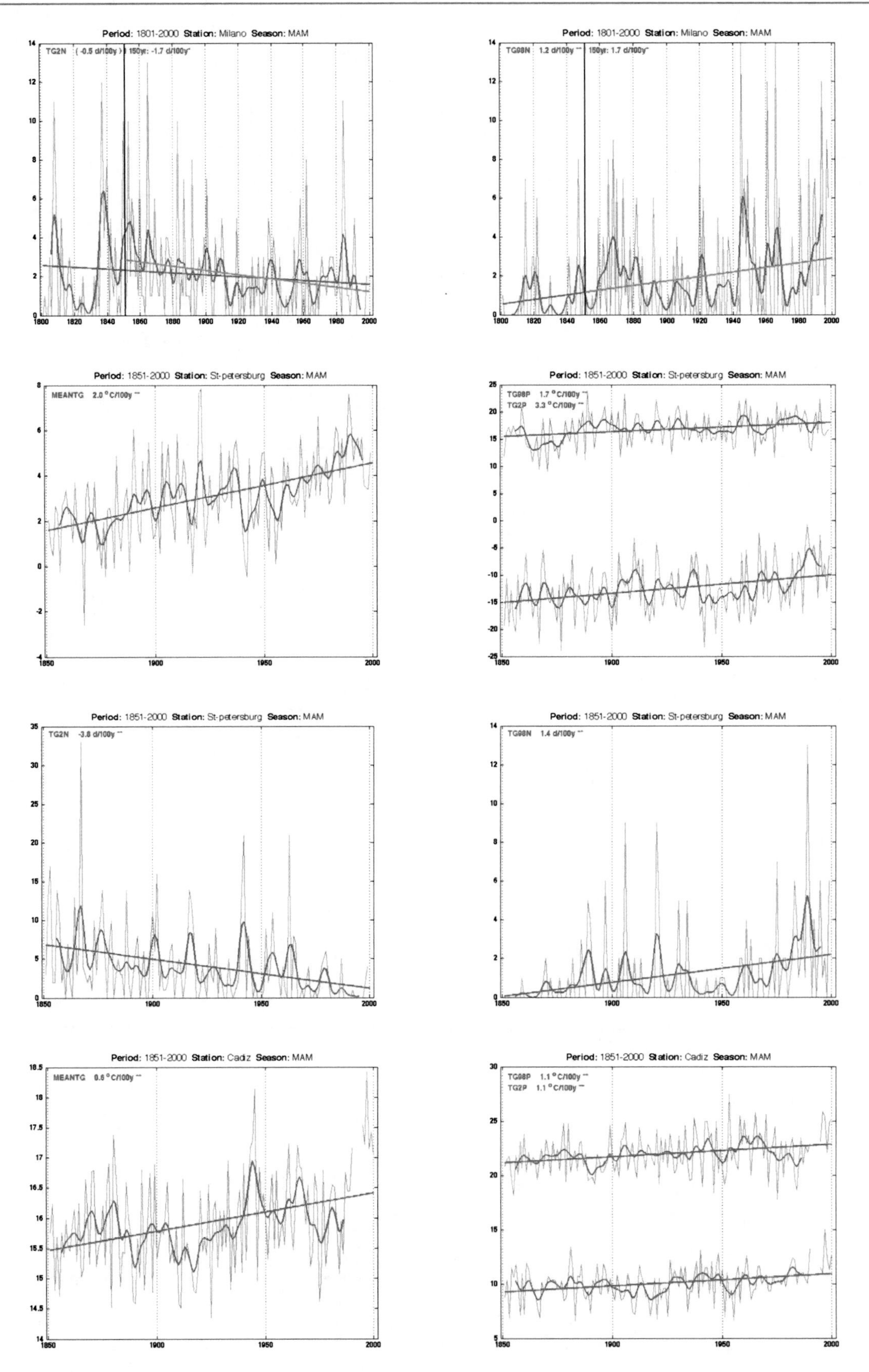

Fig. 3.32 1851–2000 MAM Tmean Milano

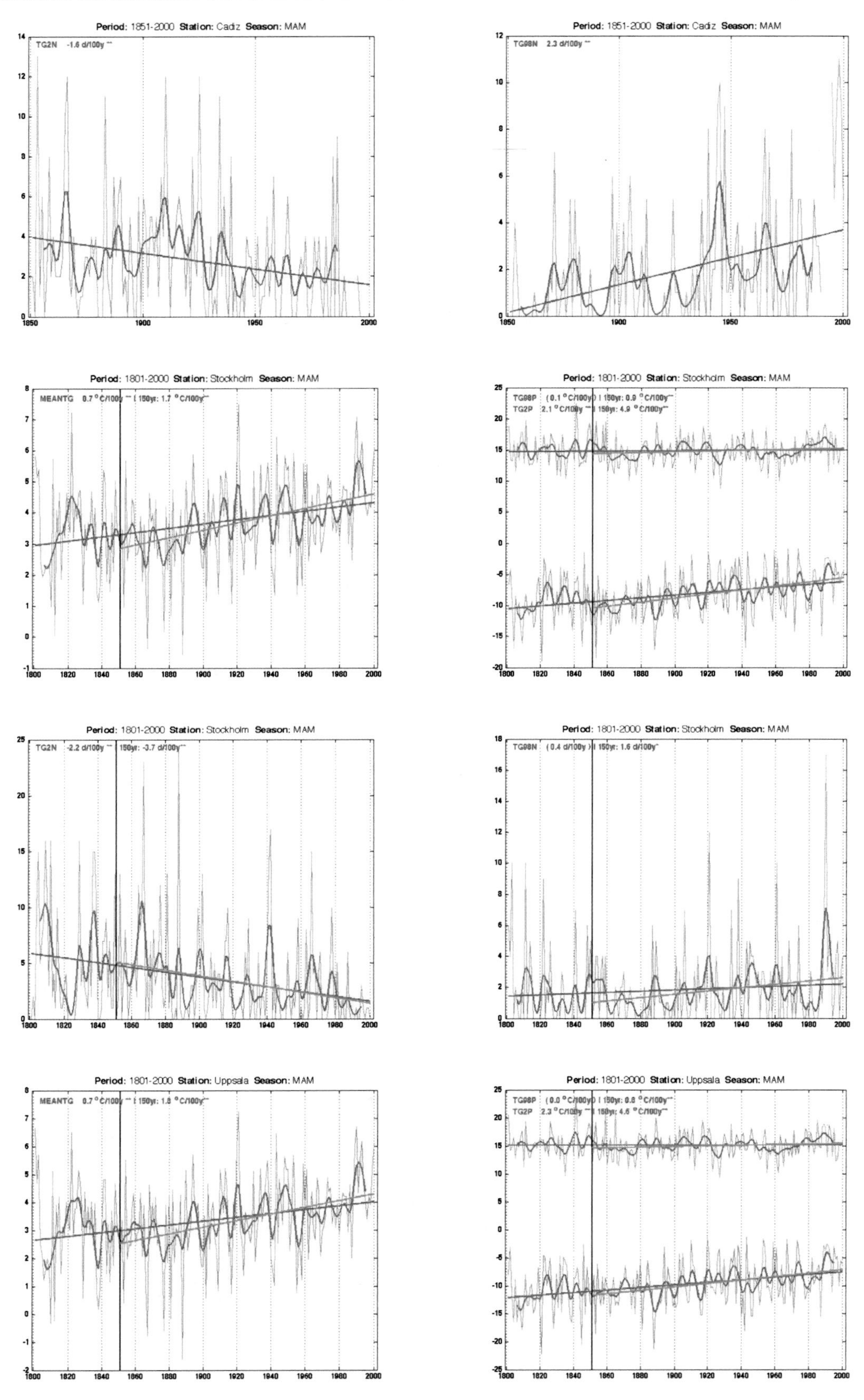

Fig. 3.33 1851–2000 MAM Tmean Cadiz

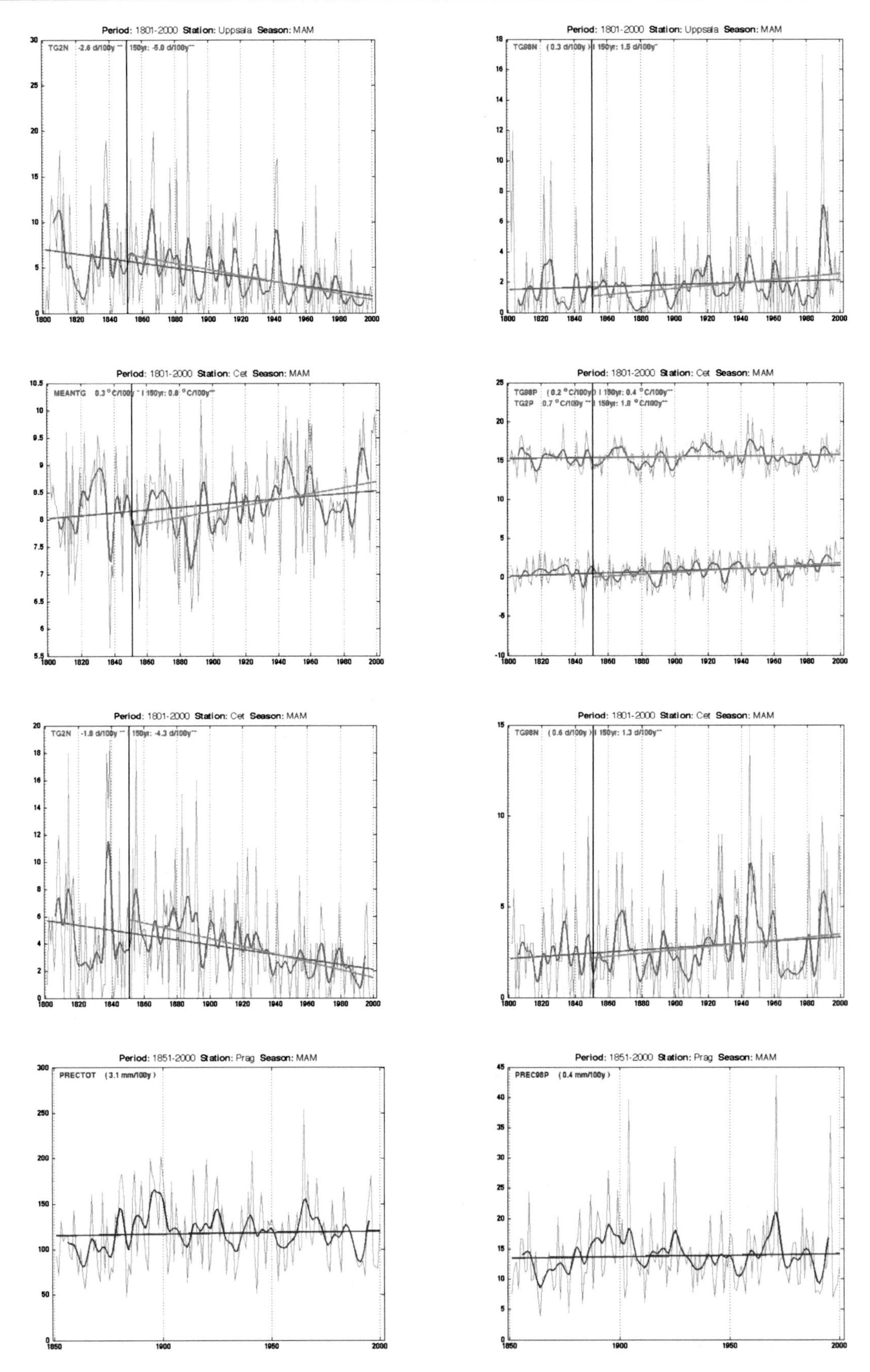

Fig. 3.34 1851–2000 MAM Tmean Uppsala

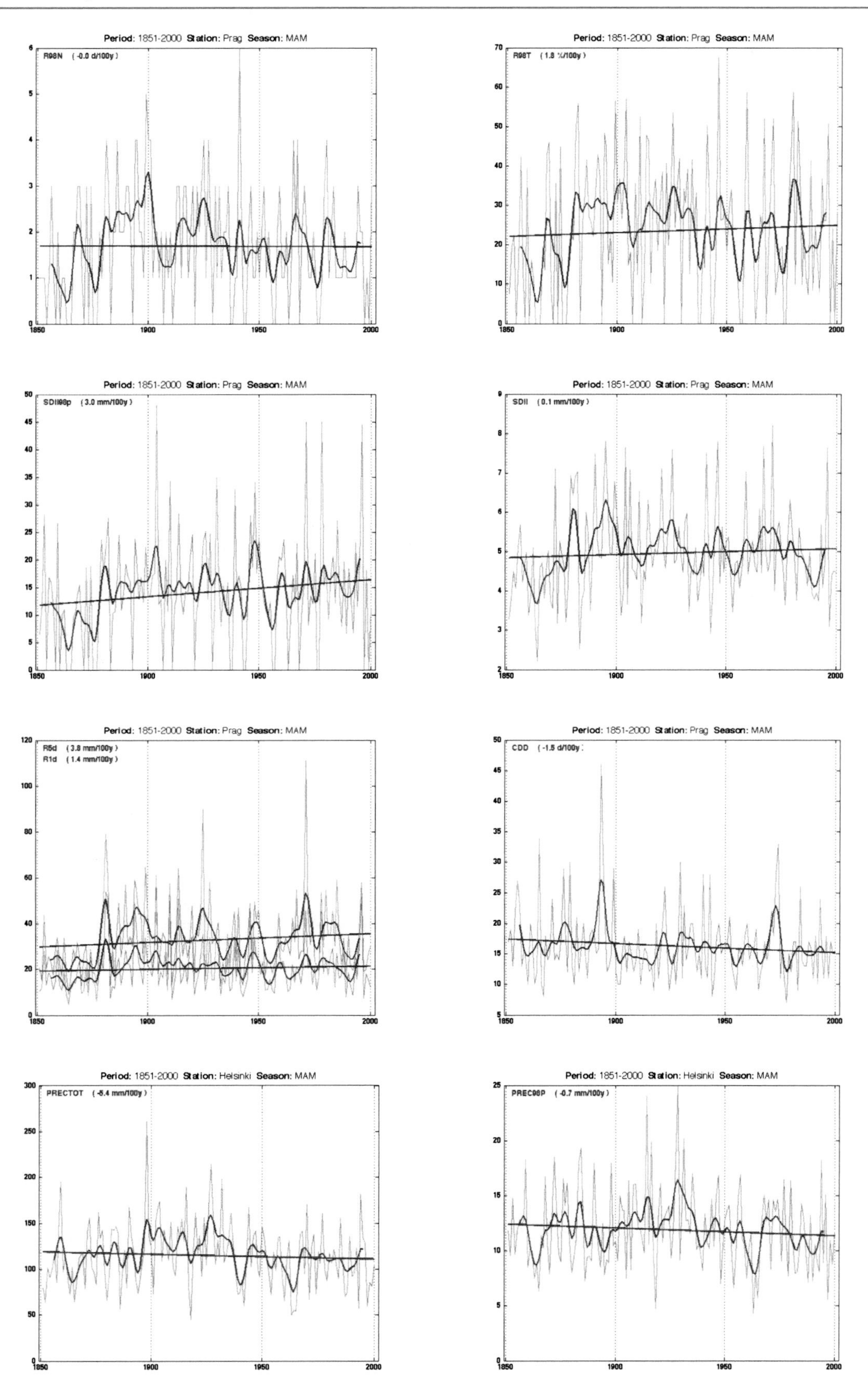

Fig. 3.35 1851–2000 MAM Prec Prag

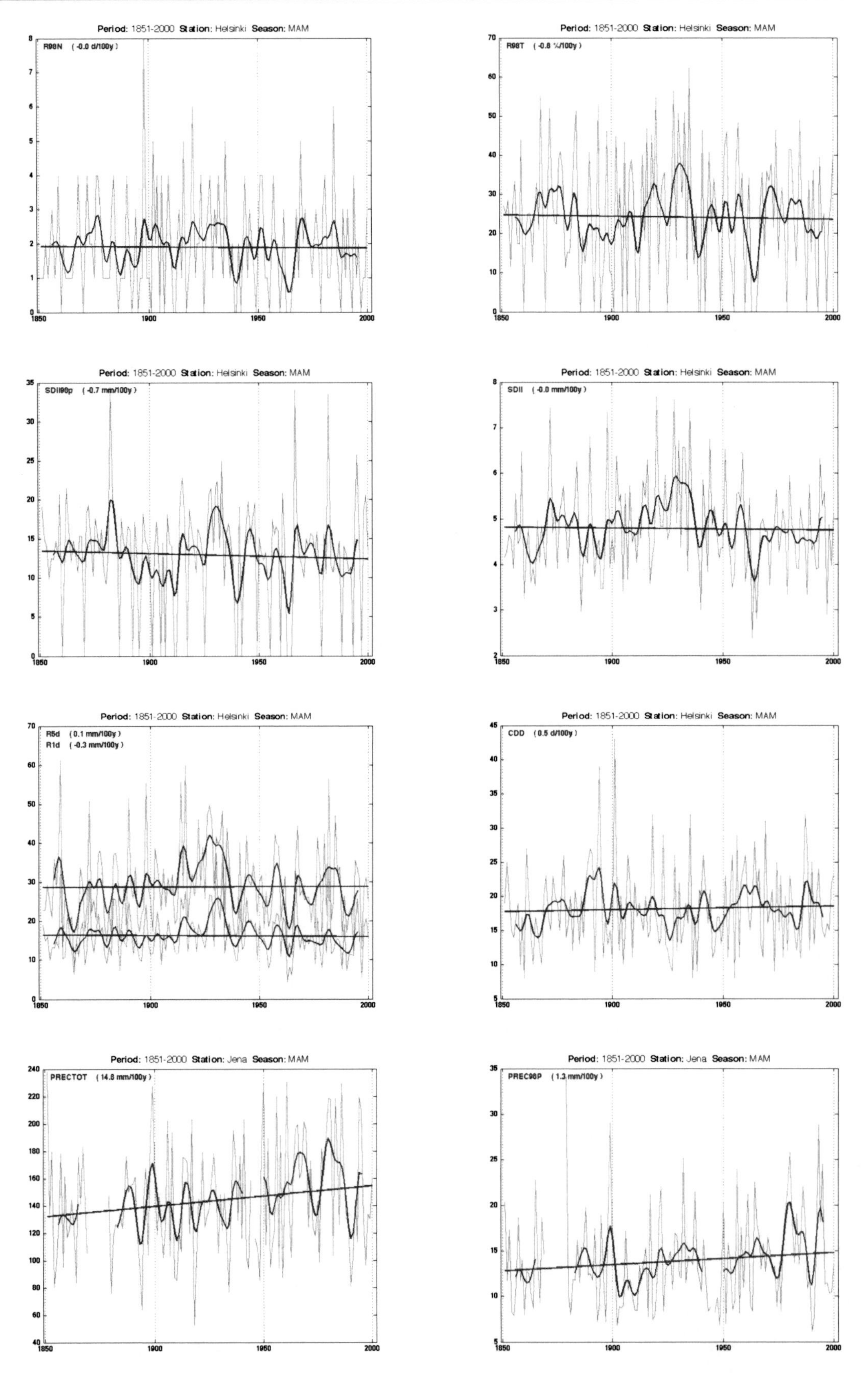

Fig. 3.36 1851–2000 MAM Prec Helsinki

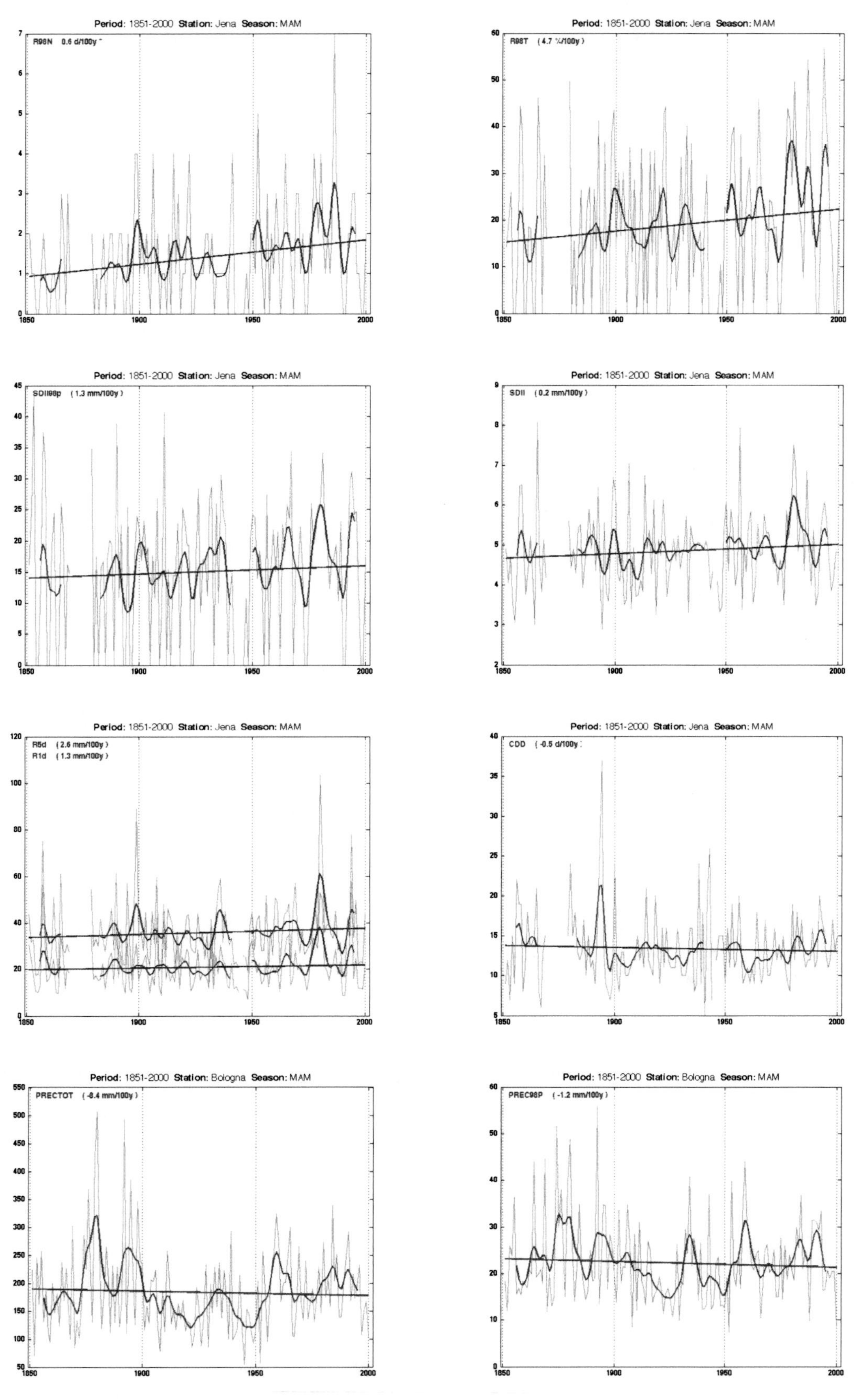

Fig. 3.37 1851–2000 MAM Prec Jena

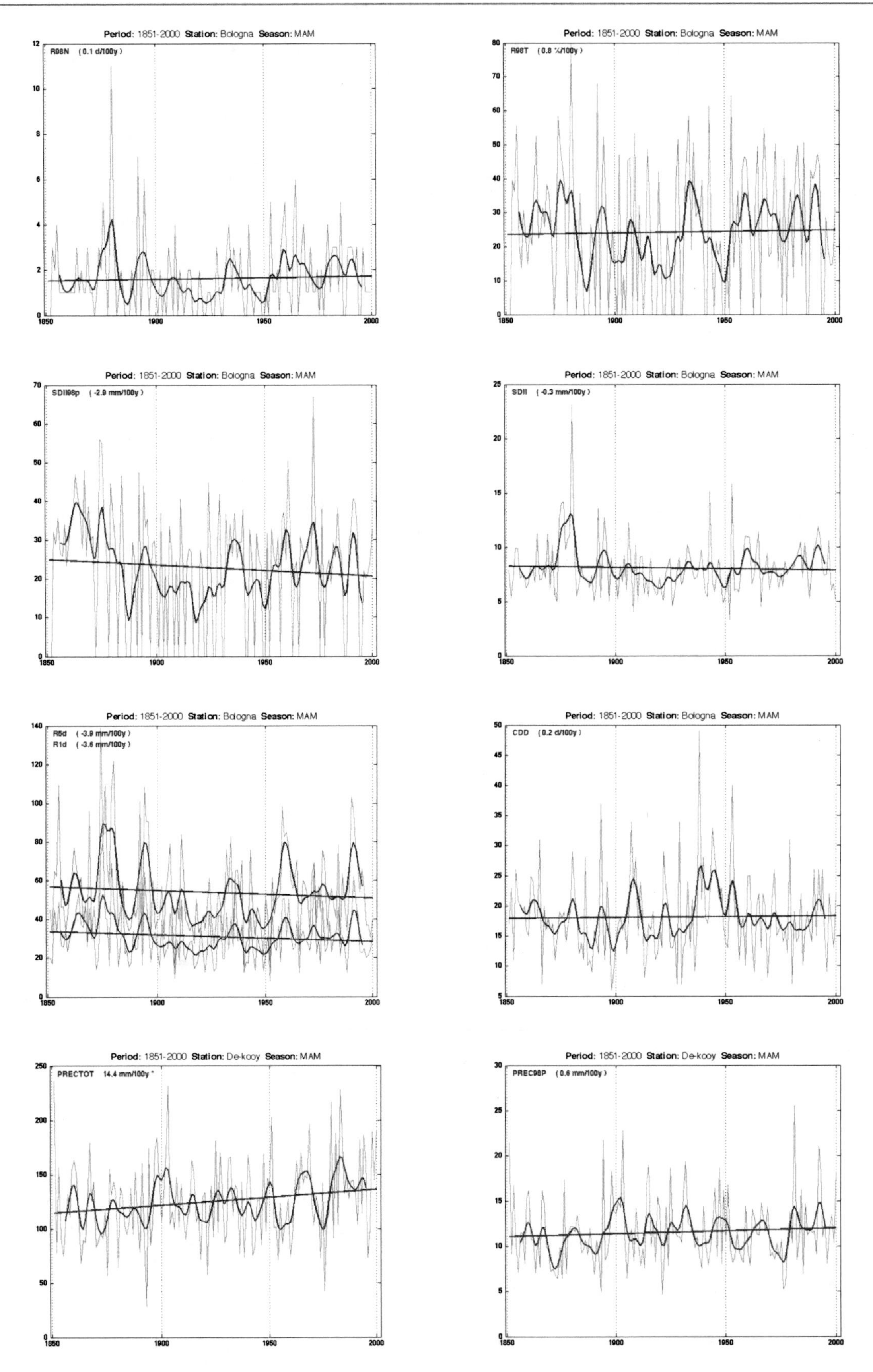

Fig. 3.38 1851–2000 MAM Prec Bologna

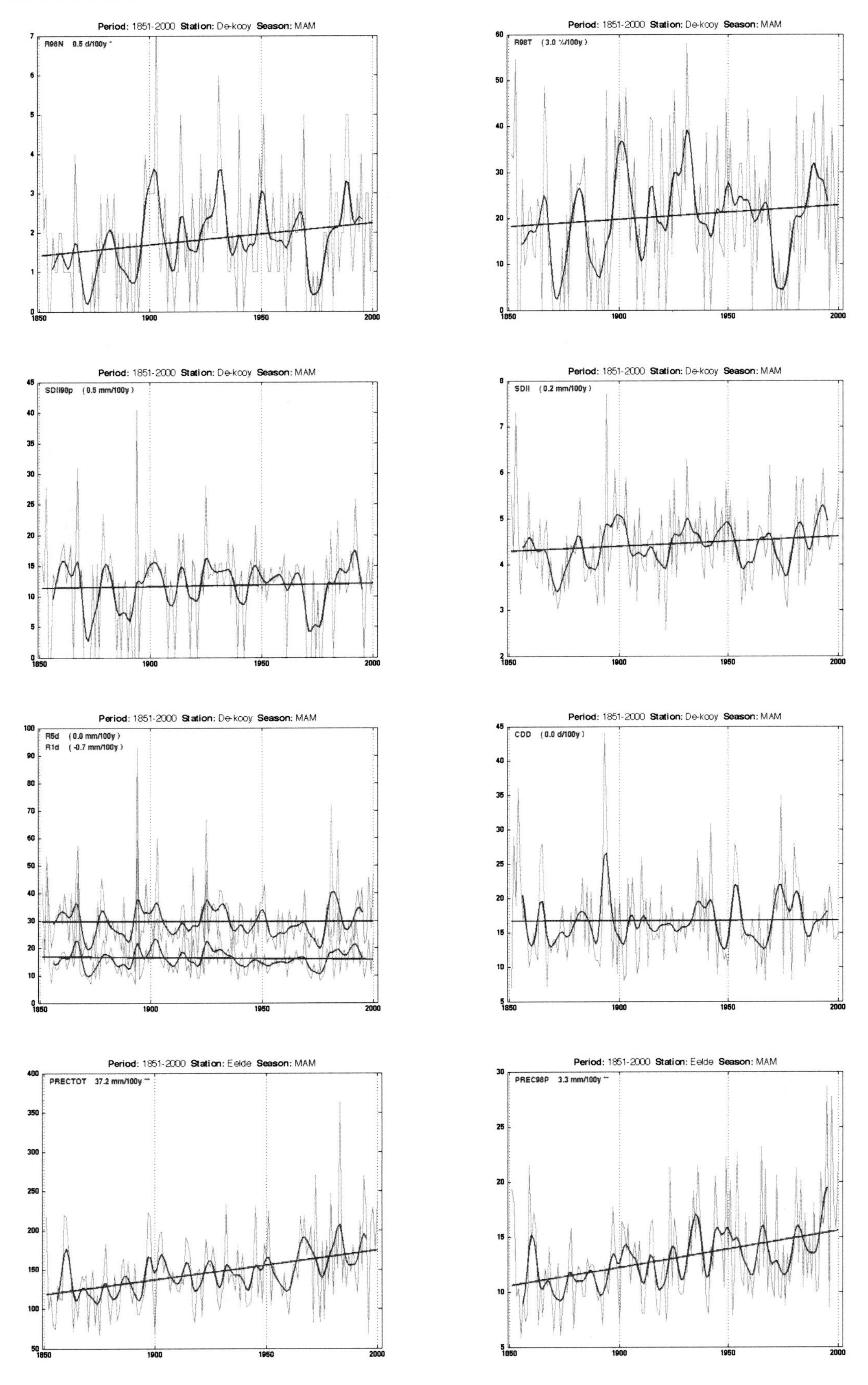

Fig. 3.39 1851–2000 MAM Prec De-kooy

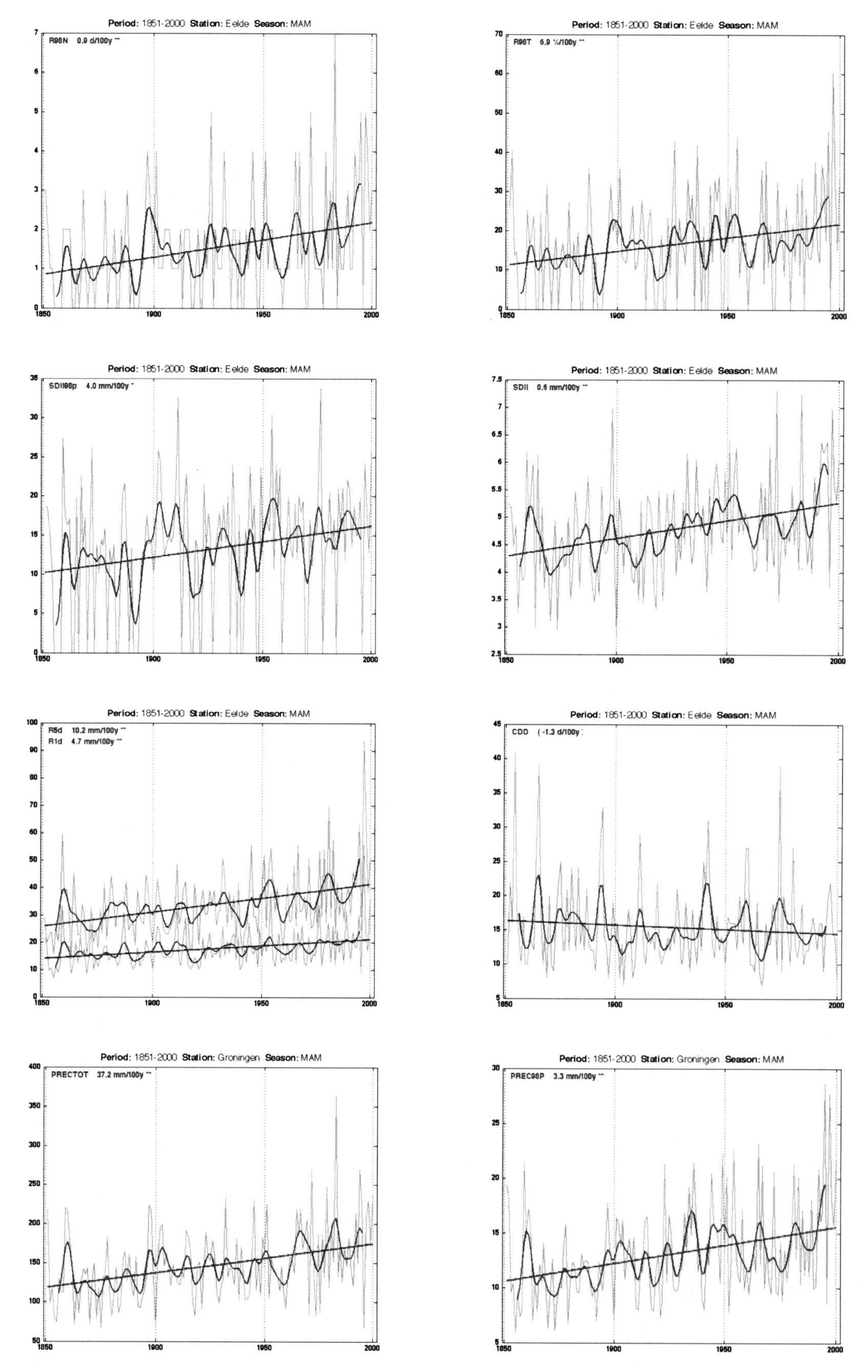

Fig. 3.40 1851–2000 MAM Prec Eelde

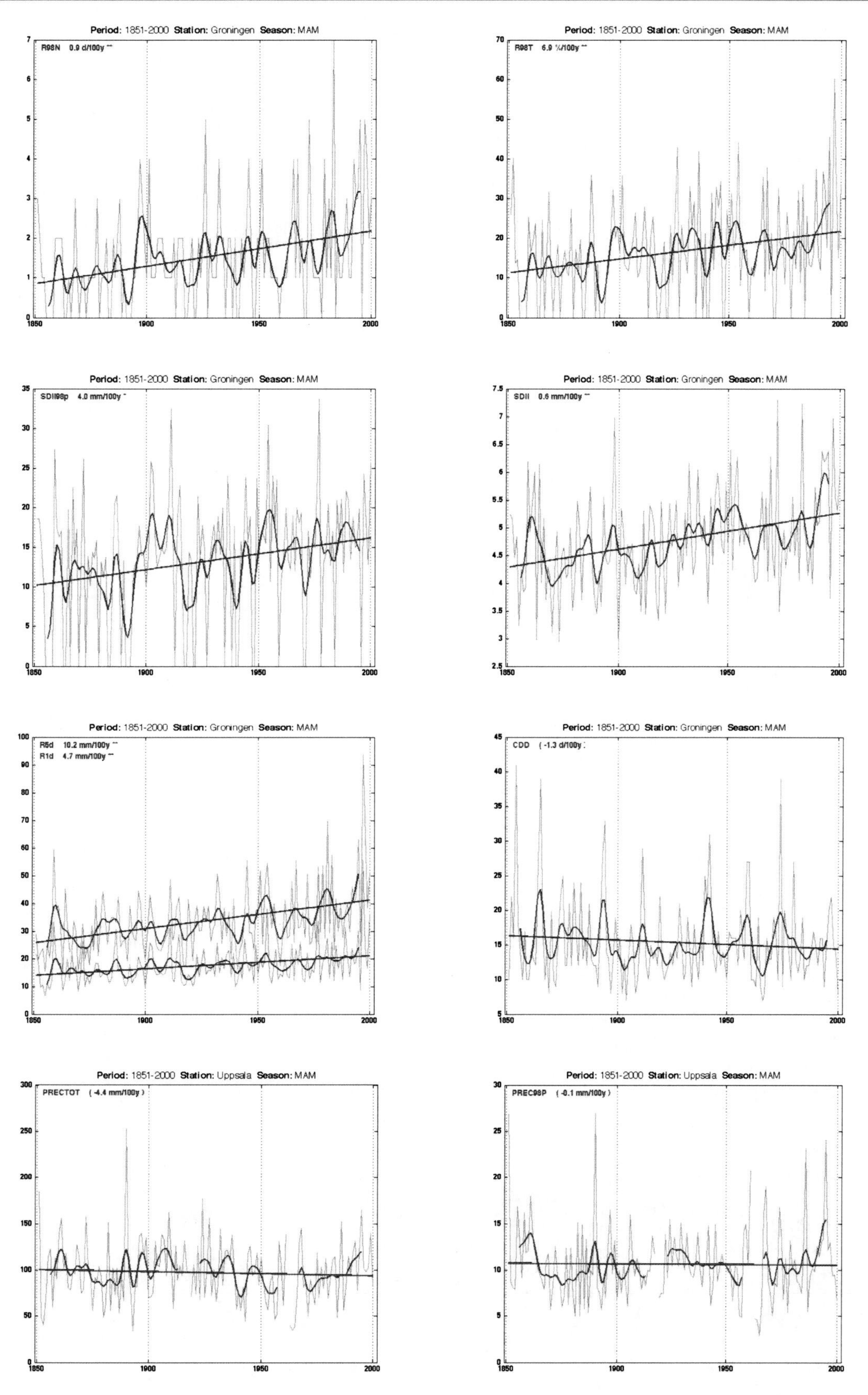

Fig. 3.41 1851–2000 MAM Prec Groningen

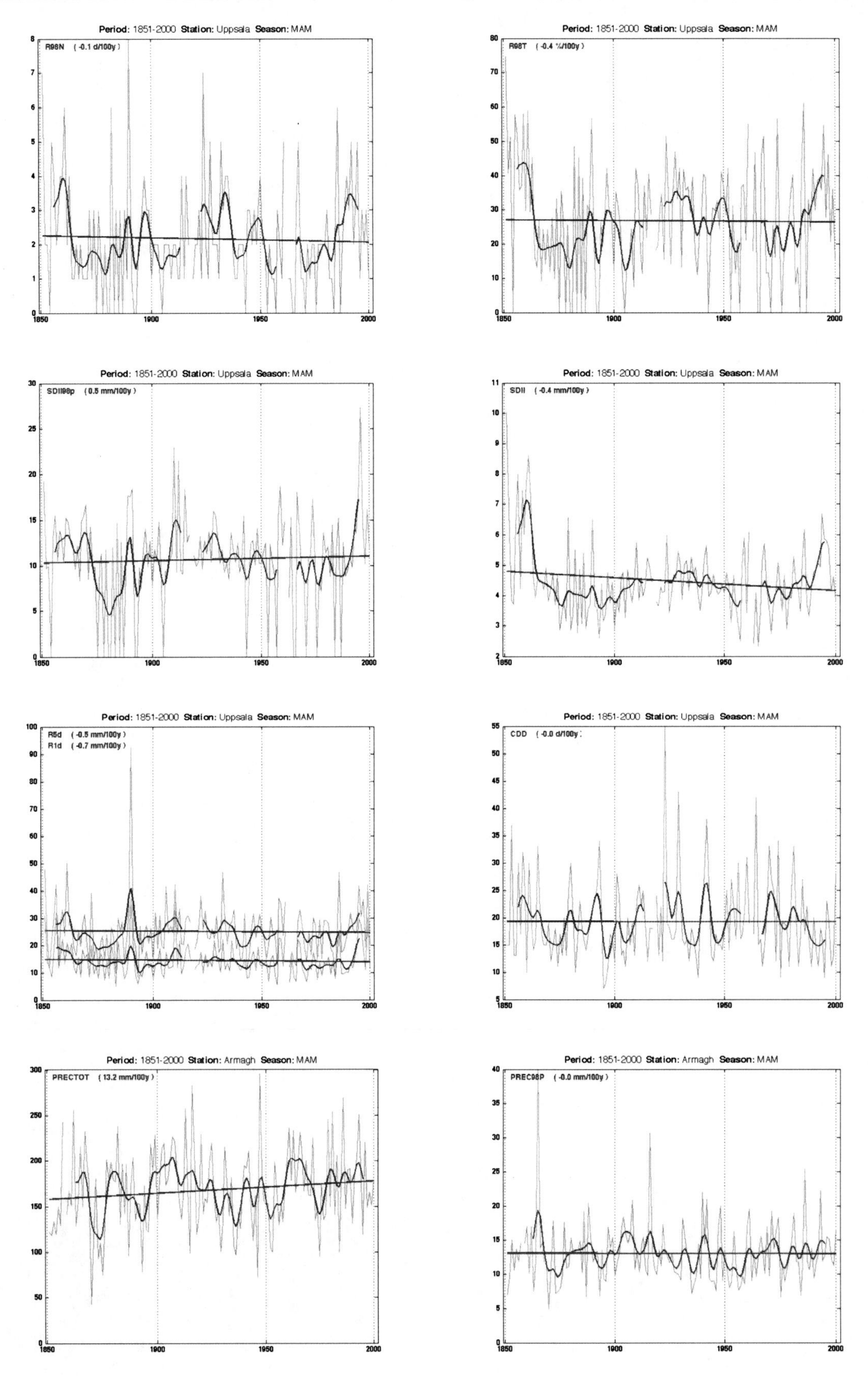

Fig. 3.42 1851–2000 MAM Prec Uppsala

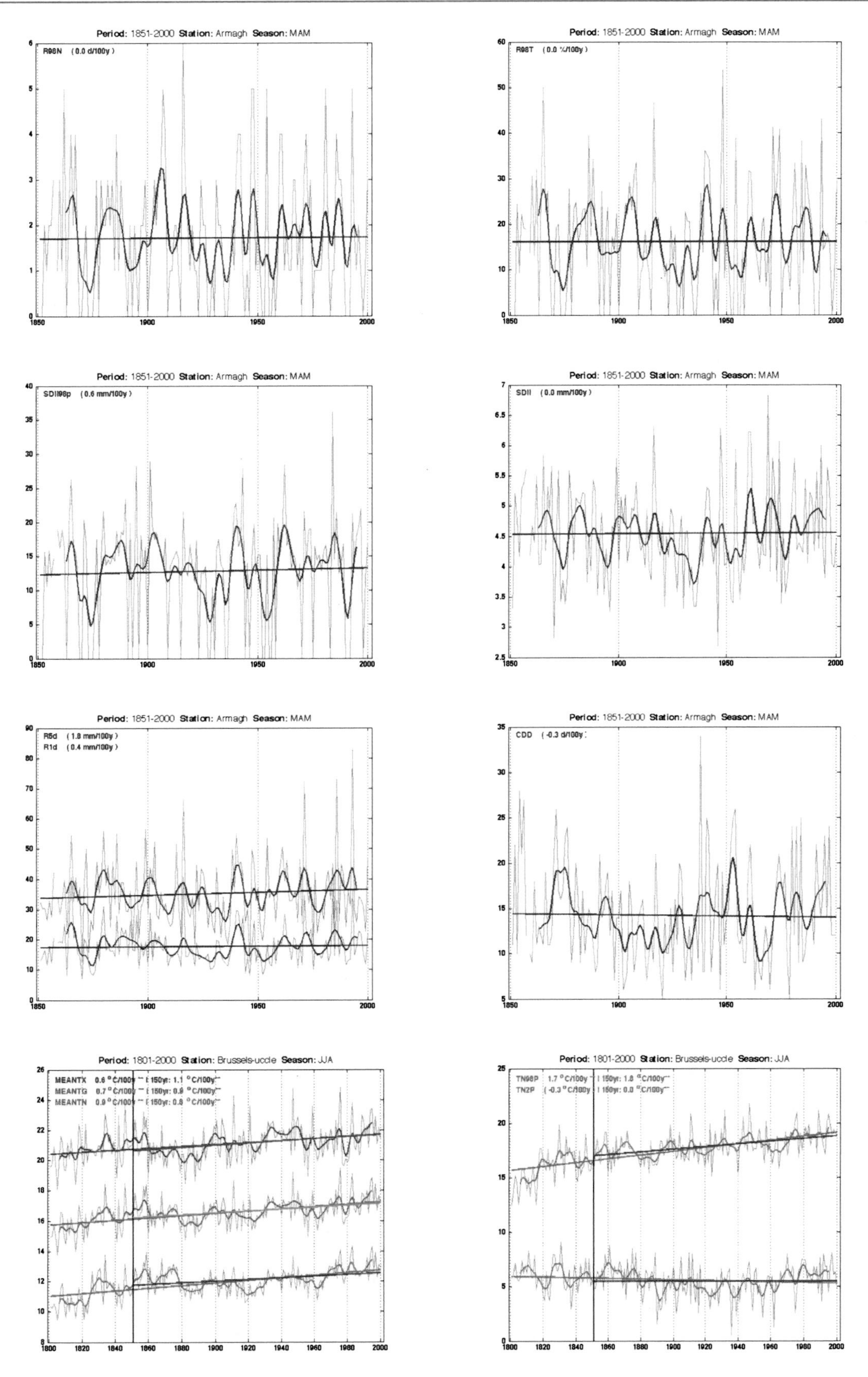

Fig. 3.43 1851–2000 MAM Prec Armagh

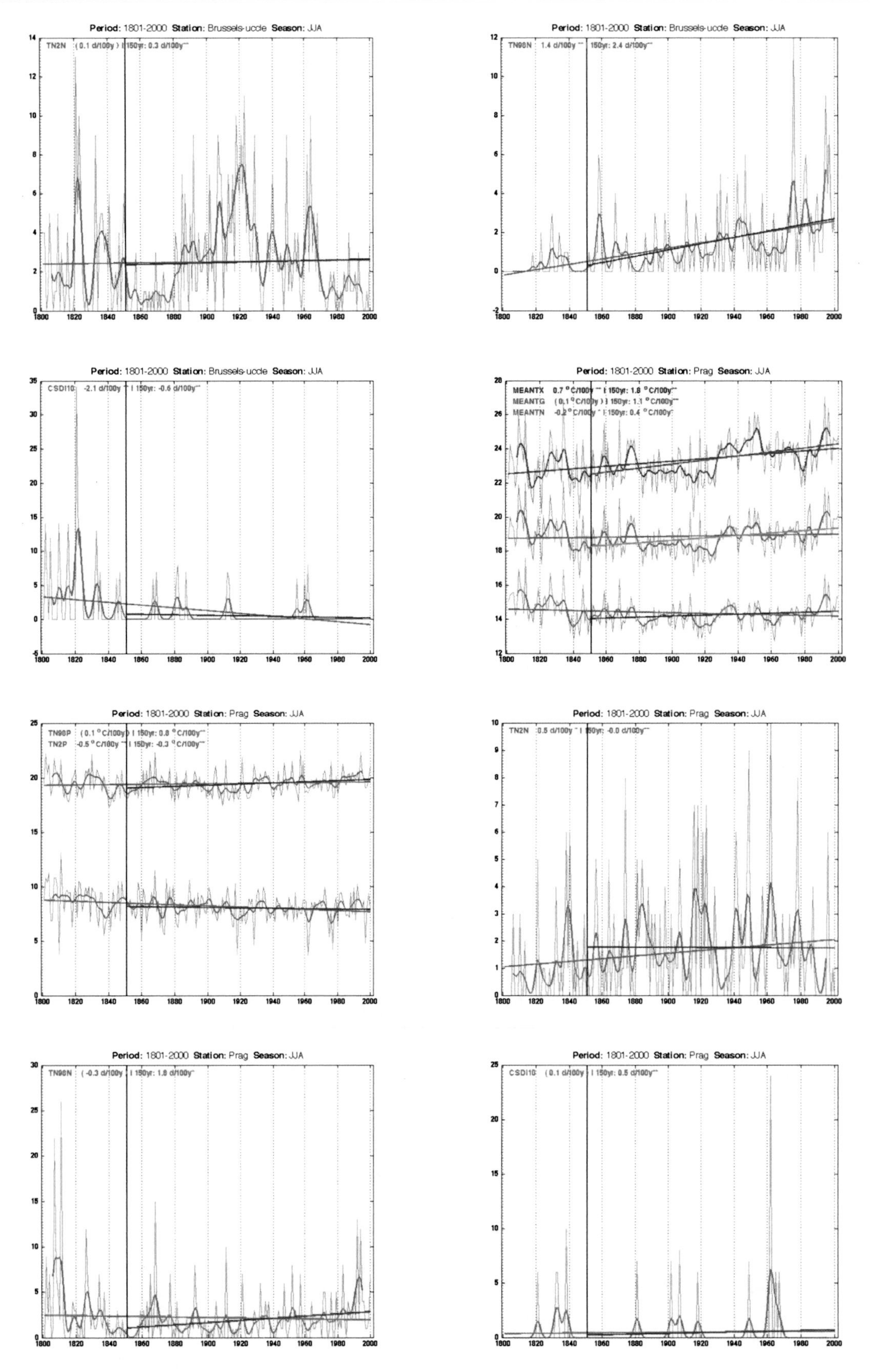

Fig. 3.44 1851–2000 JJA Tmin Brussels-uccle

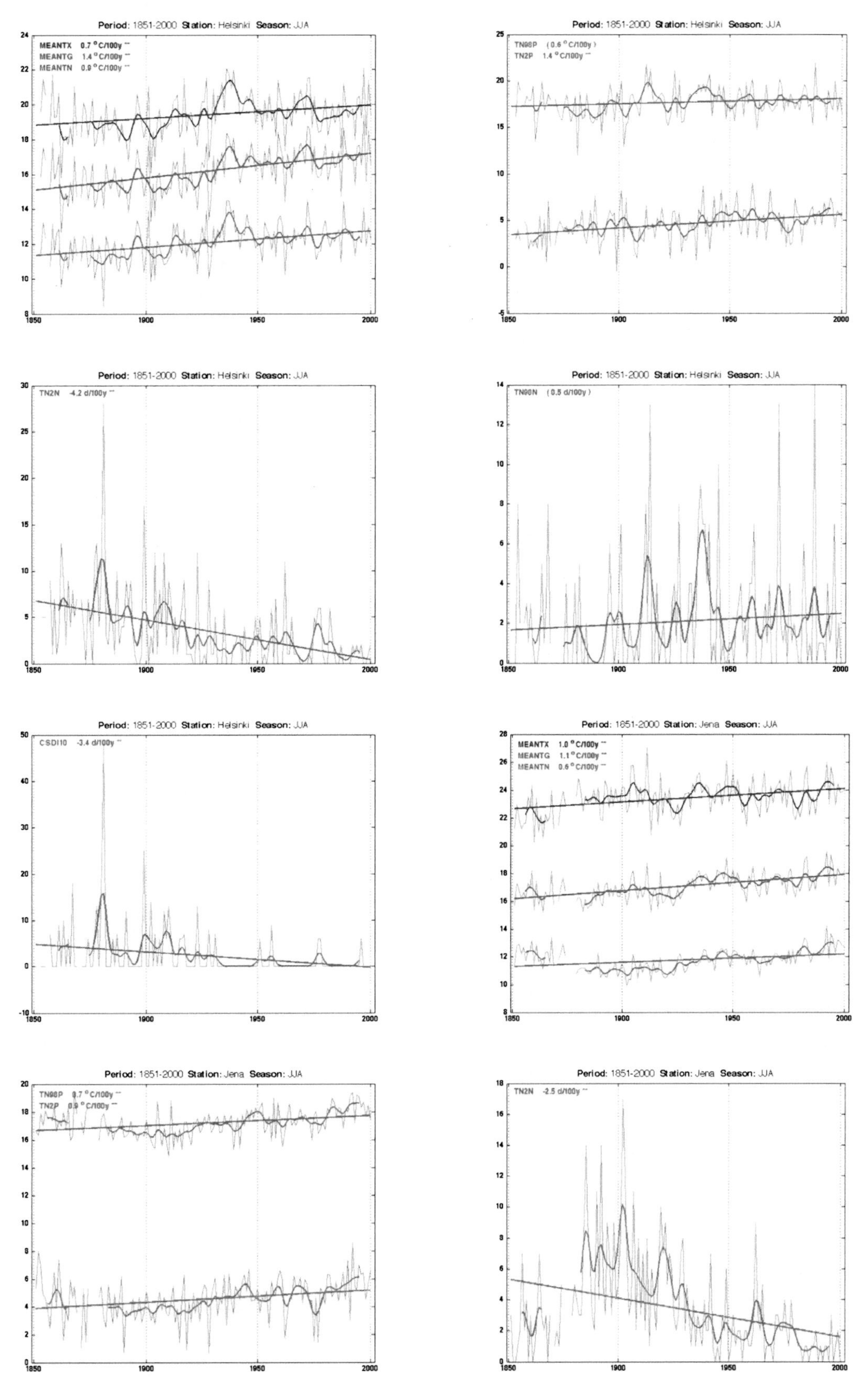

Fig. 3.45 1851–2000 JJA Tmin Helsinki

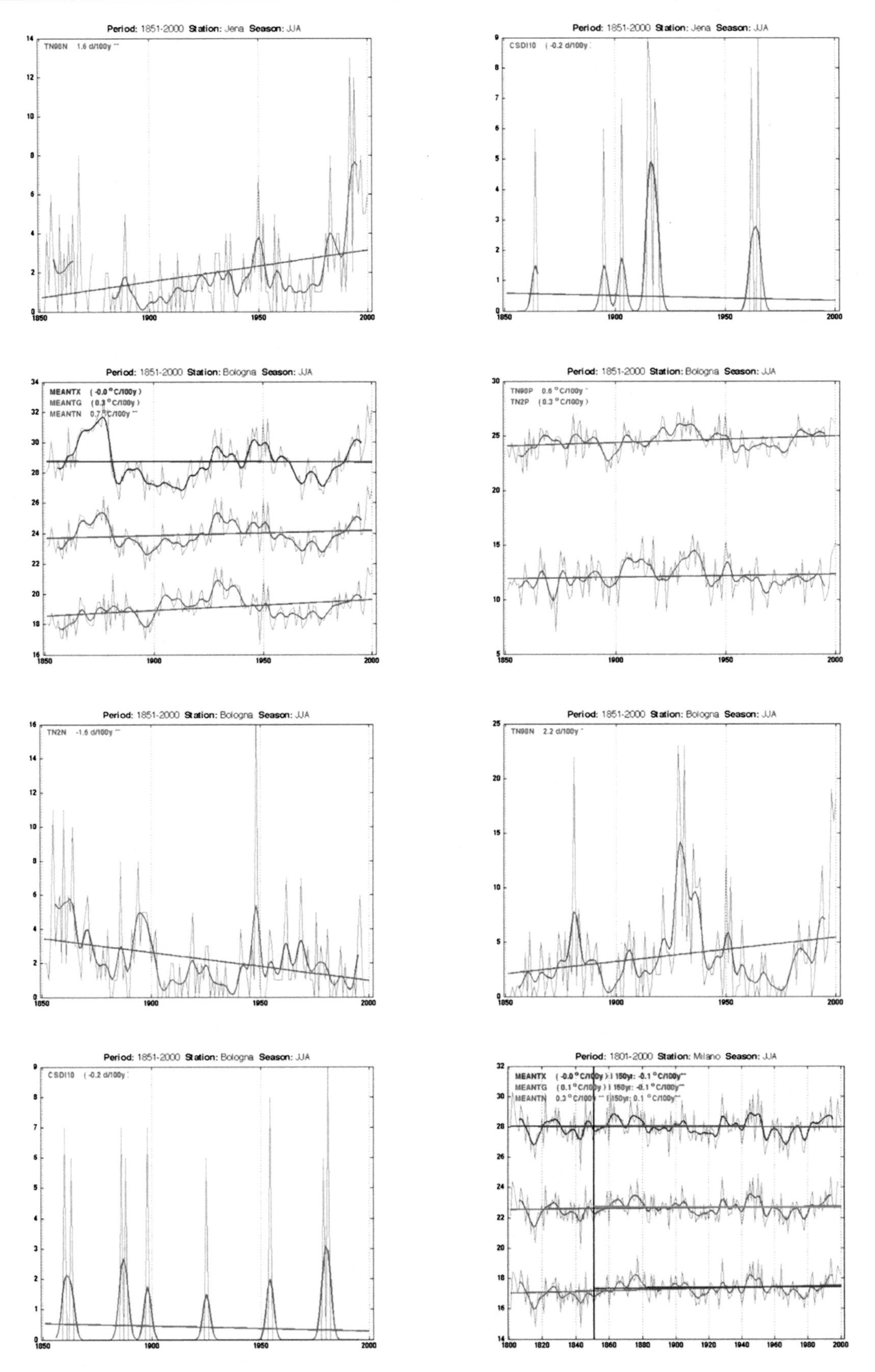

Fig. 3.46 1851–2000 JJA Tmin Jena

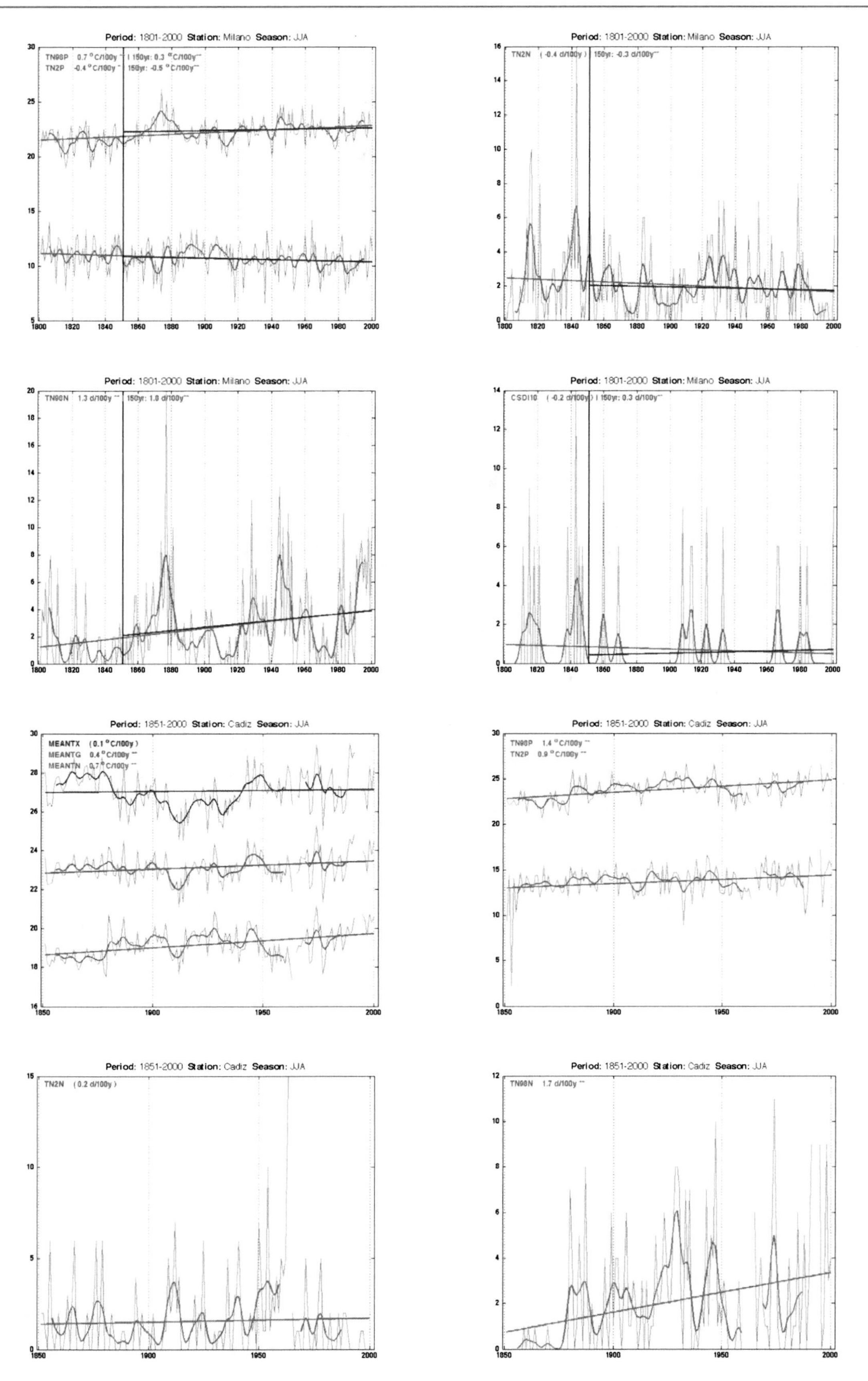

Fig. 3.47 1851–2000 JJA Tmin Milano

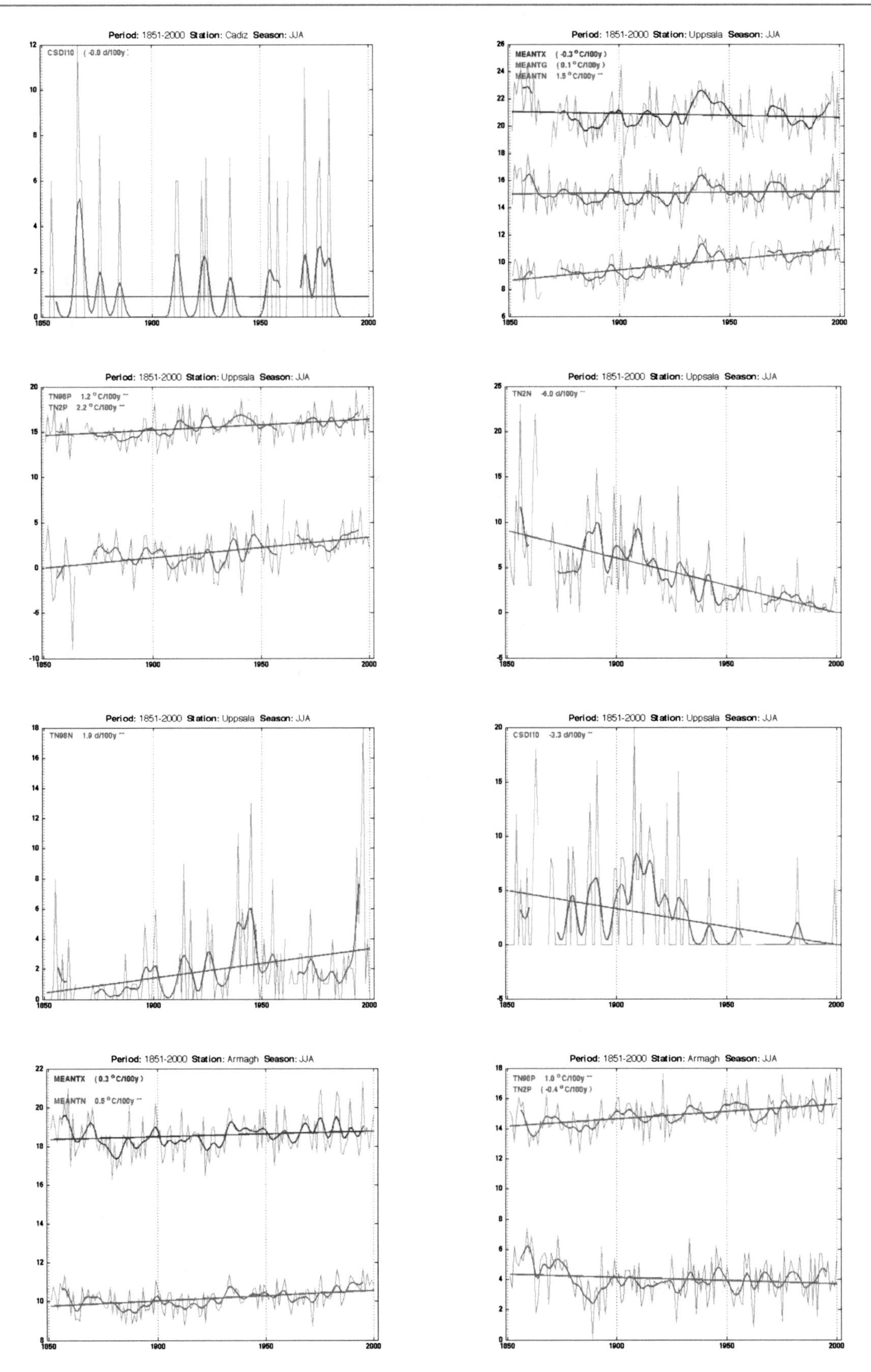

Fig. 3.48 1851–2000 JJA Tmin Cadiz

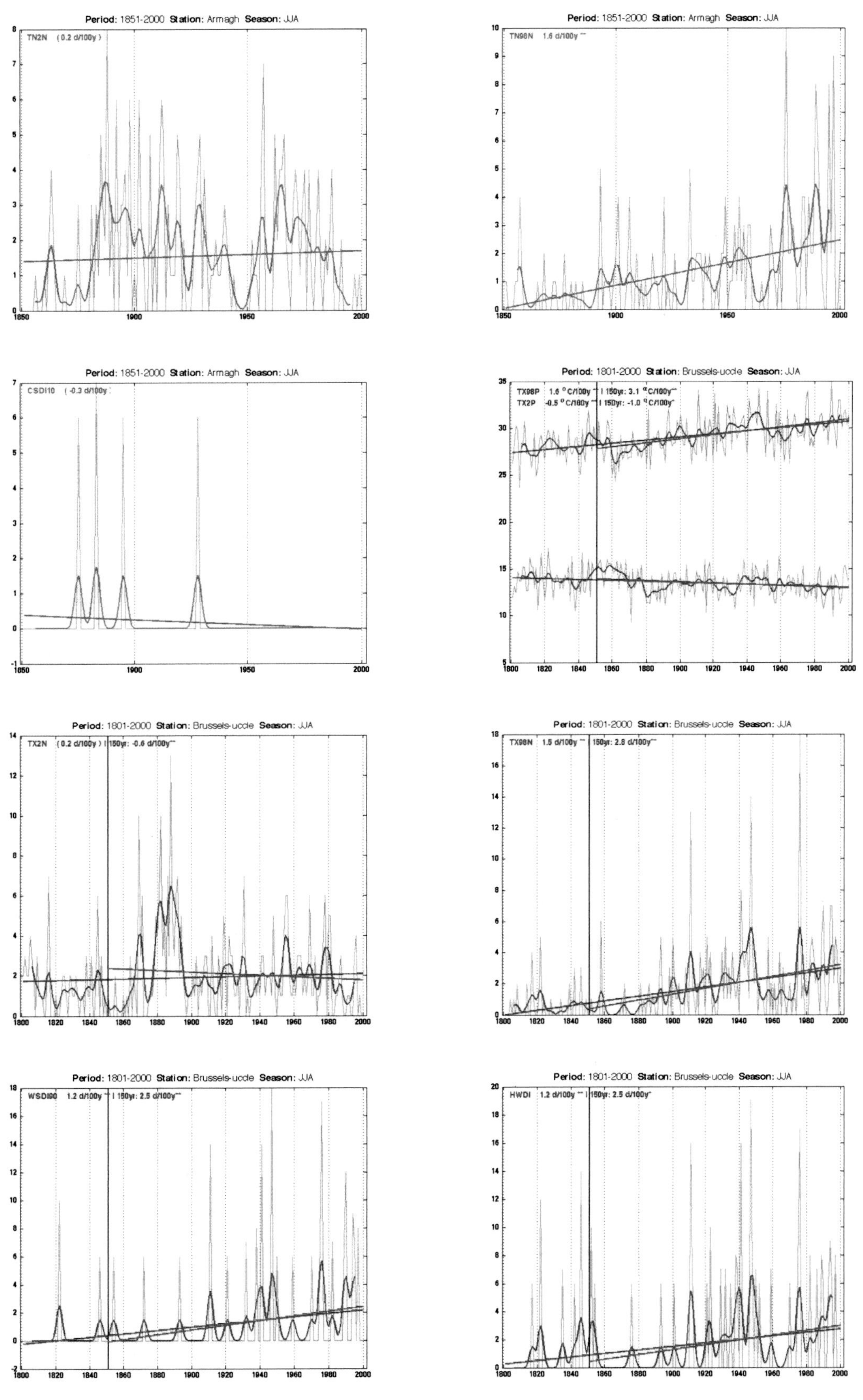

Fig. 3.49 1851–2000 JJA Tmin Armagh

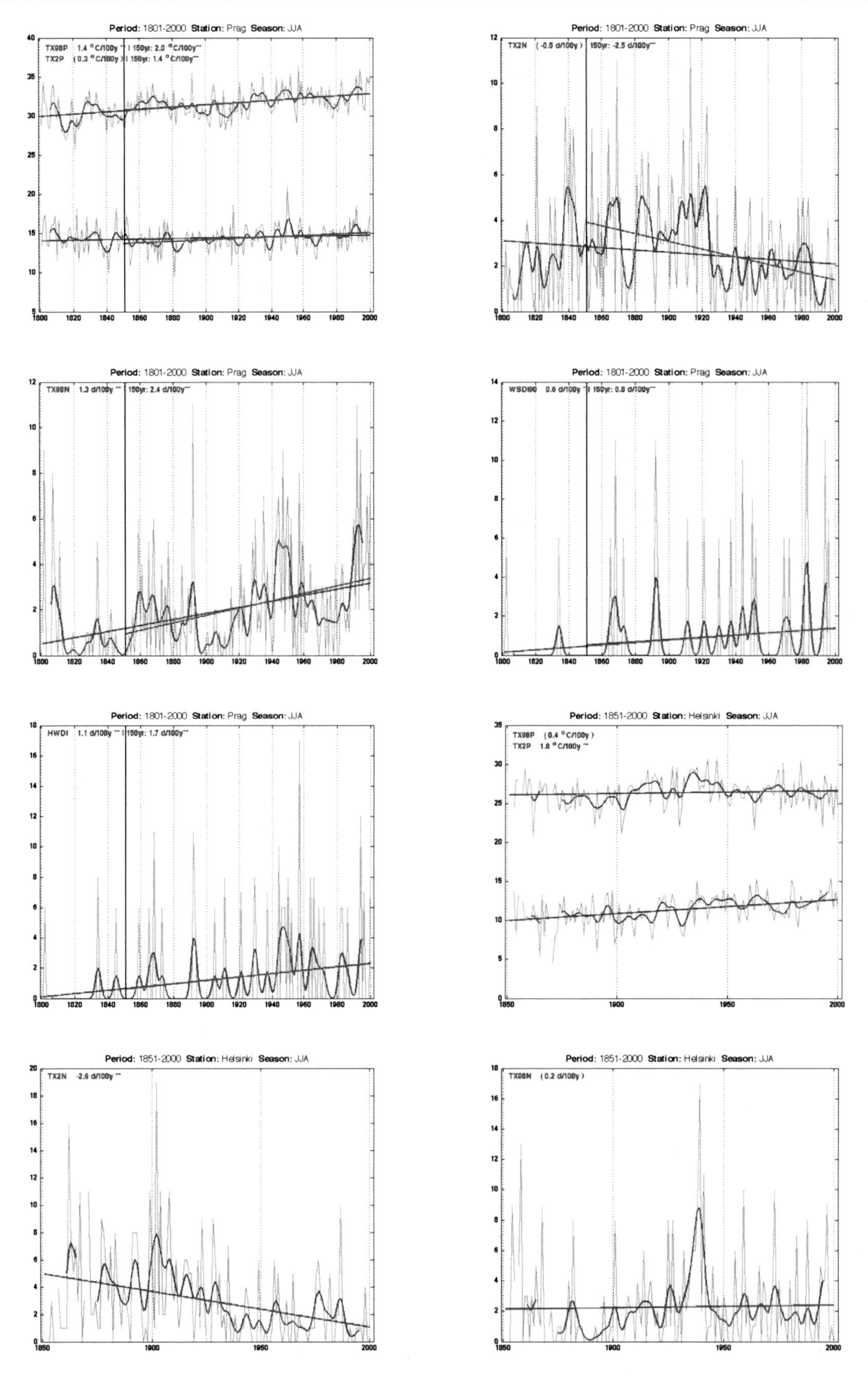

Fig. 3.50 1851–2000 JJA Tmax Prag

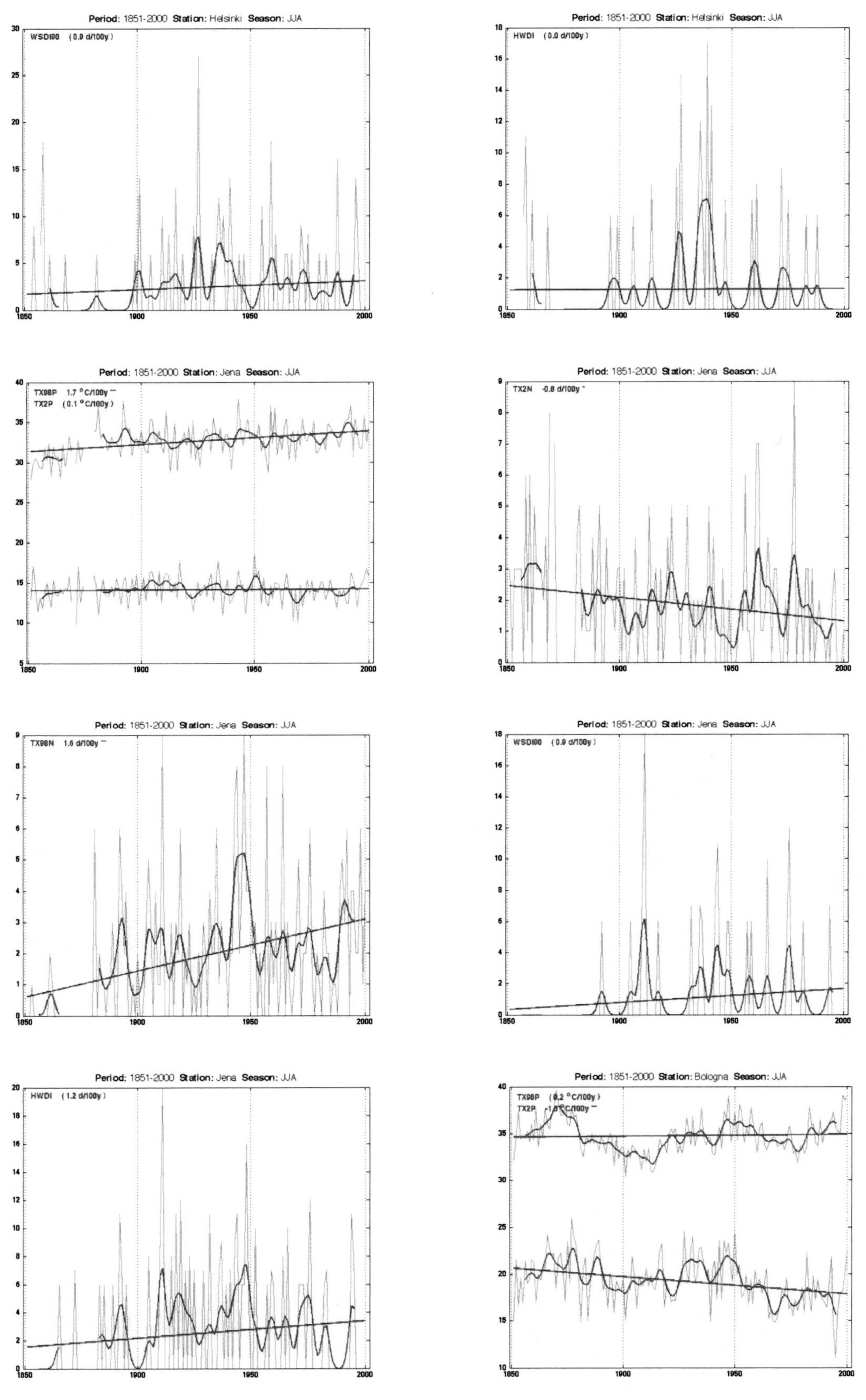

Fig. 3.51 1851–2000 JJA Tmax Helsinki

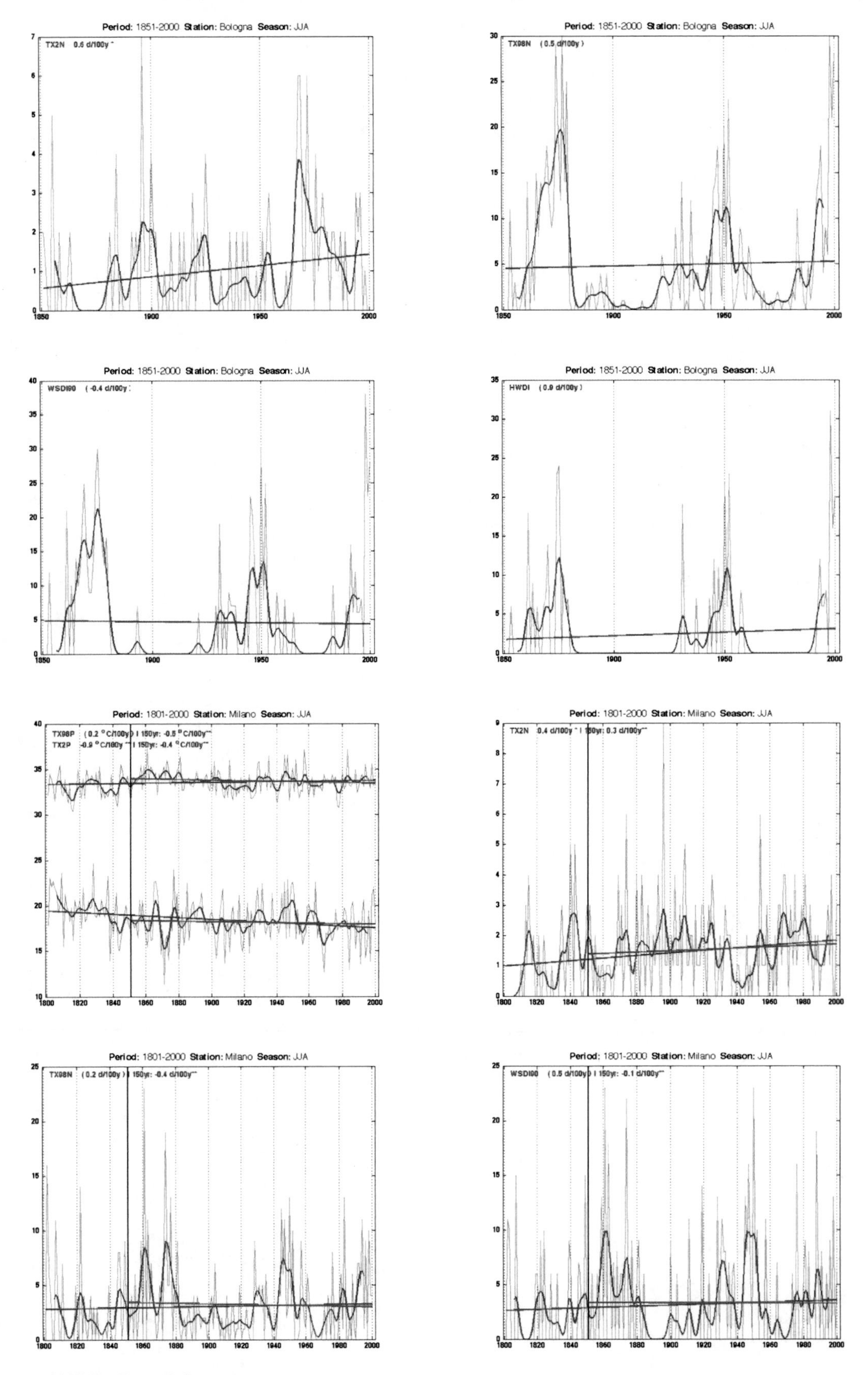

Fig. 3.52 1851–2000 JJA Tmax Bologna

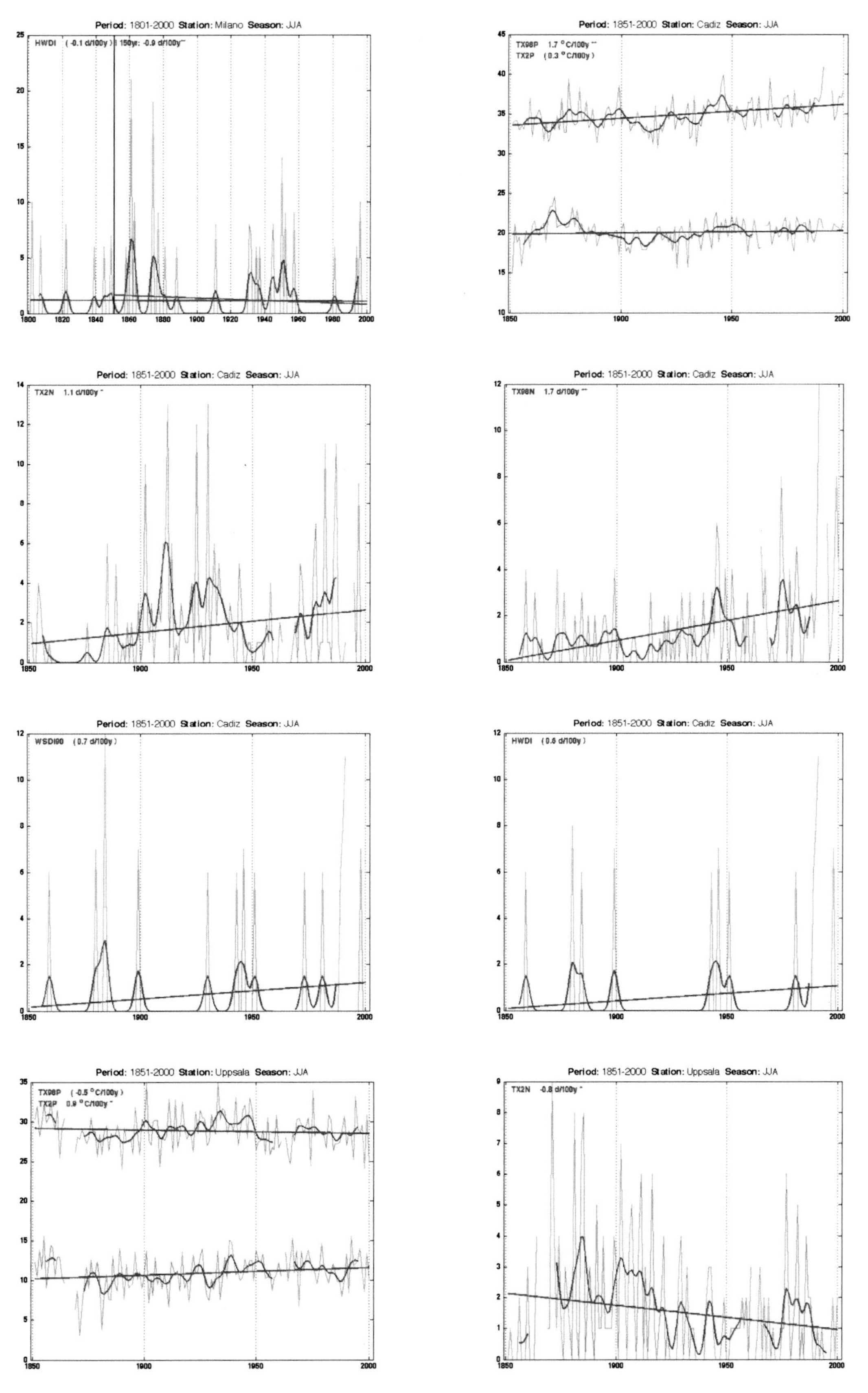

Fig. 3.53 1851–2000 JJA Tmax Milano

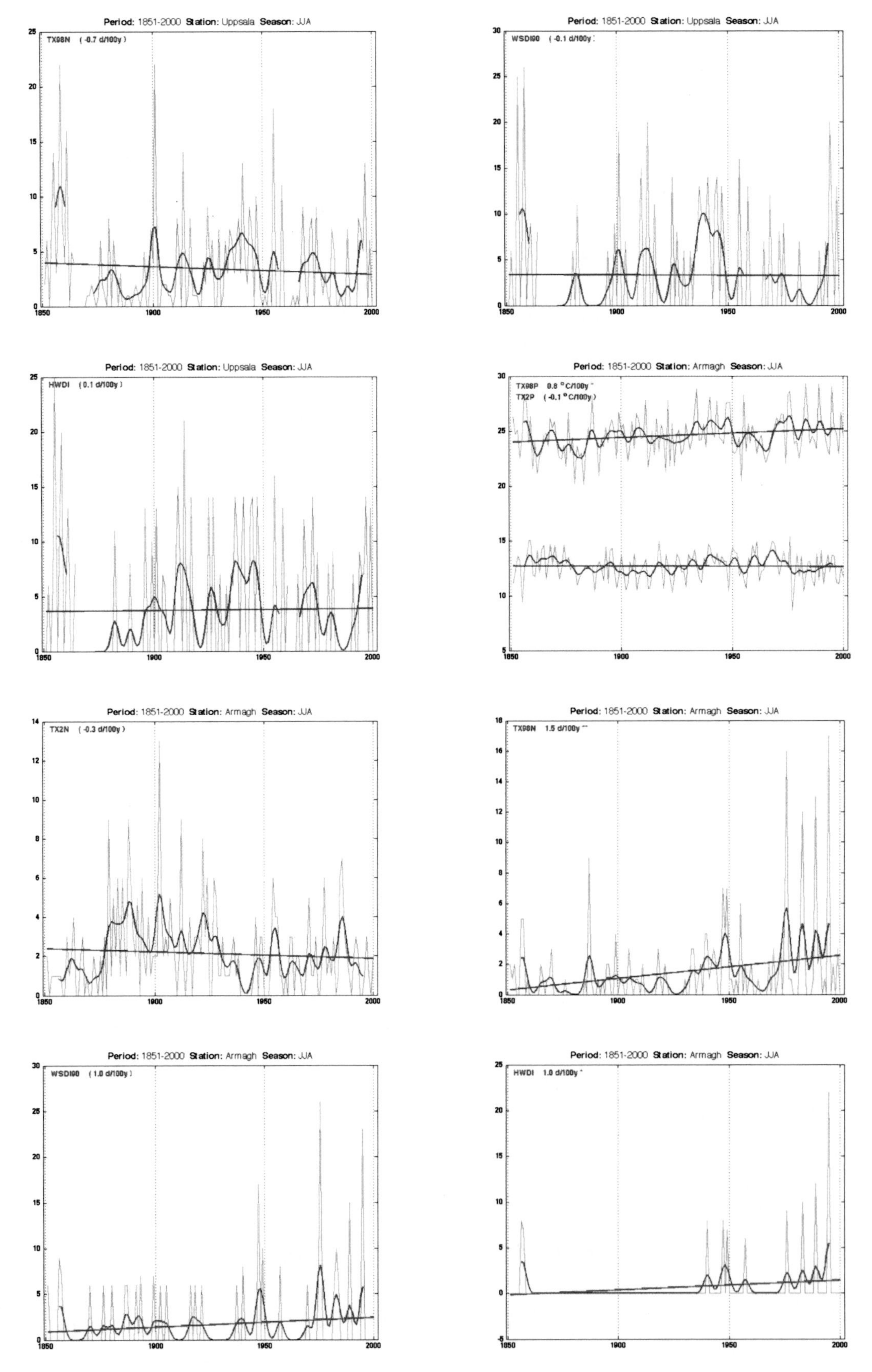

Fig. 3.54 1851–2000 JJA Tmax Uppsala

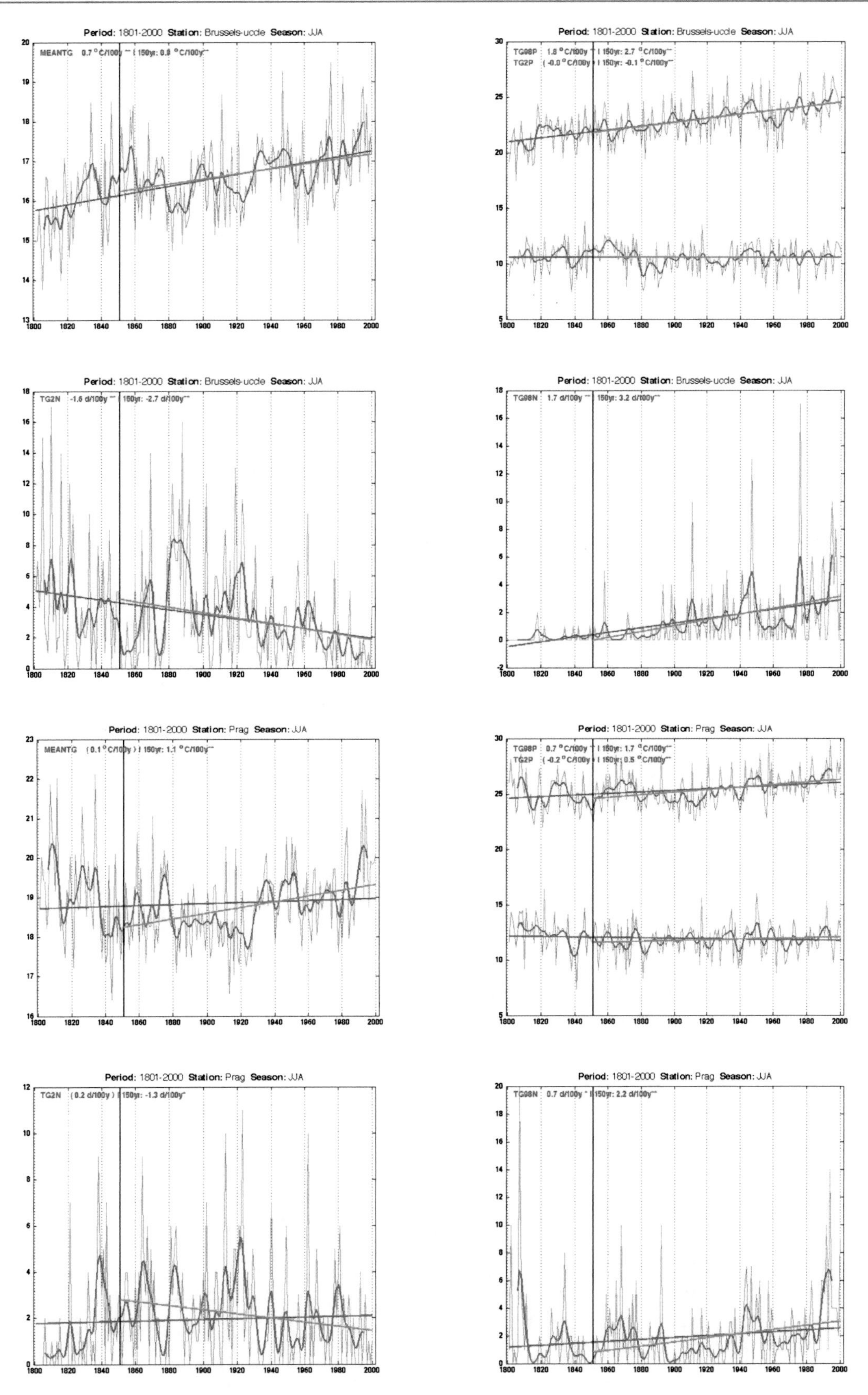

Fig. 3.55 1851–2000 JJA Tmean Brussels-uccle

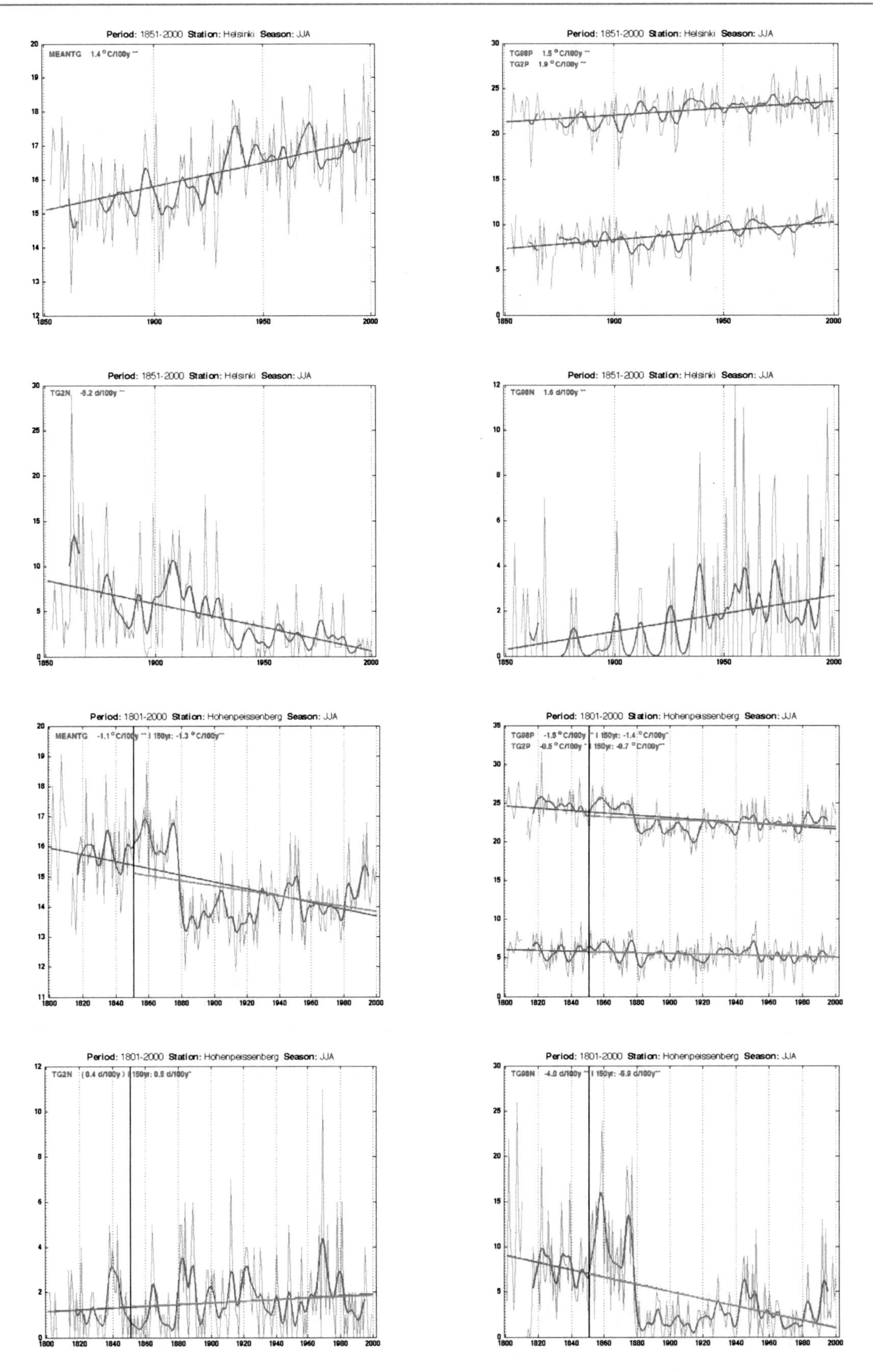

Fig. 3.56 1851–2000 JJA Tmean Helsinki

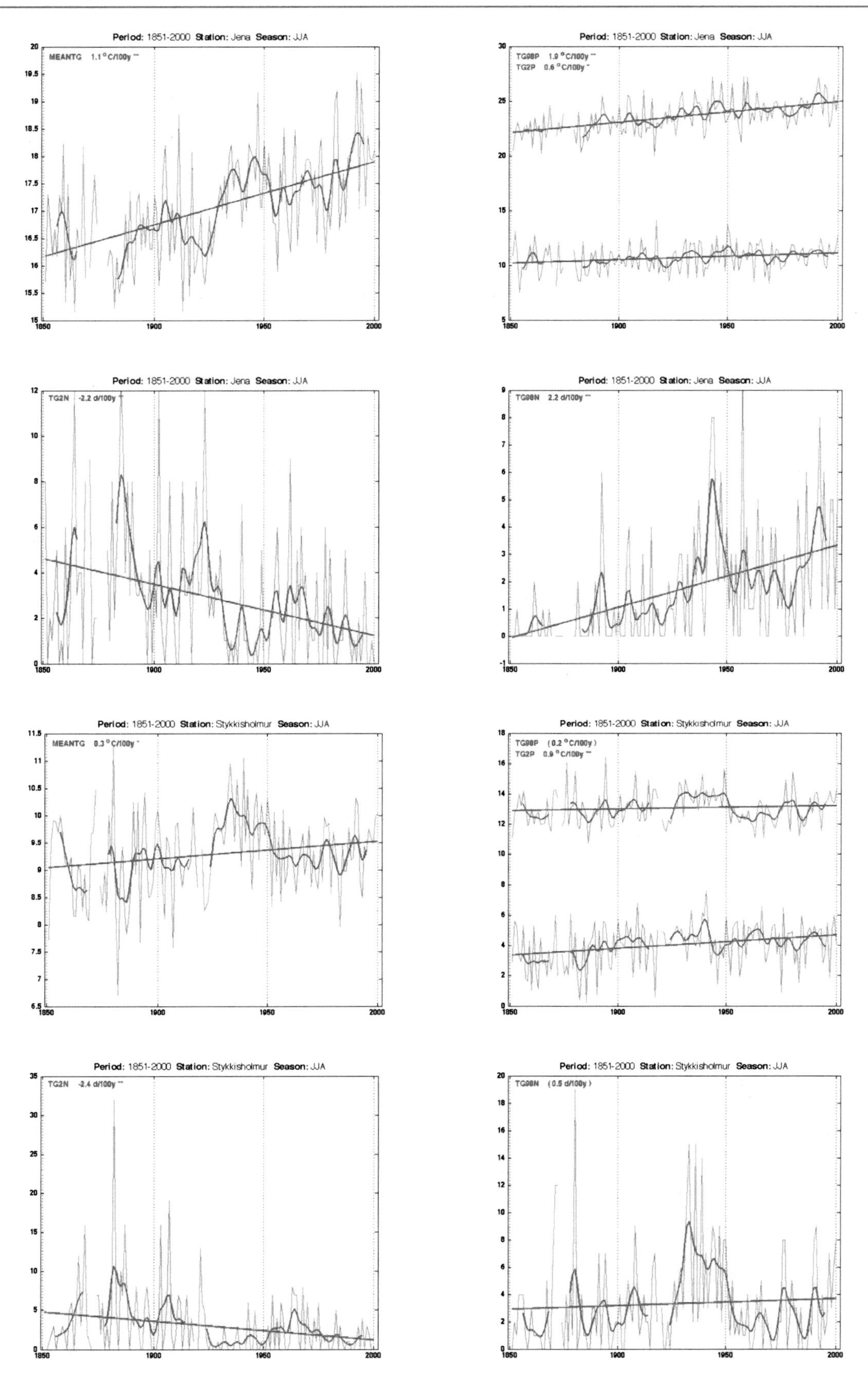

Fig. 3.57 1851–2000 JJA Tmean Jena

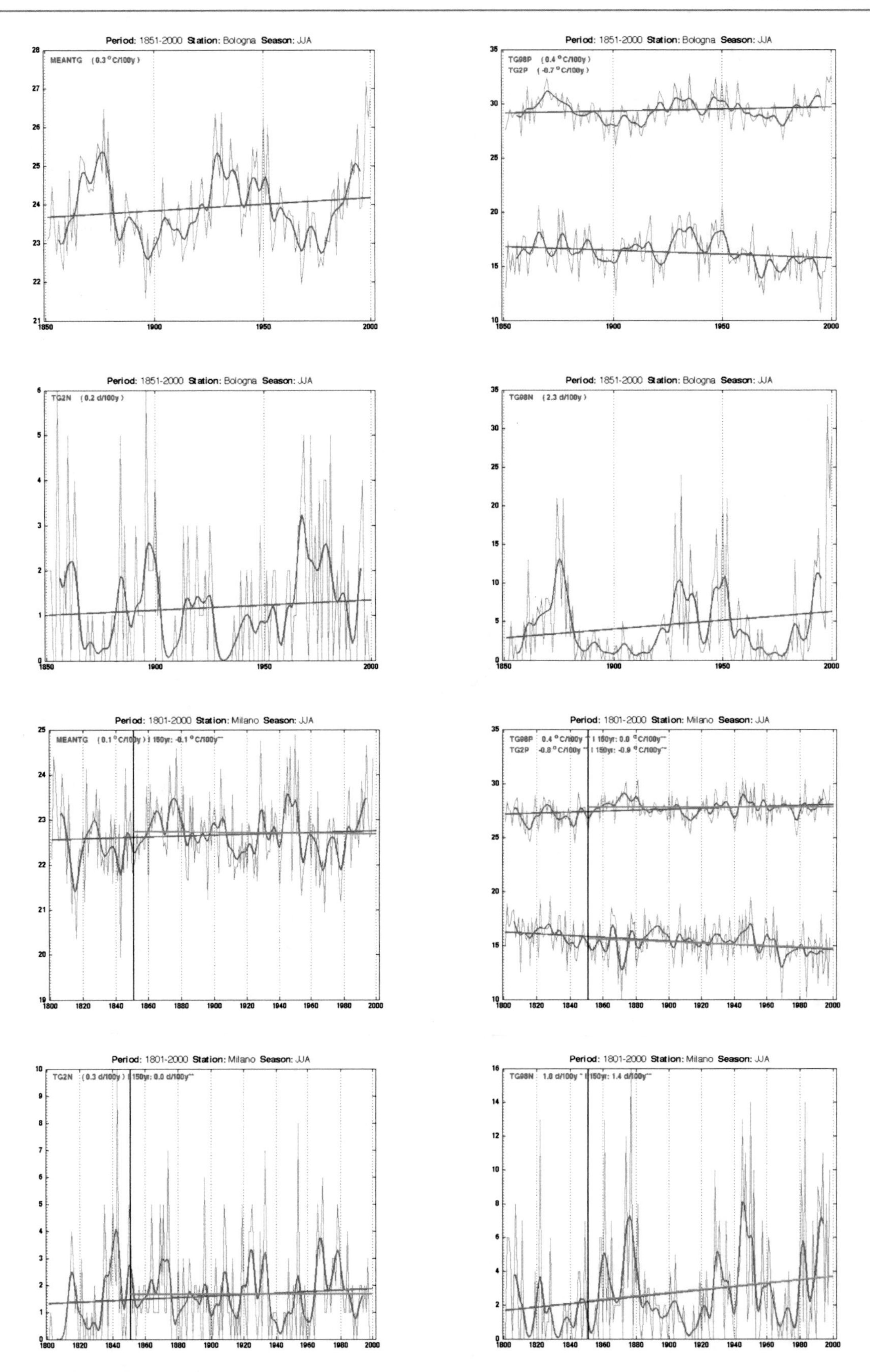

Fig. 3.58 1851–2000 JJA Tmean Bologna

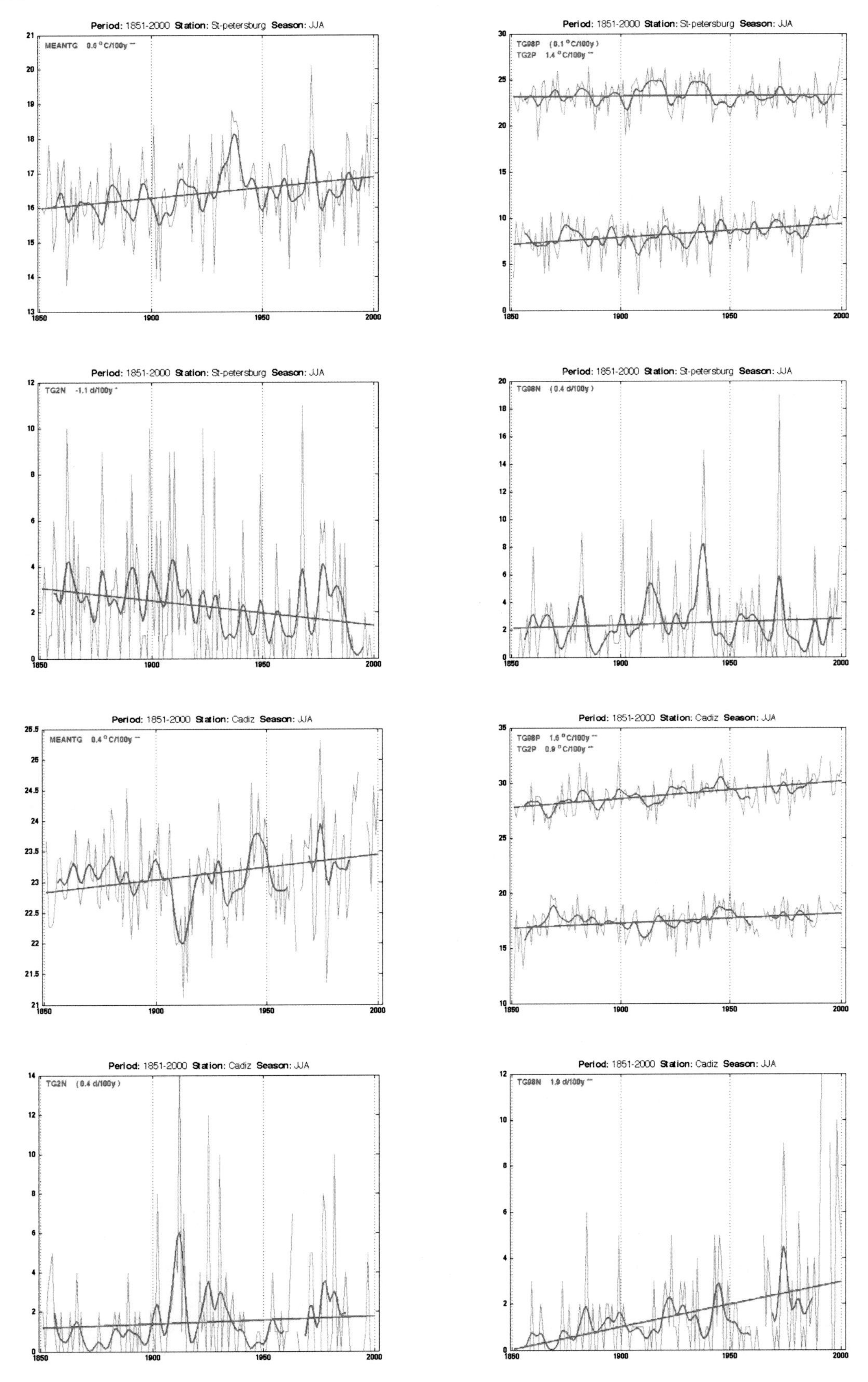

Fig. 3.59 1851–2000 JJA Tmean St-petersburg

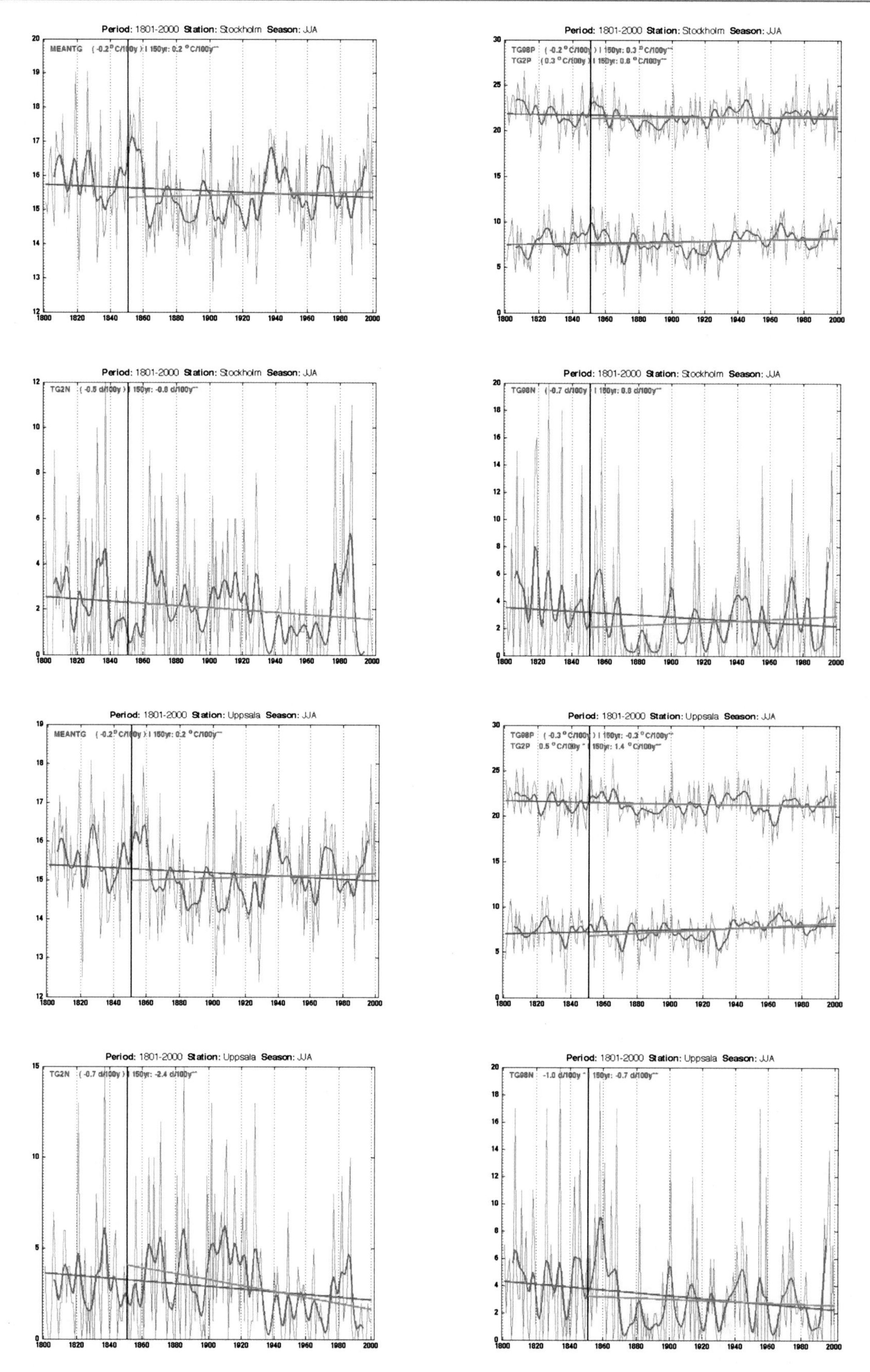

Fig. 3.60 1851–2000 JJA Tmean Stockholm

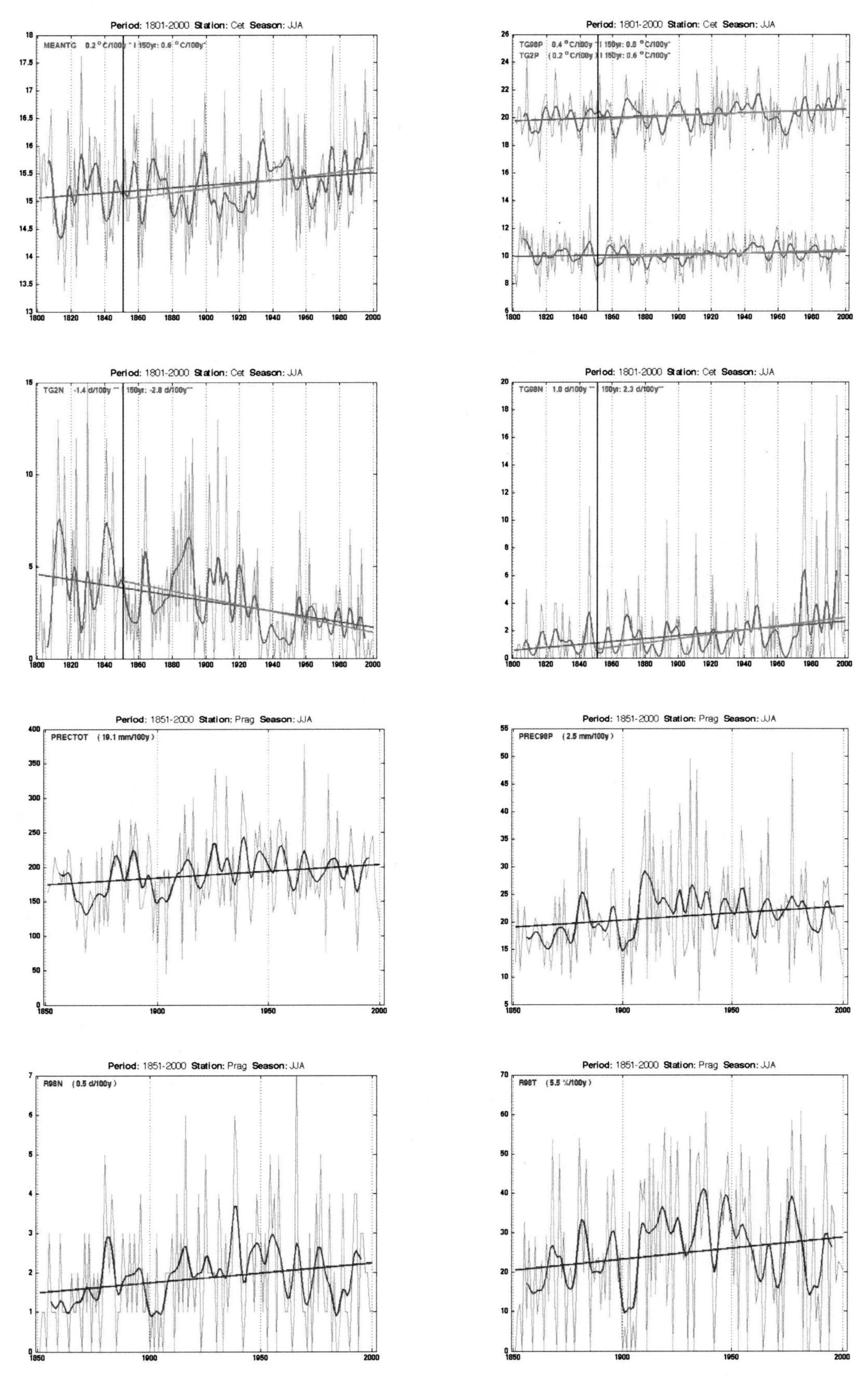

Fig. 3.61 1851–2000 JJA Tmean Cet

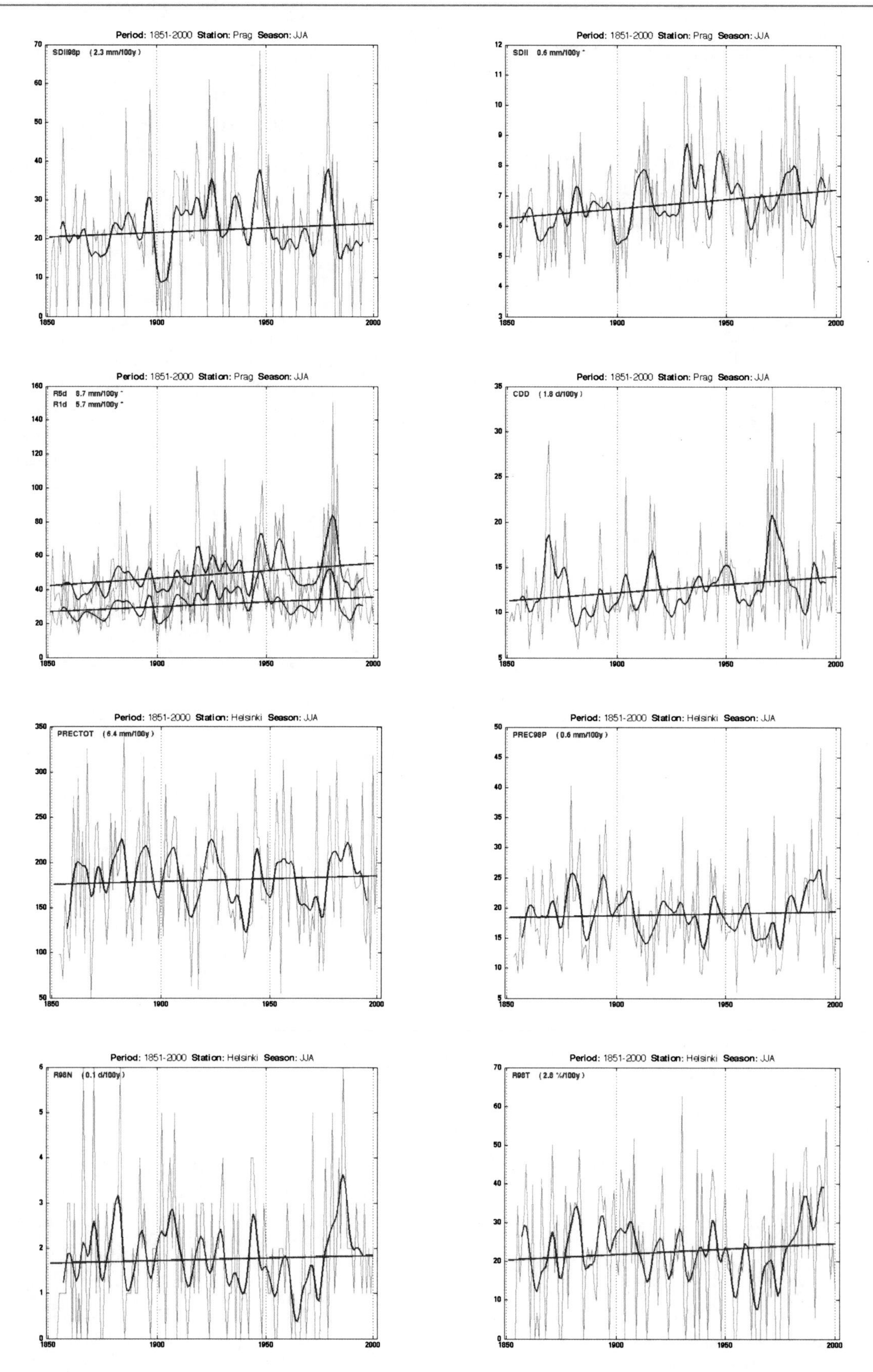

Fig. 3.62 1851–2000 JJA Prec Prag

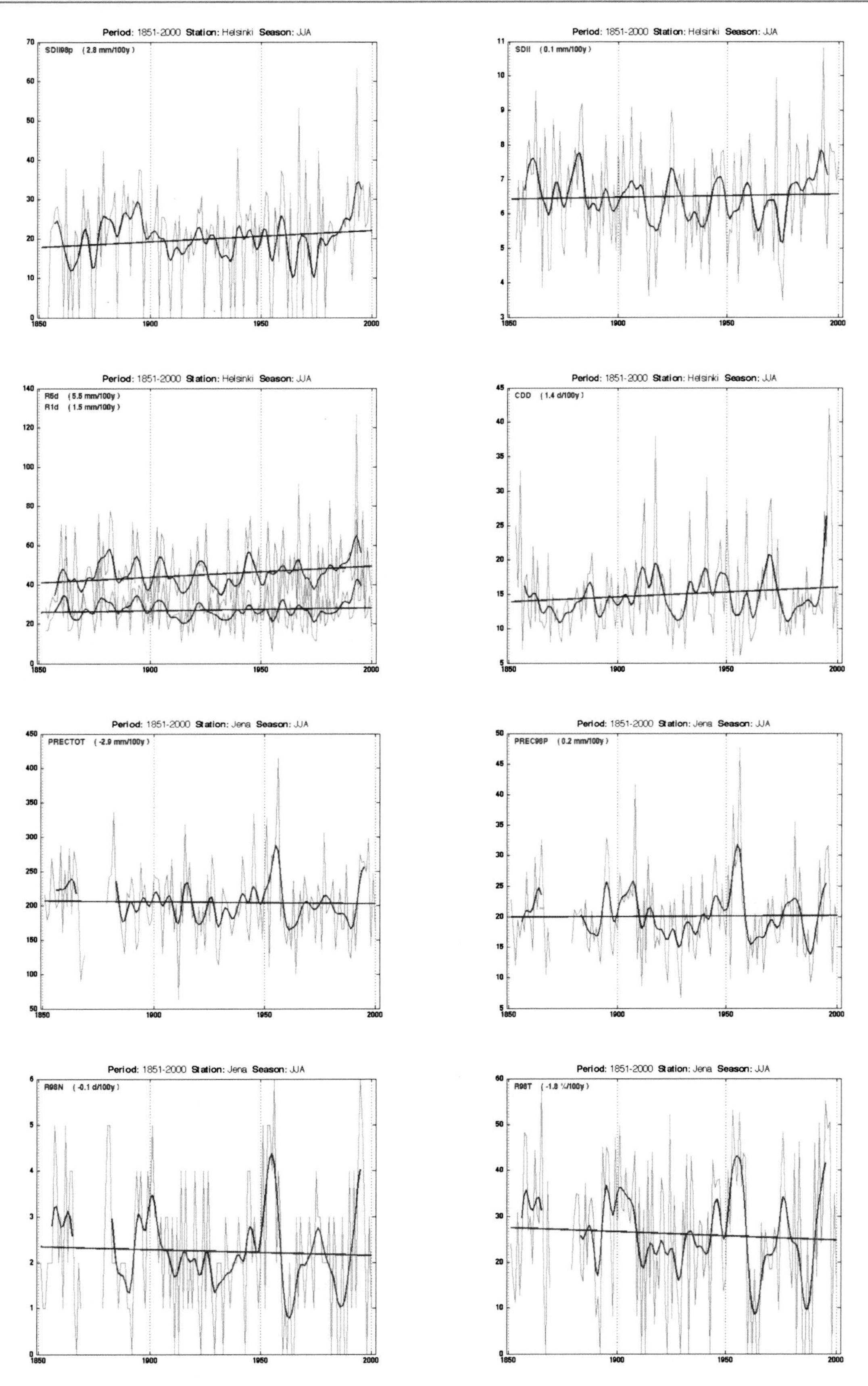

Fig. 3.63 1851–2000 JJA Prec Helsinki

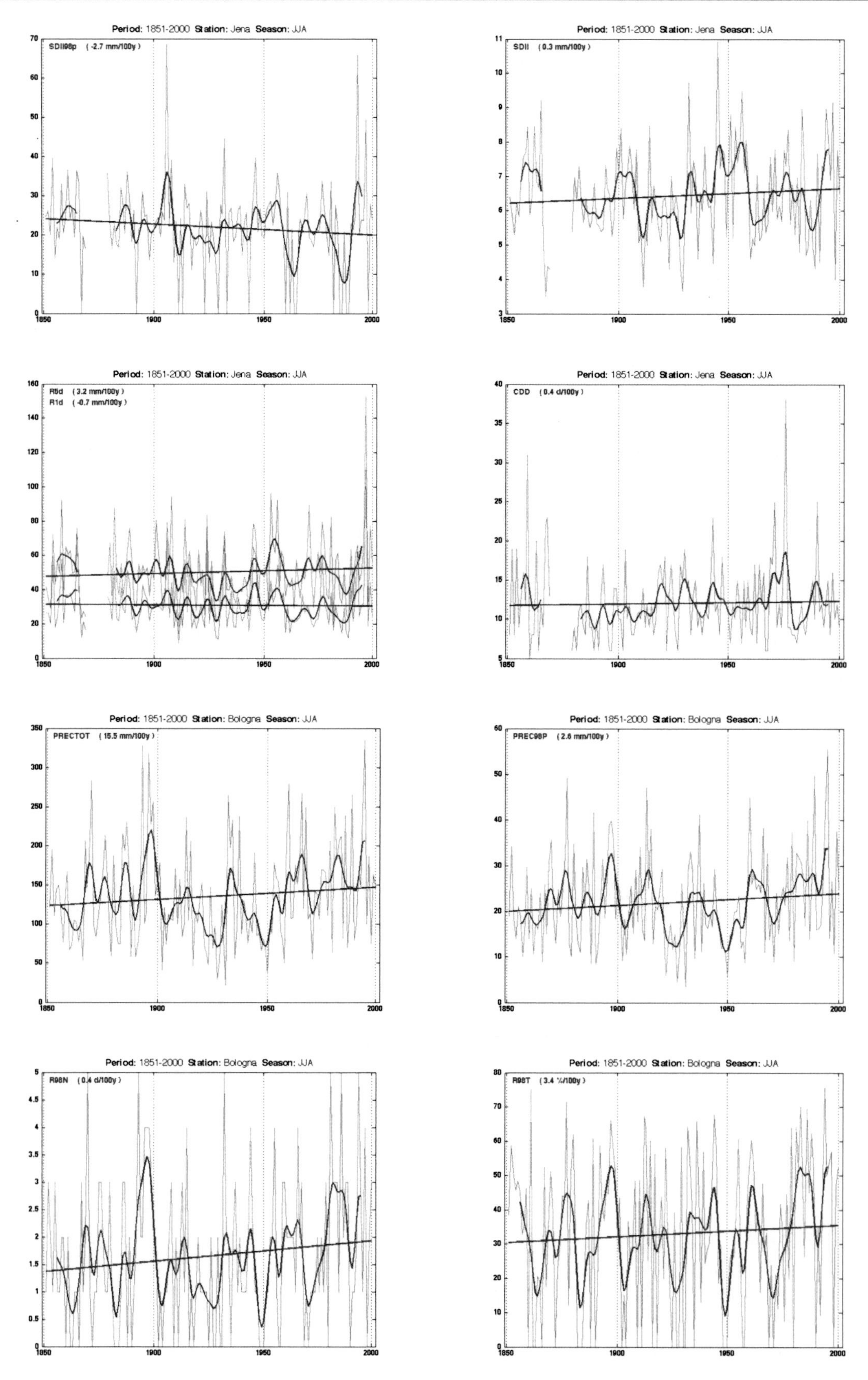

Fig. 3.64 1851–2000 JJA Prec Jena

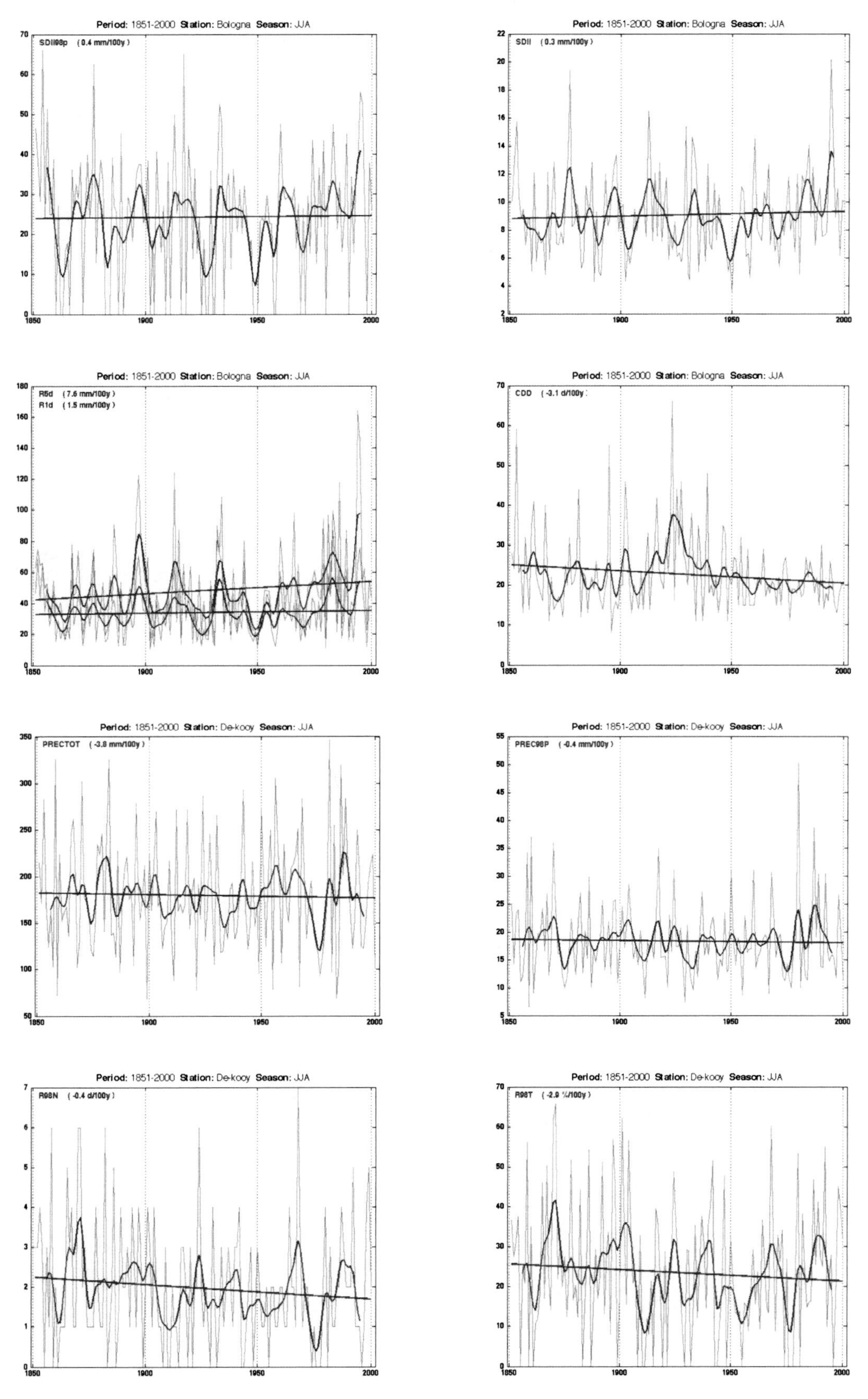

Fig. 3.65 1851–2000 JJA Prec Bologna

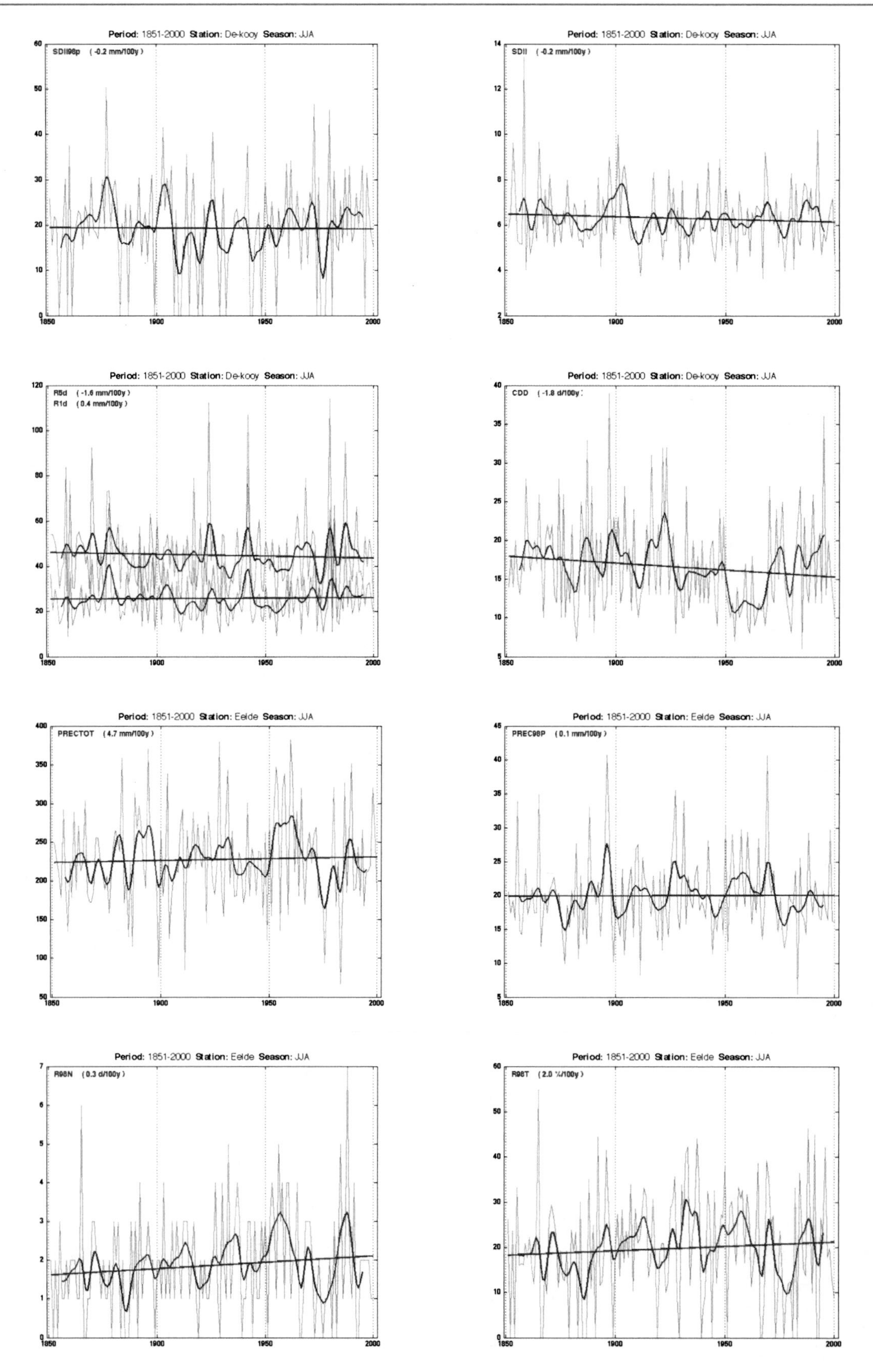

Fig. 3.66 1851–2000 JJA Prec De-kooy

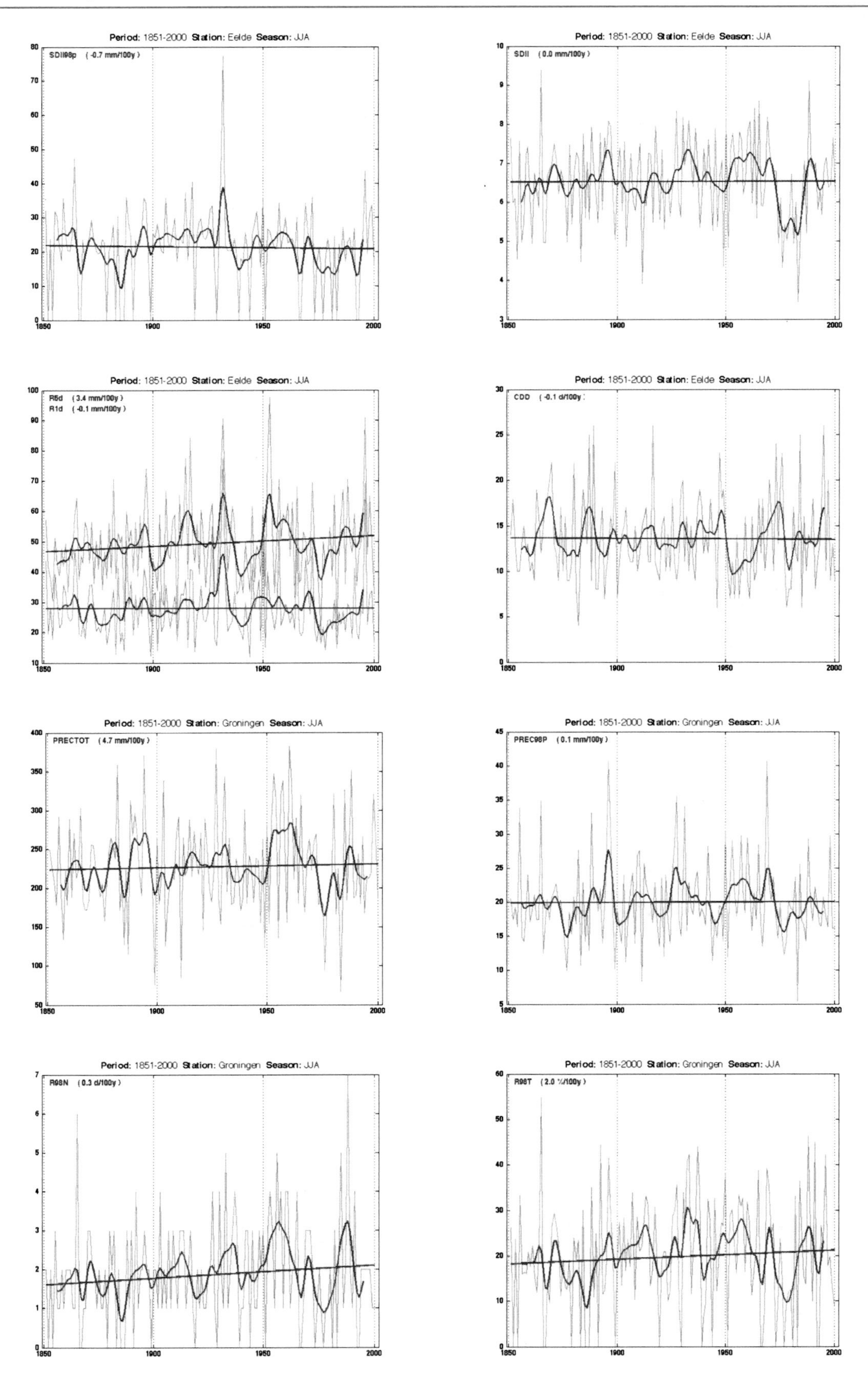

Fig. 3.67 1851–2000 JJA Prec Eelde

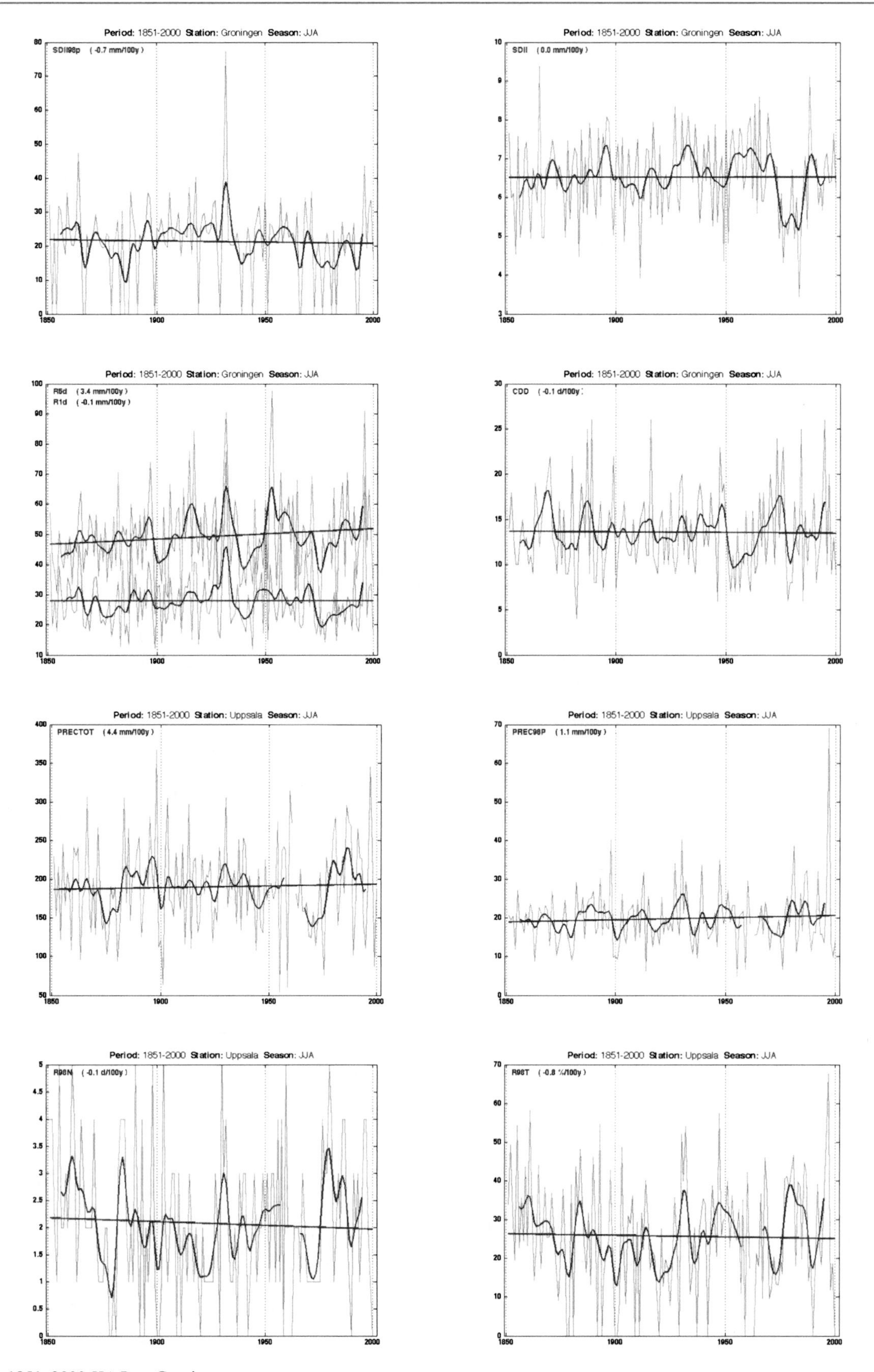

Fig. 3.68 1851–2000 JJA Prec Groningen

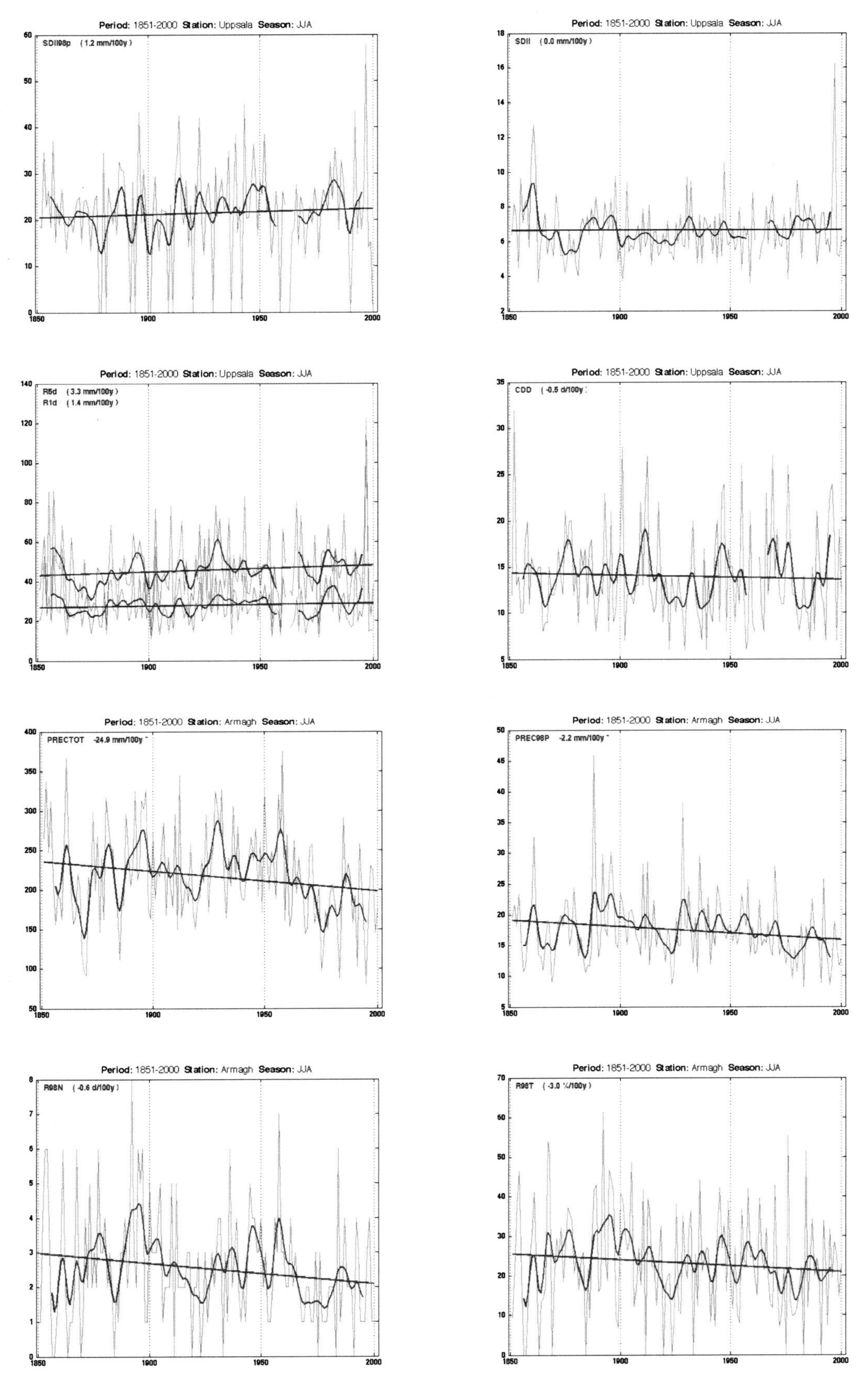

Fig. 3.69 1851–2000 JJA Prec Uppsala

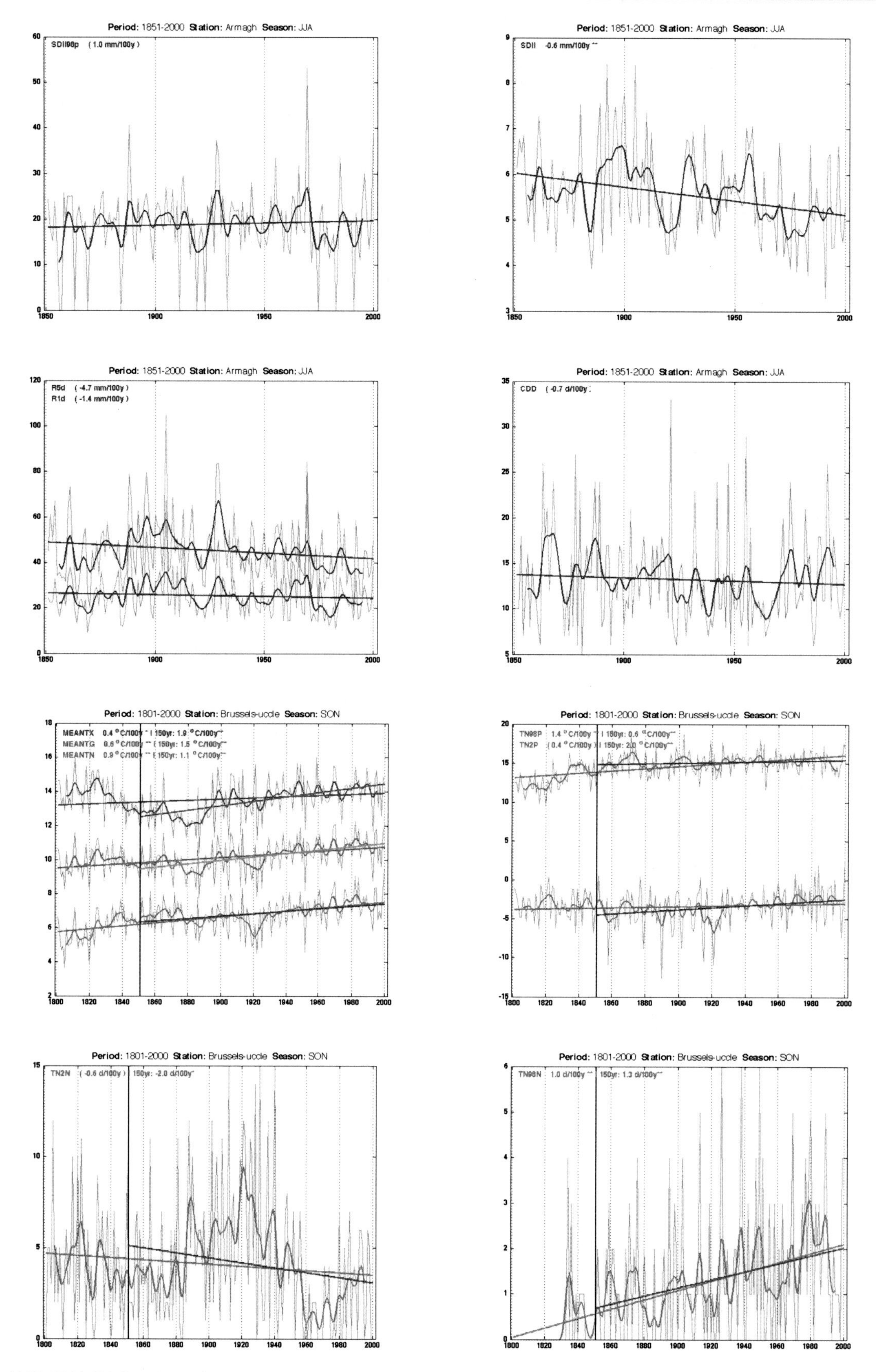

Fig. 3.70 1851–2000 JJA Prec Armagh

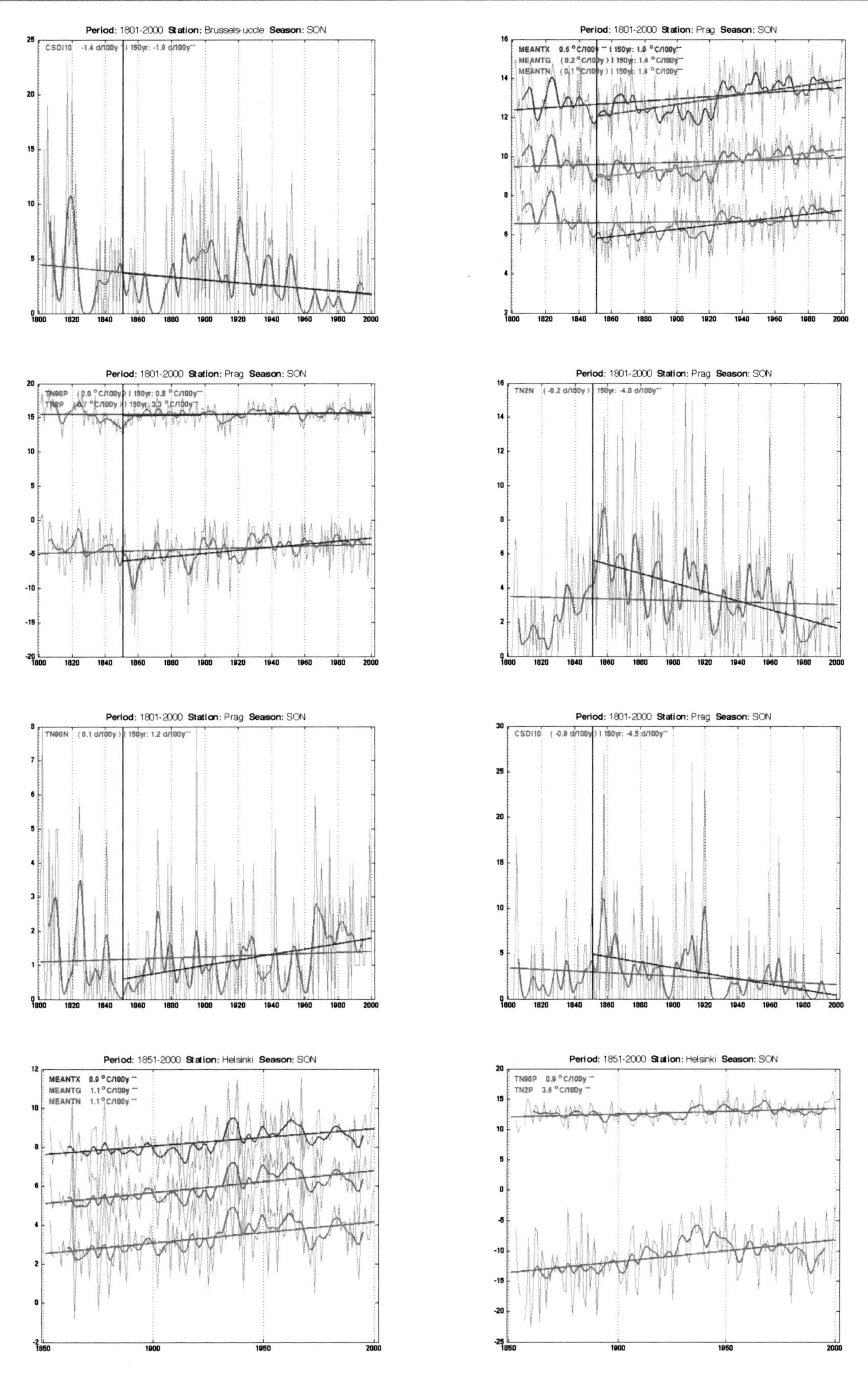

Fig. 3.71 1851–2000 SON Tmin Brussels-uccle

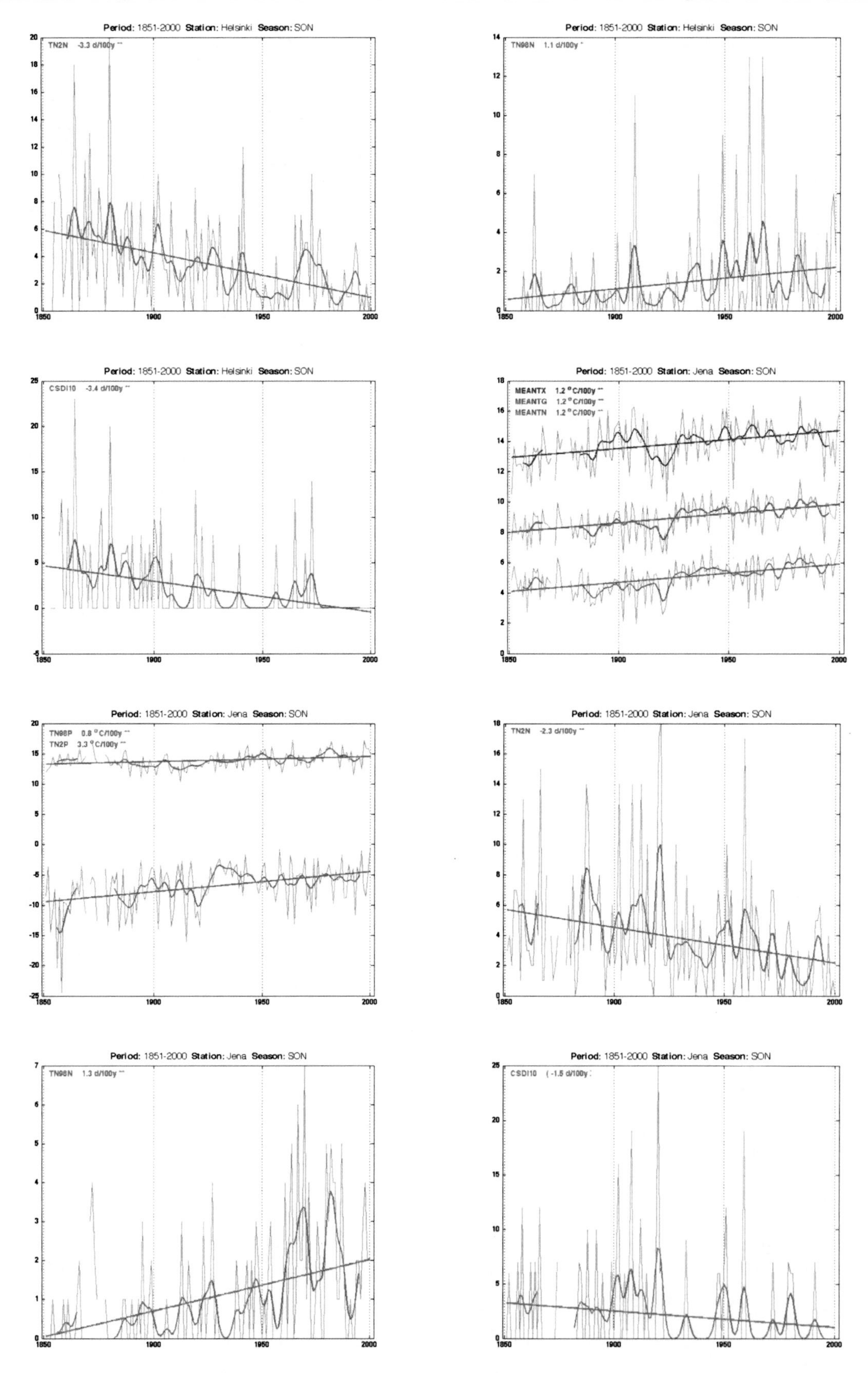

Fig. 3.72 1851–2000 SON Tmin Helsinki

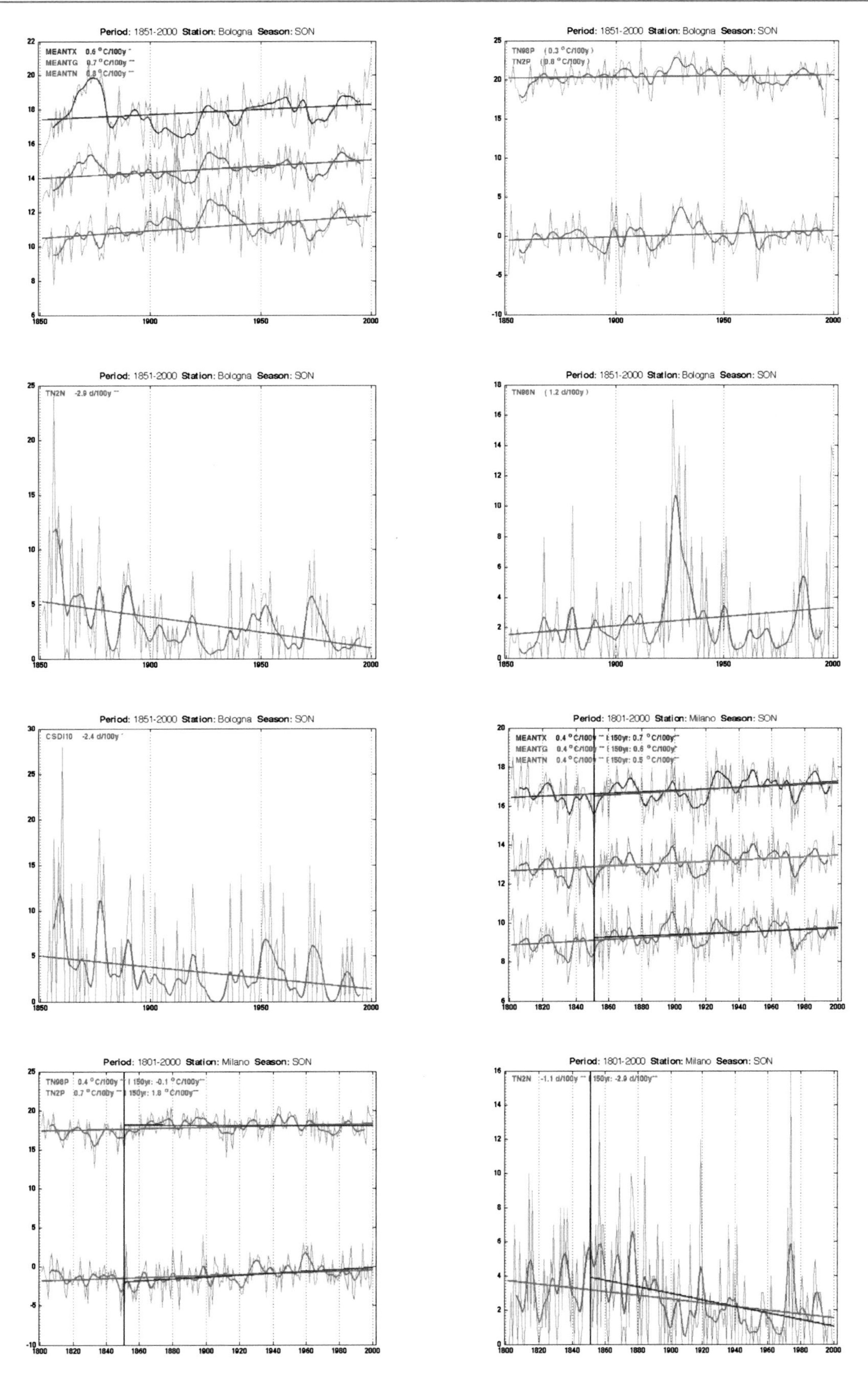

Fig. 3.73 1851–2000 SON Tmin Bologna

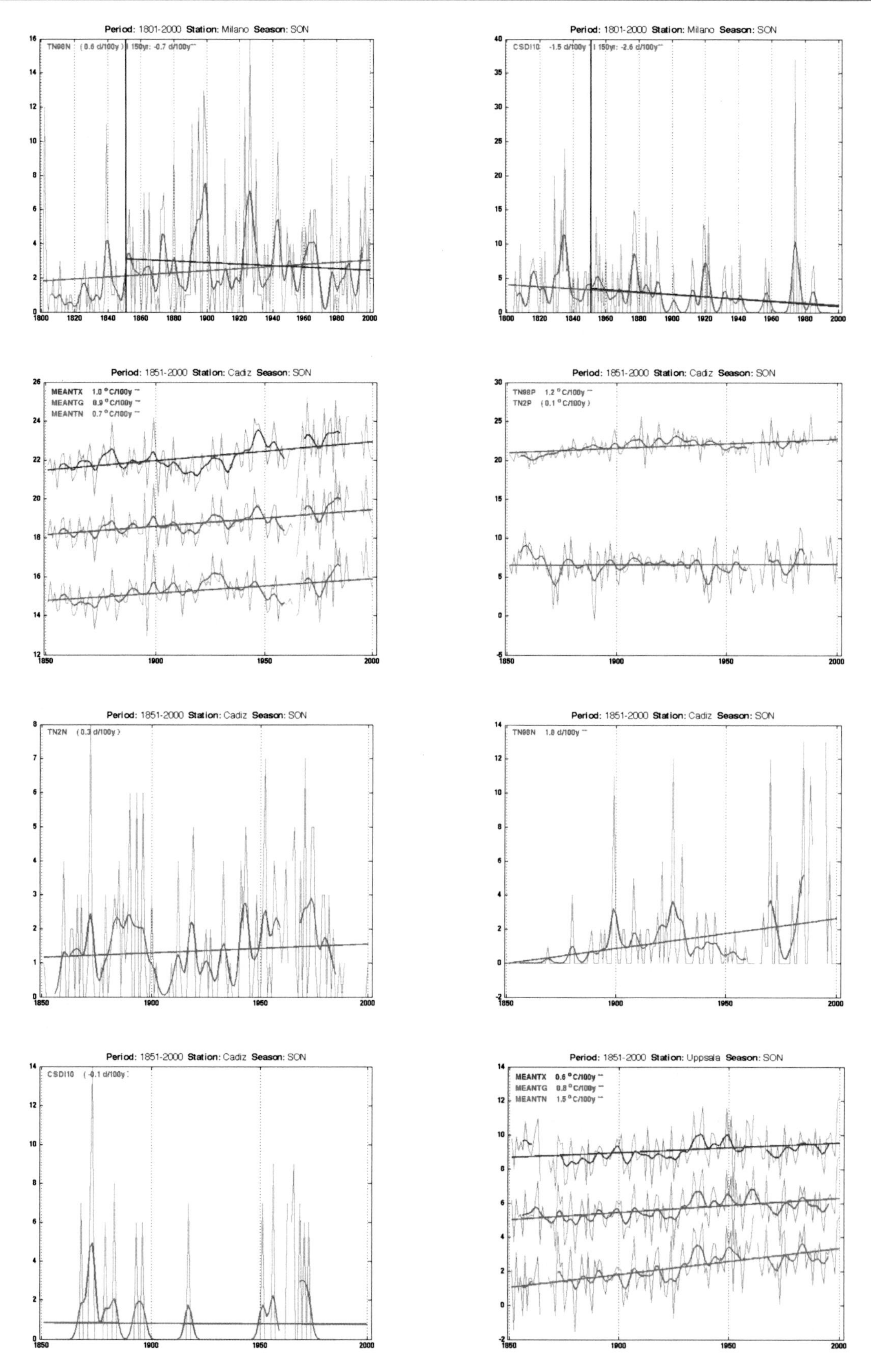

Fig. 3.74 1851–2000 SON Tmin Milano

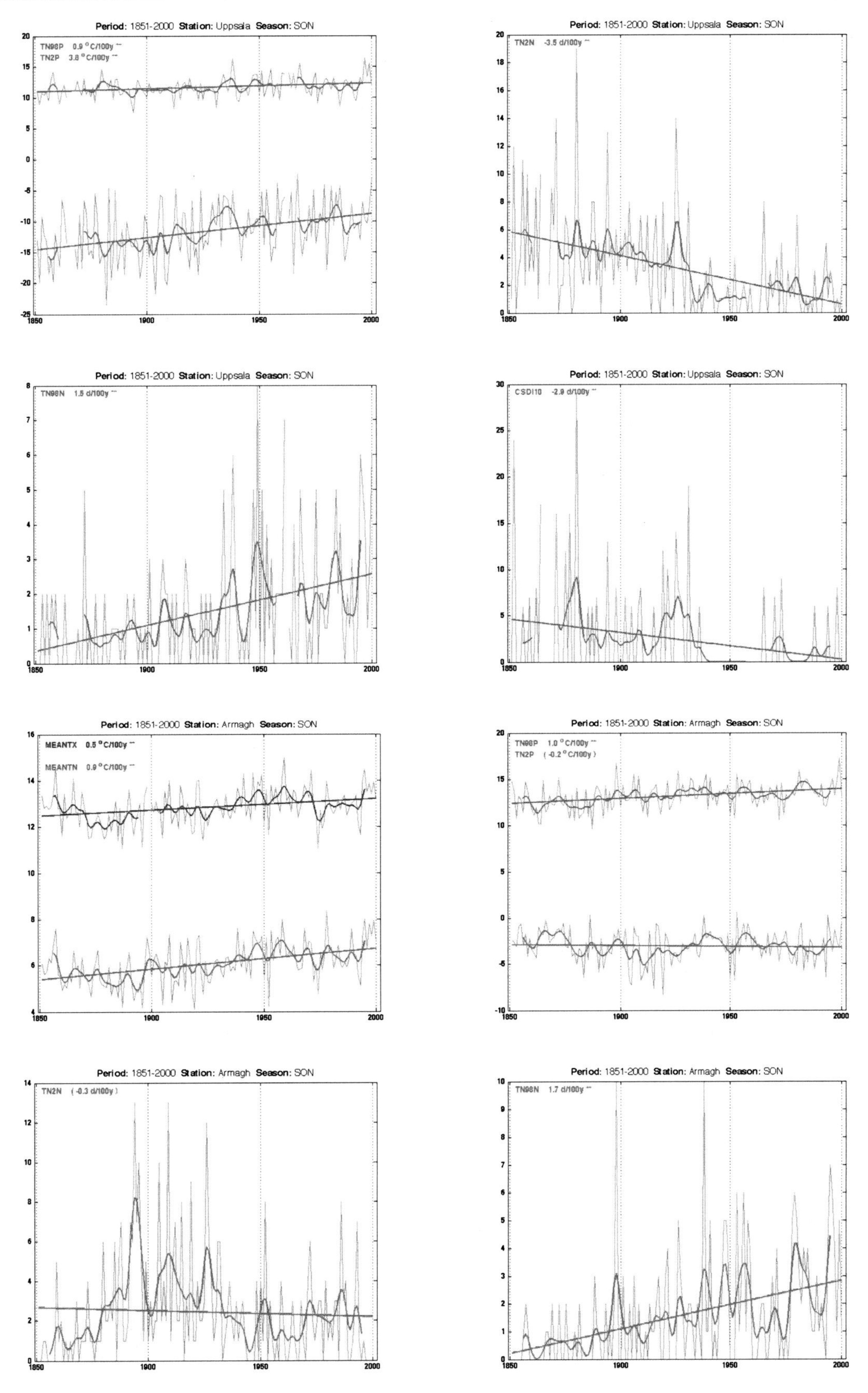

Fig. 3.75 1851–2000 SON Tmin Uppsala

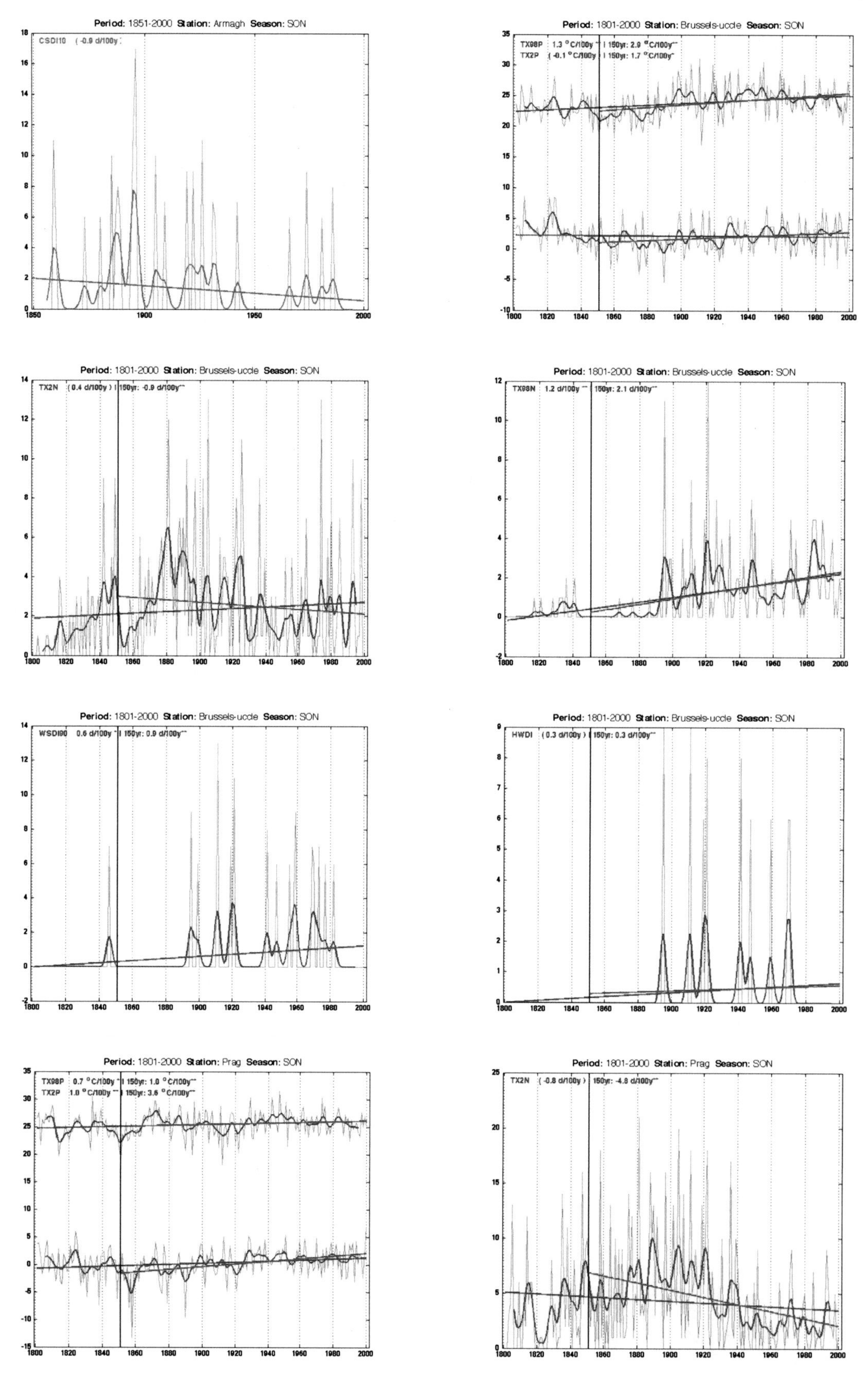

Fig. 3.76 1851–2000 SON Tmin Armagh

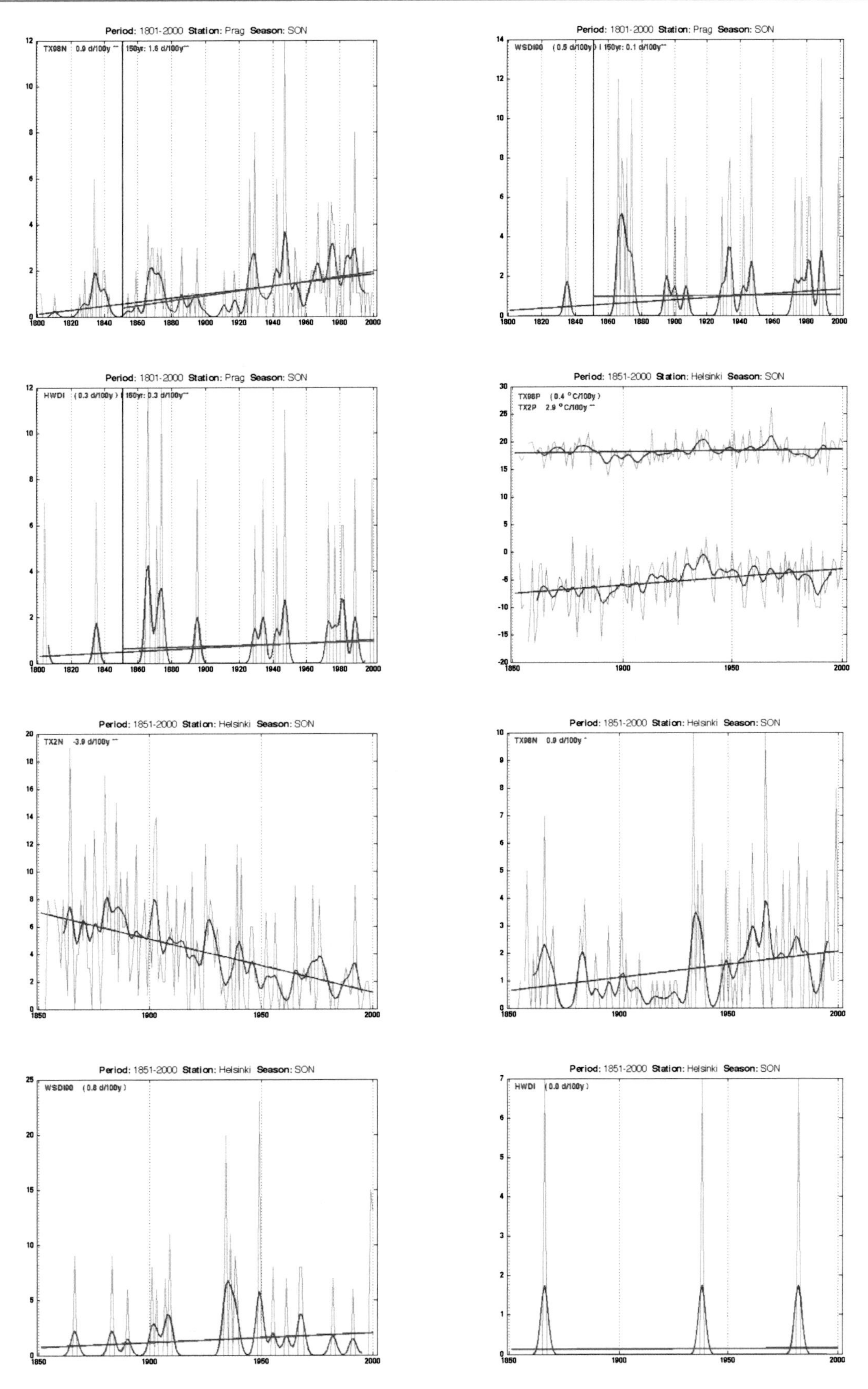

Fig. 3.77 1851–2000 SON Tmax Prag

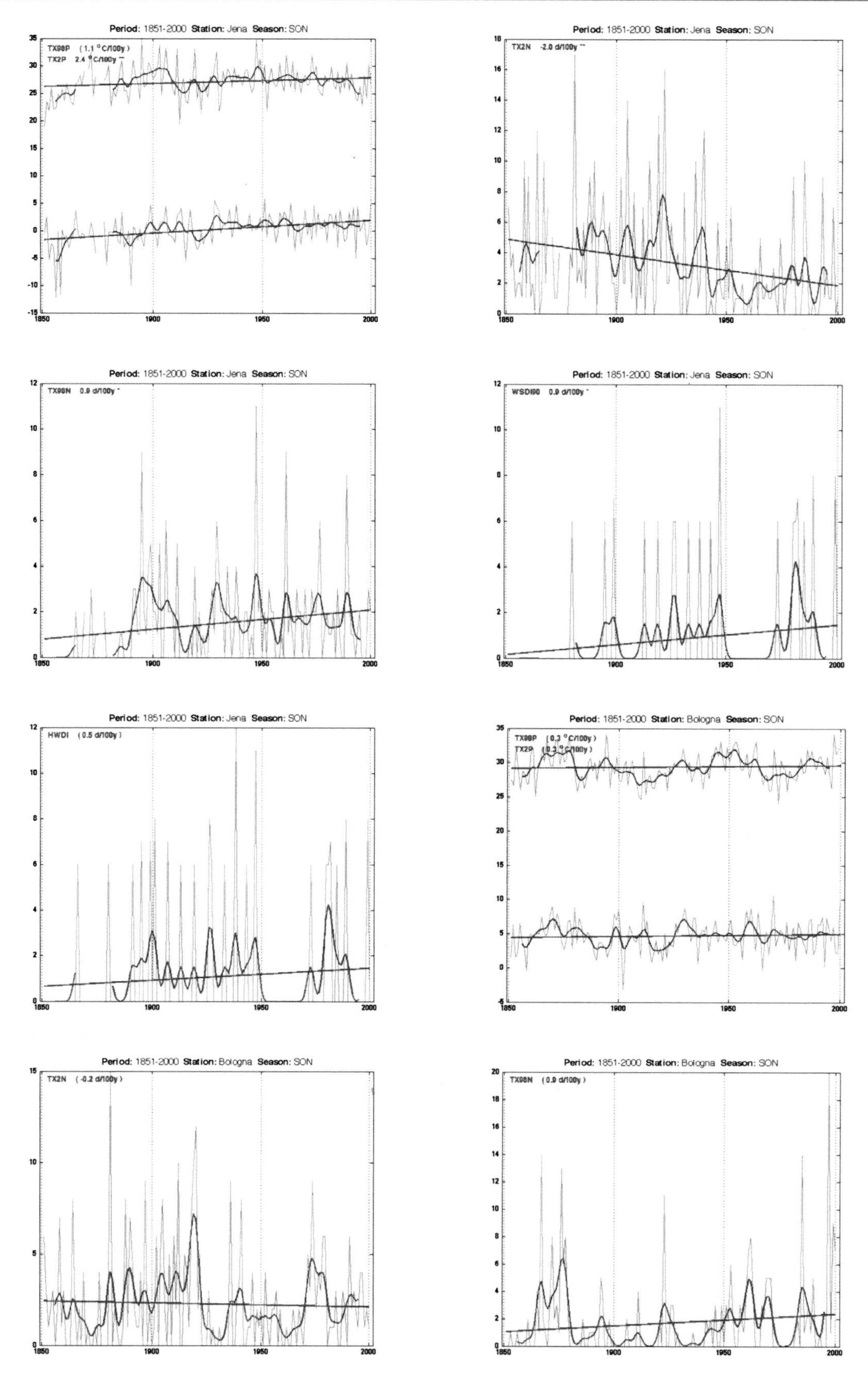

Fig. 3.78 1851–2000 SON Tmax Jena

Fig. 3.79 1851–2000 SON Tmax Bologna

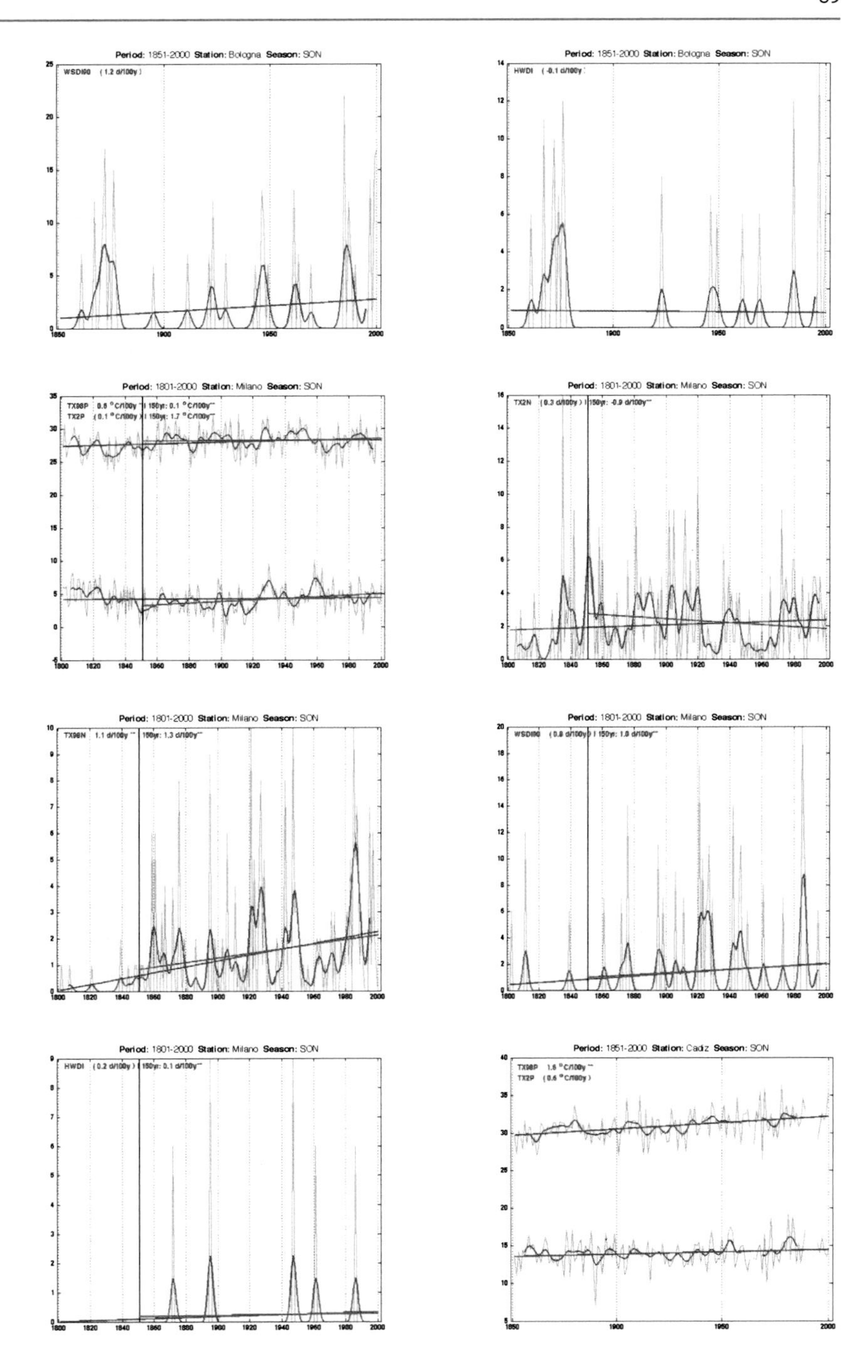

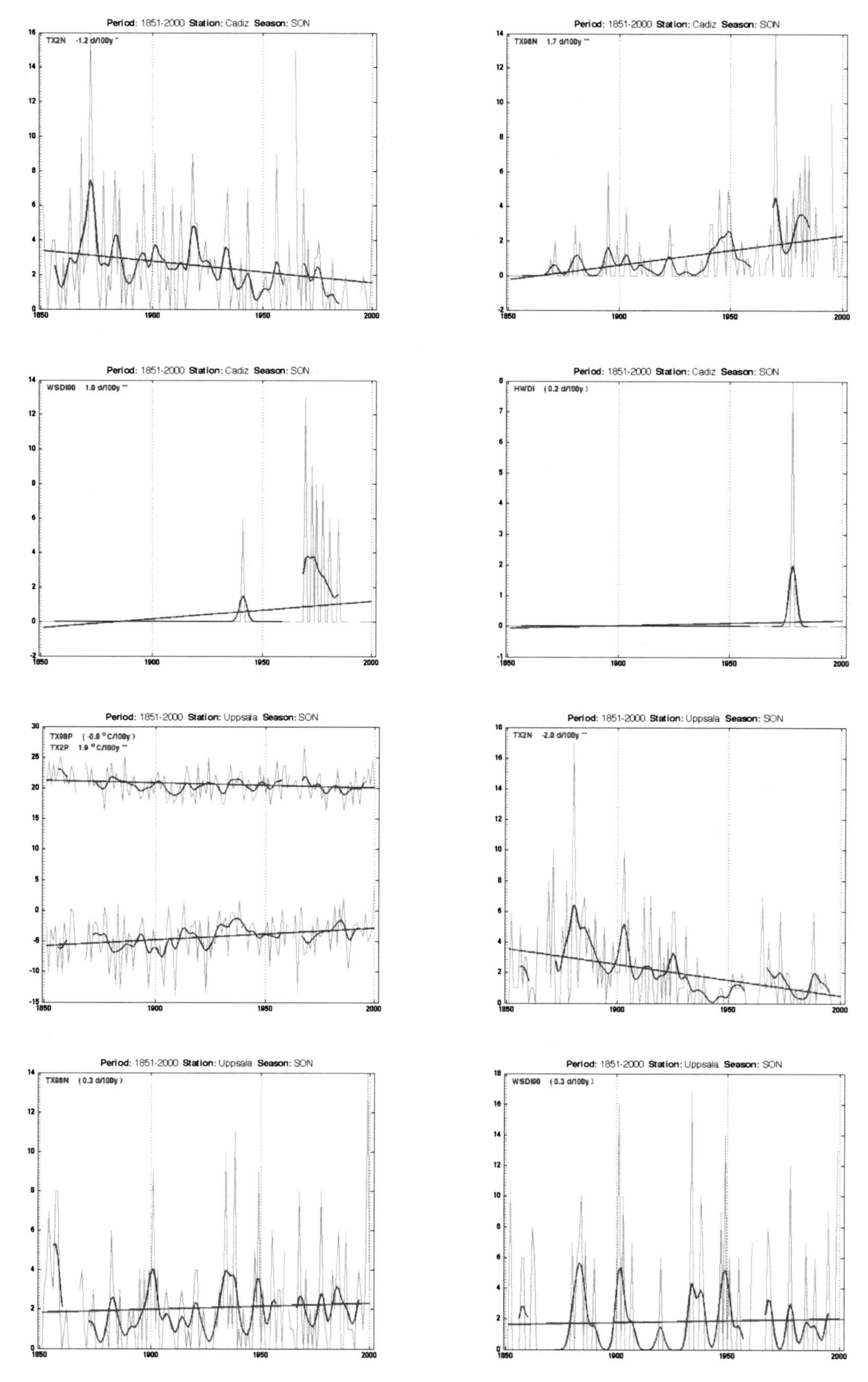

Fig. 3.80 1851–2000 SON Tmax Cadiz

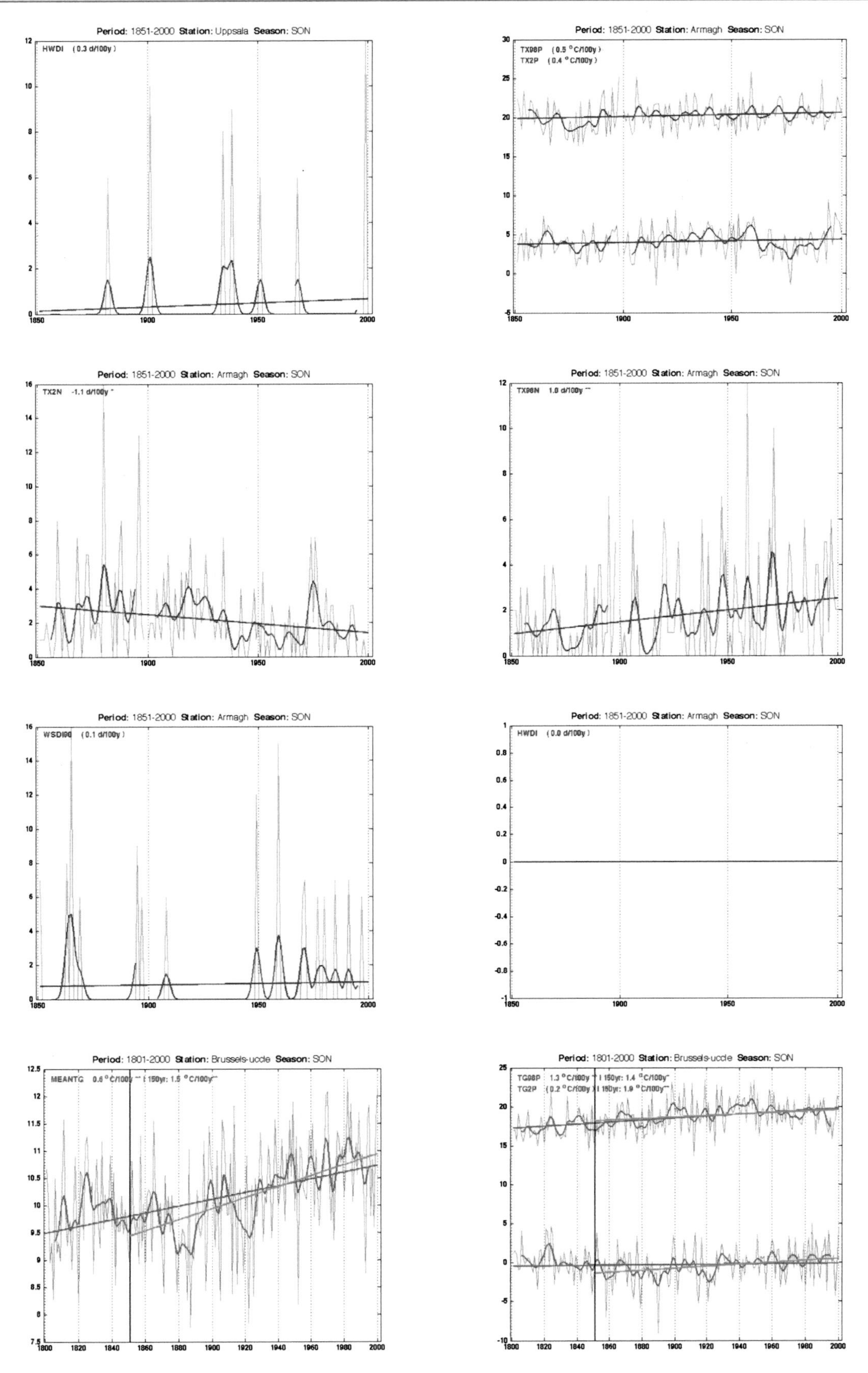

Fig. 3.81 1851–2000 SON Tmax Uppsala

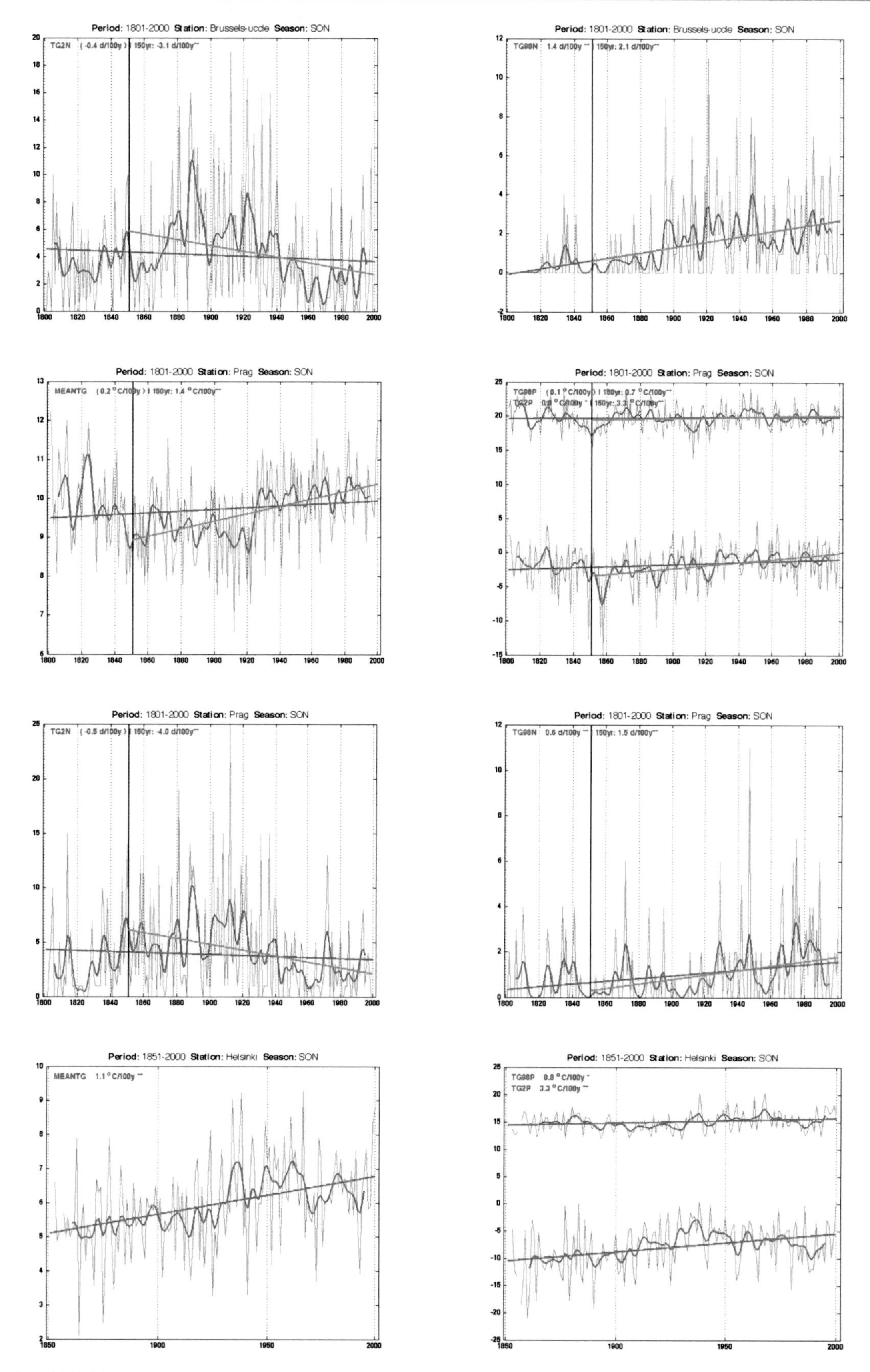

Fig. 3.82 1851–2000 SON Tmean Brussels-uccle

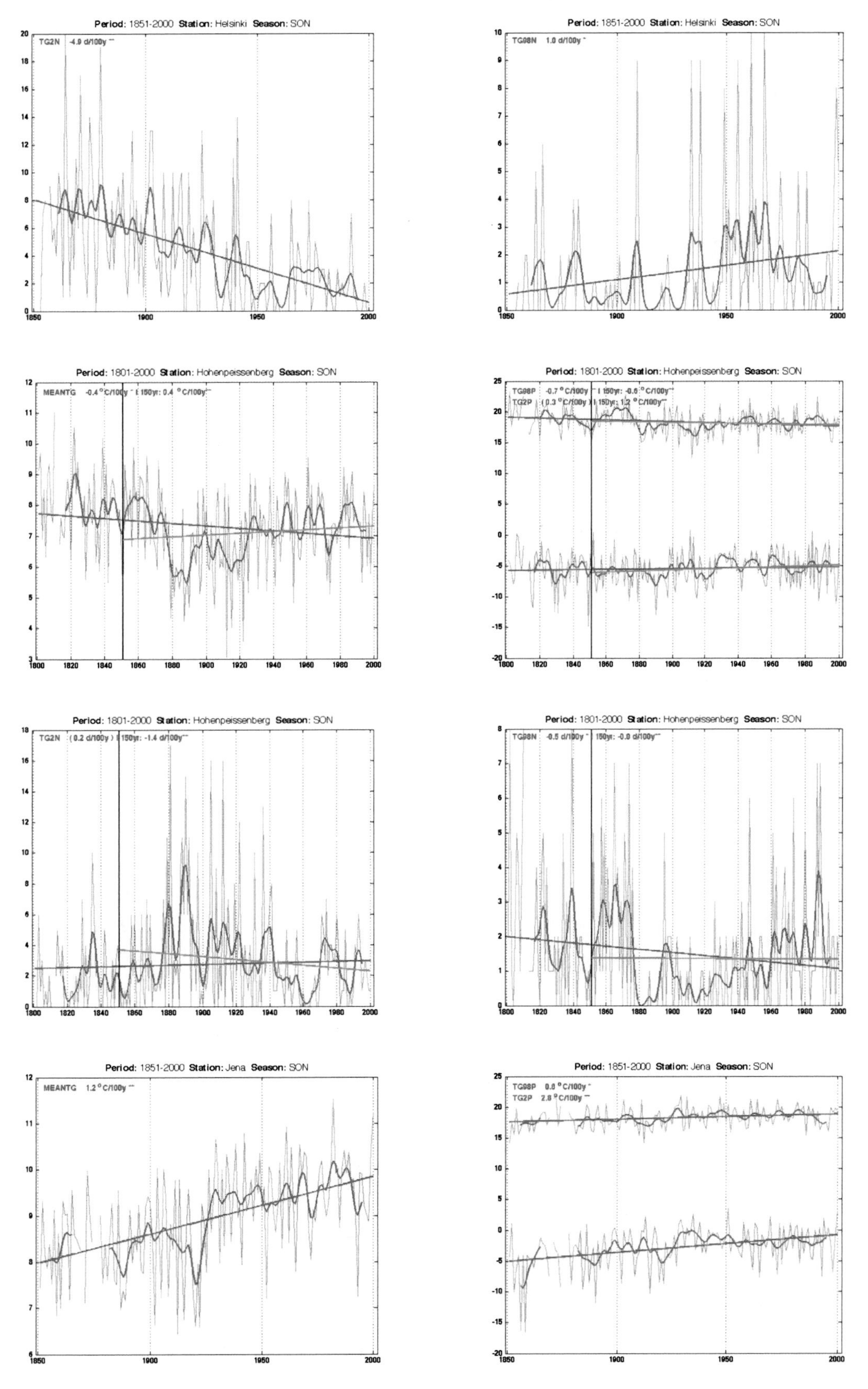

Fig. 3.83 1851–2000 SON Tmean Helsinki

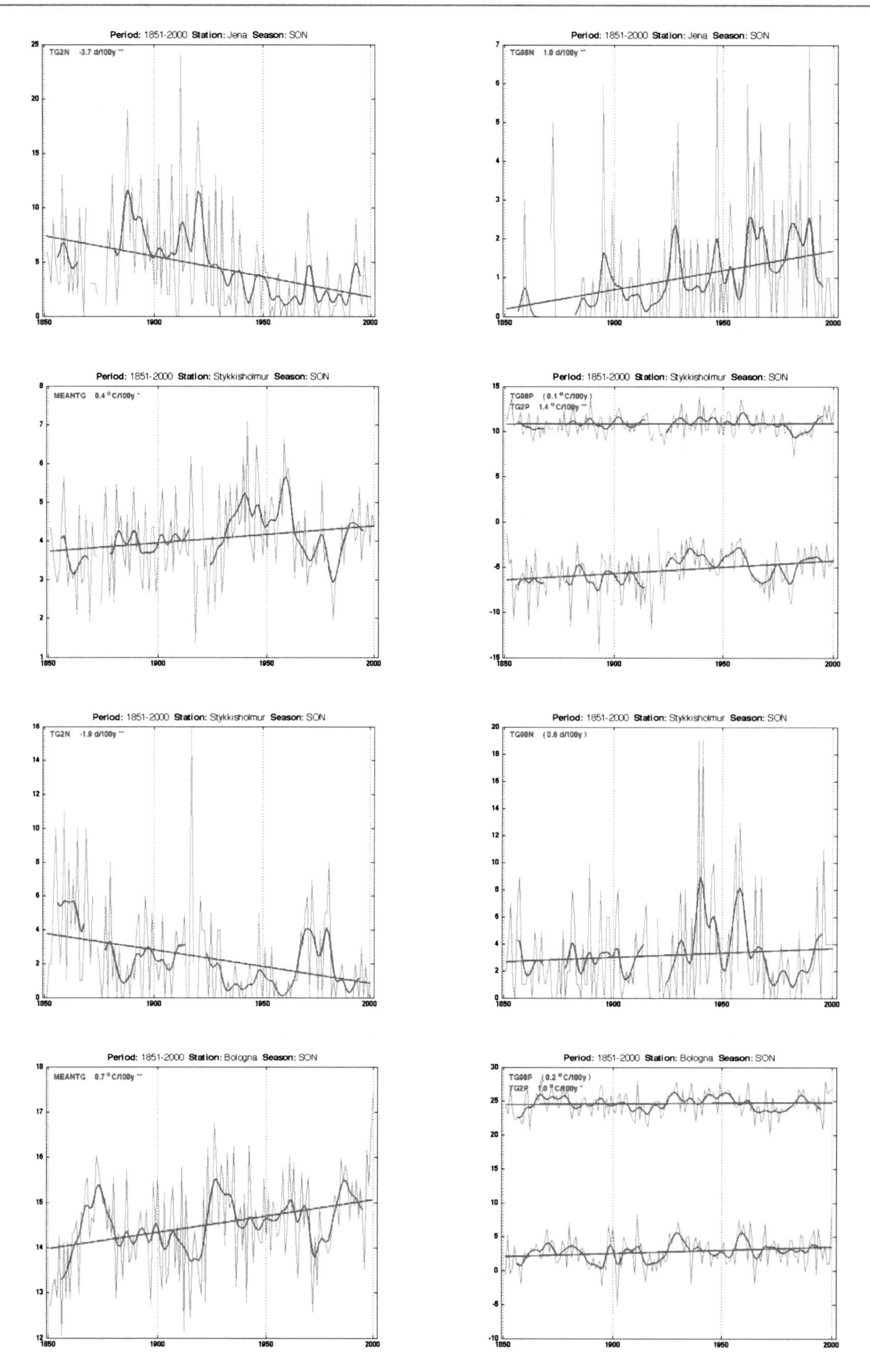

Fig. 3.84 1851–2000 SON Tmean Jena

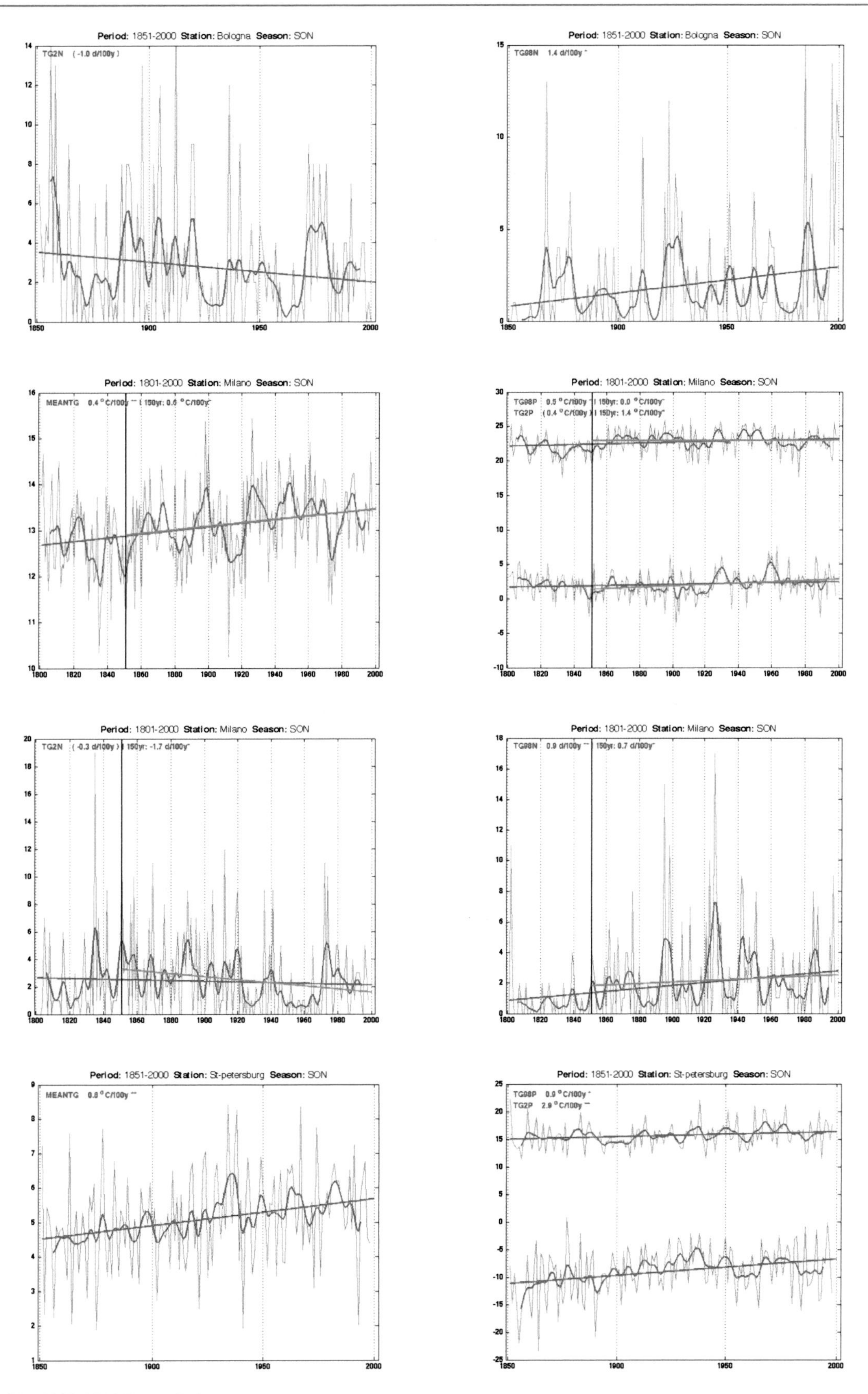

Fig. 3.85 1851–2000 SON Tmean Bologna

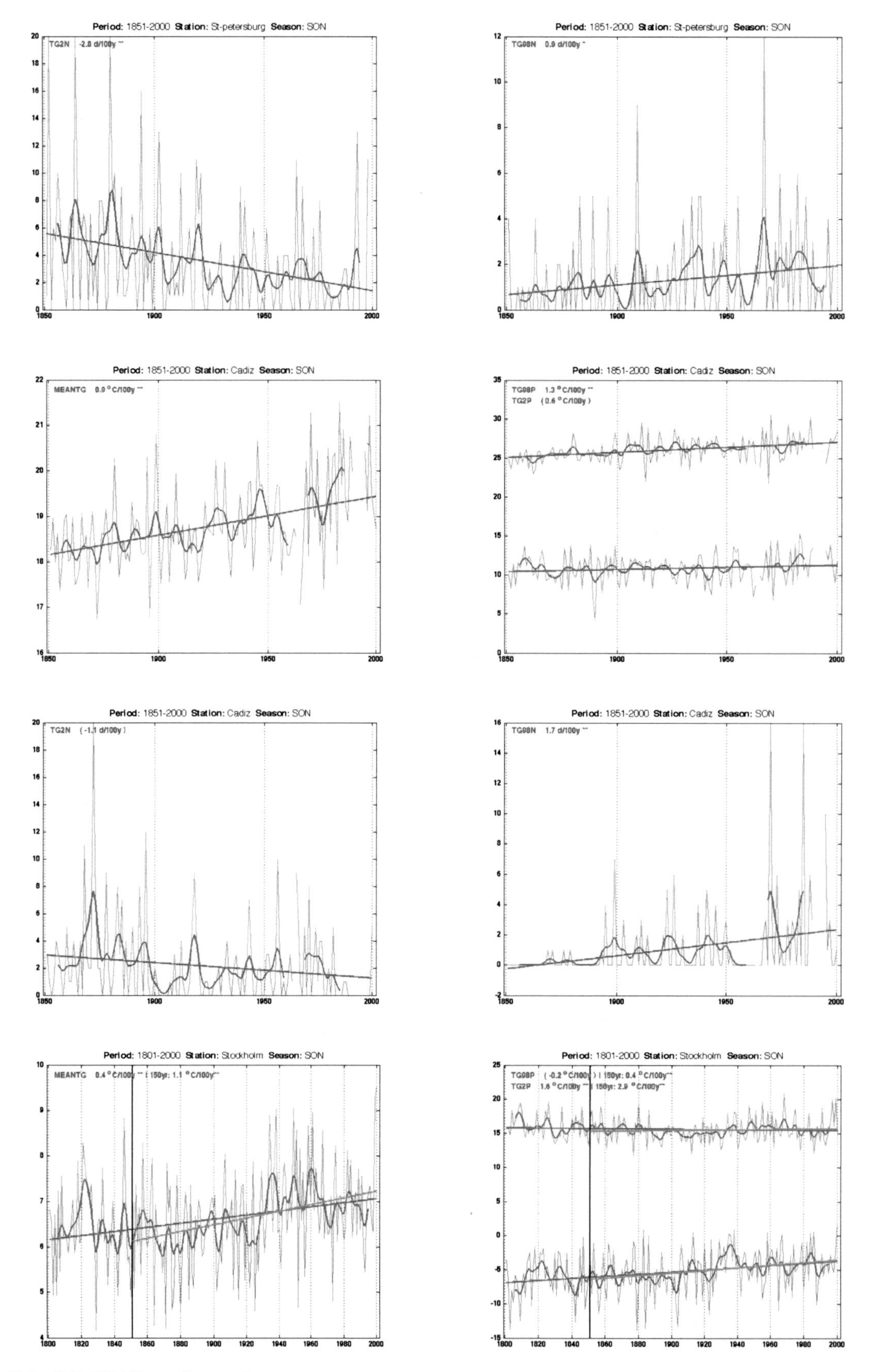

Fig. 3.86 1851–2000 SON Tmean St-petersburg

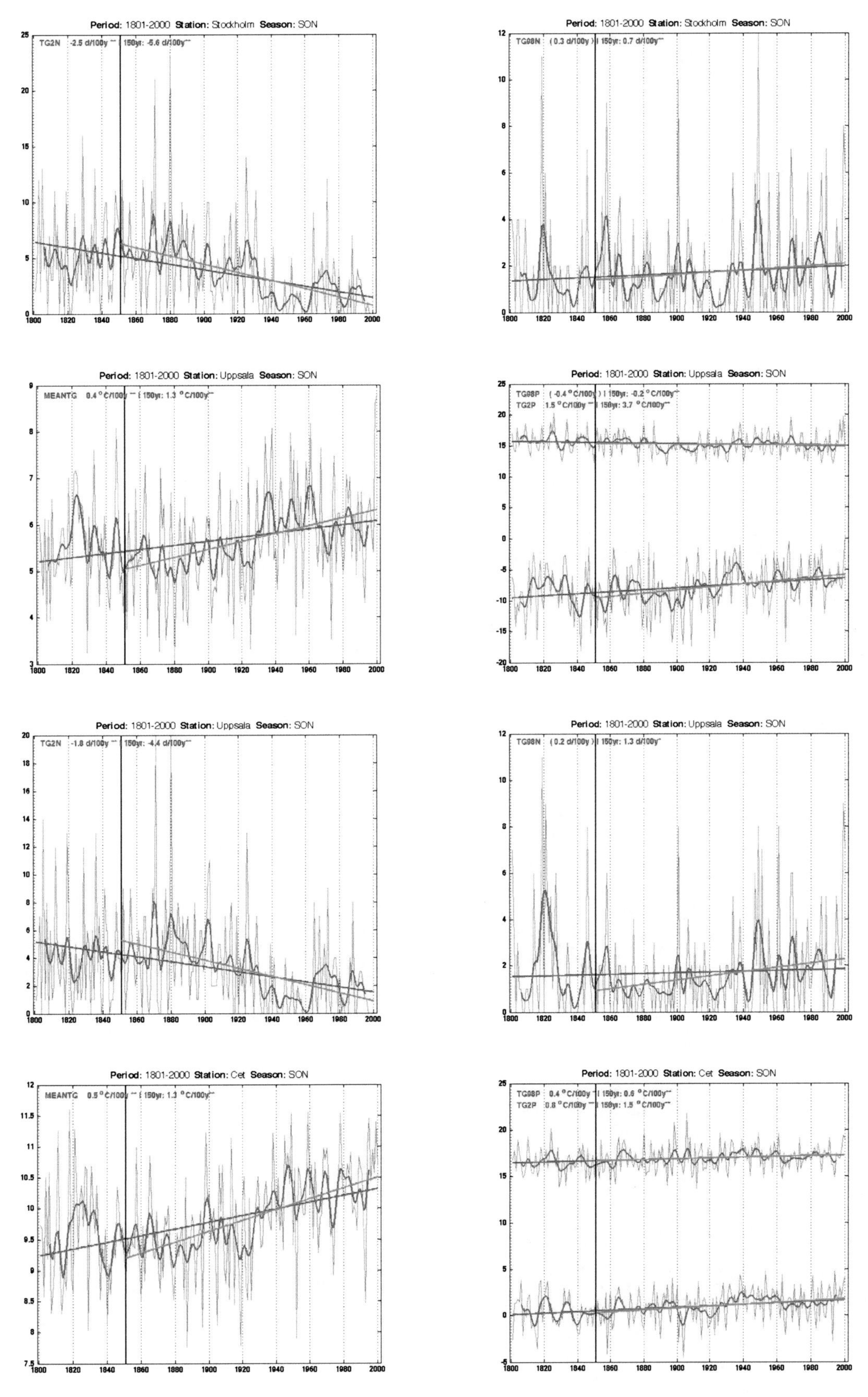

Fig. 3.87 1851–2000 SON Tmean Stockholm

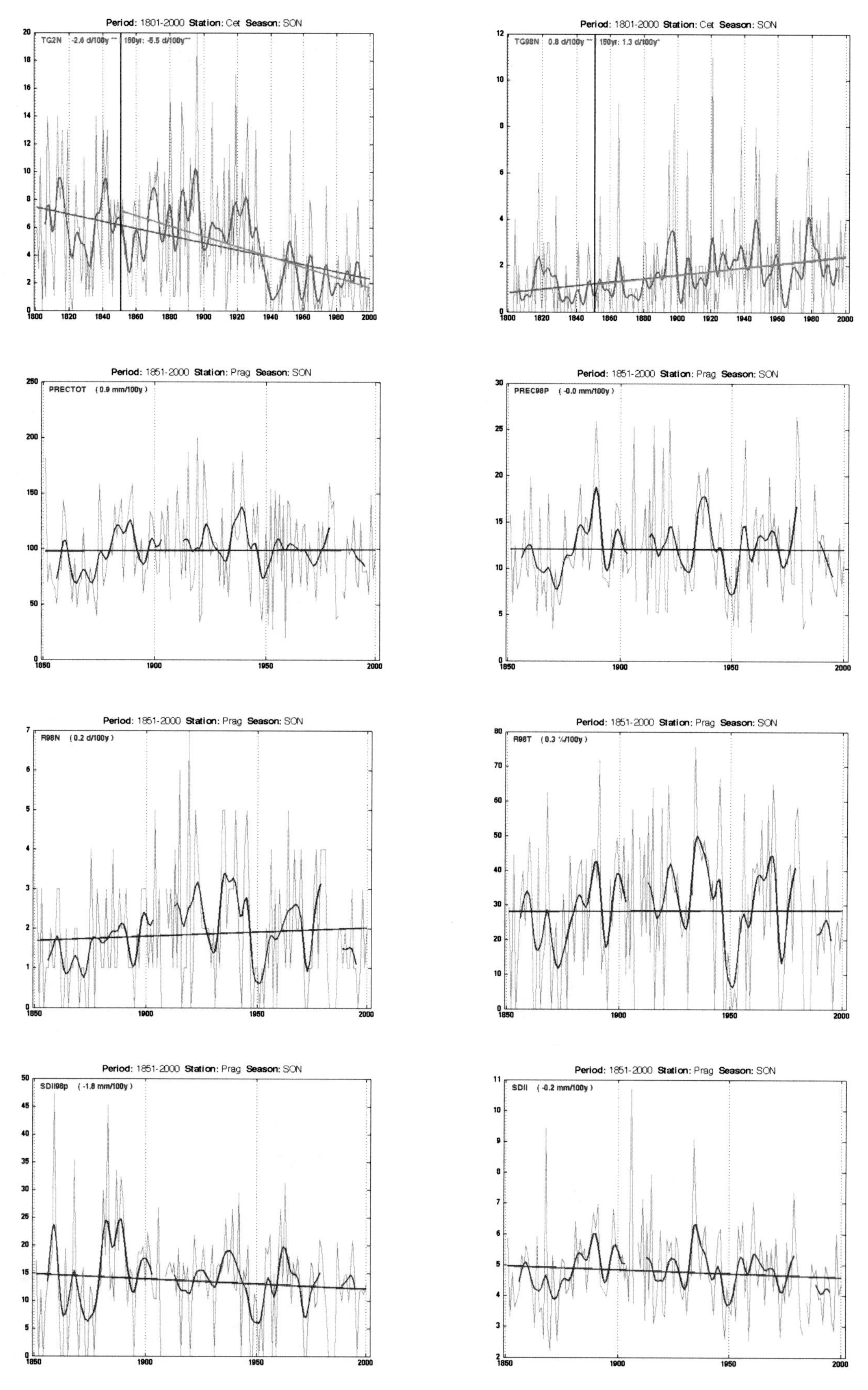

Fig. 3.88 1851–2000 SON Tmean Cet

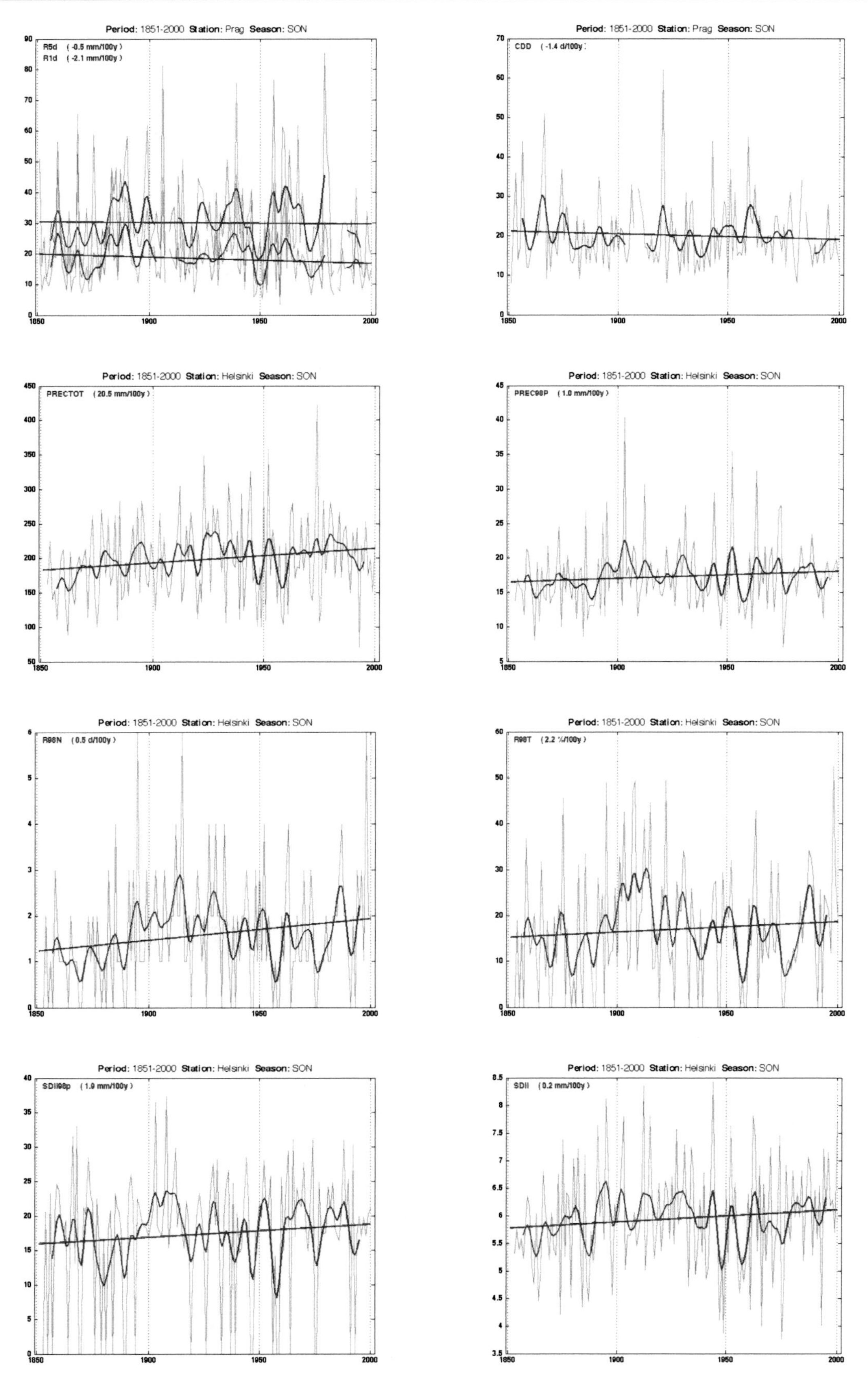

Fig. 3.89 1851–2000 SON Prec Prag

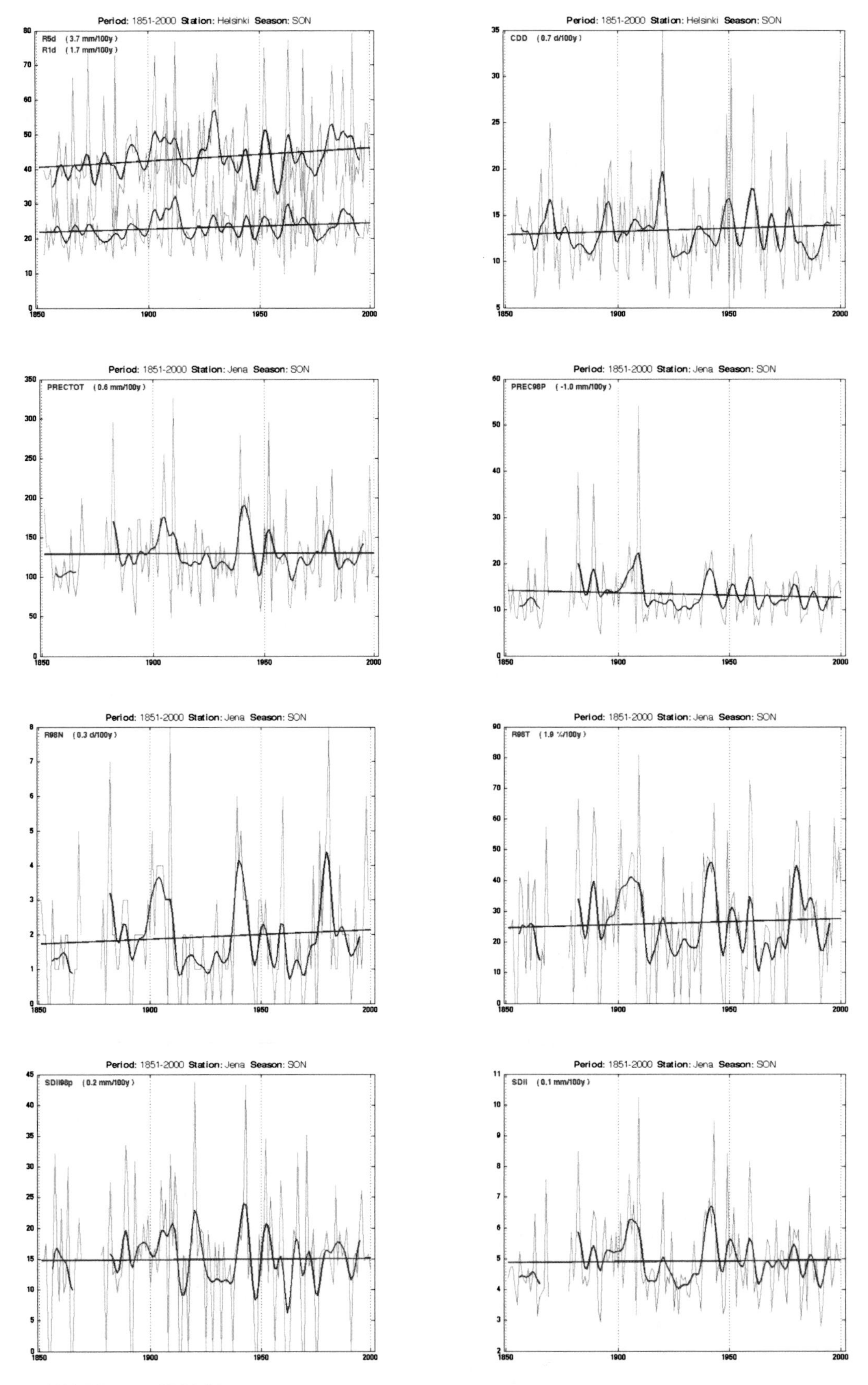

Fig. 3.90 1851–2000 SON Prec Helsinki

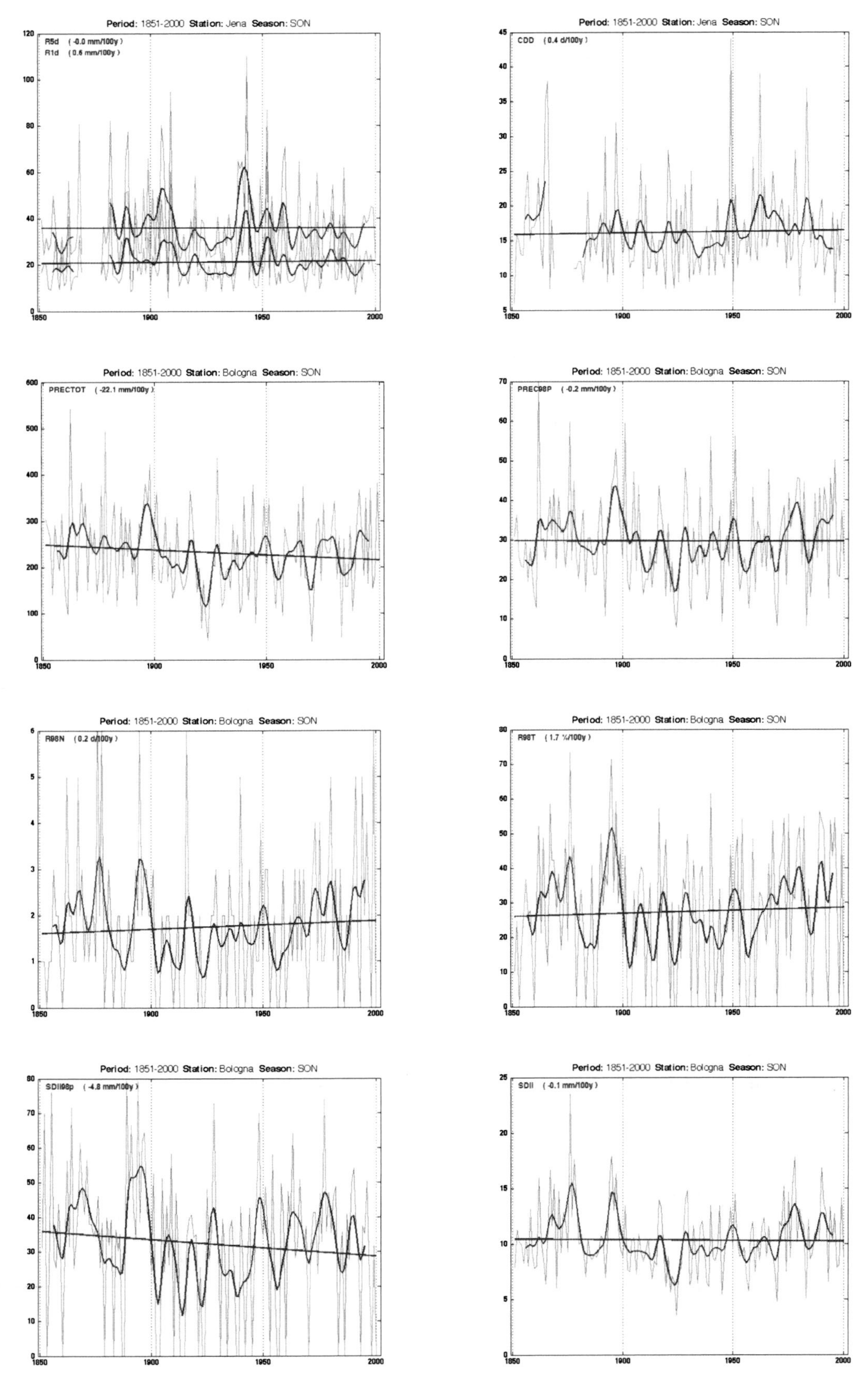

Fig. 3.91 1851–2000 SON Prec Jena

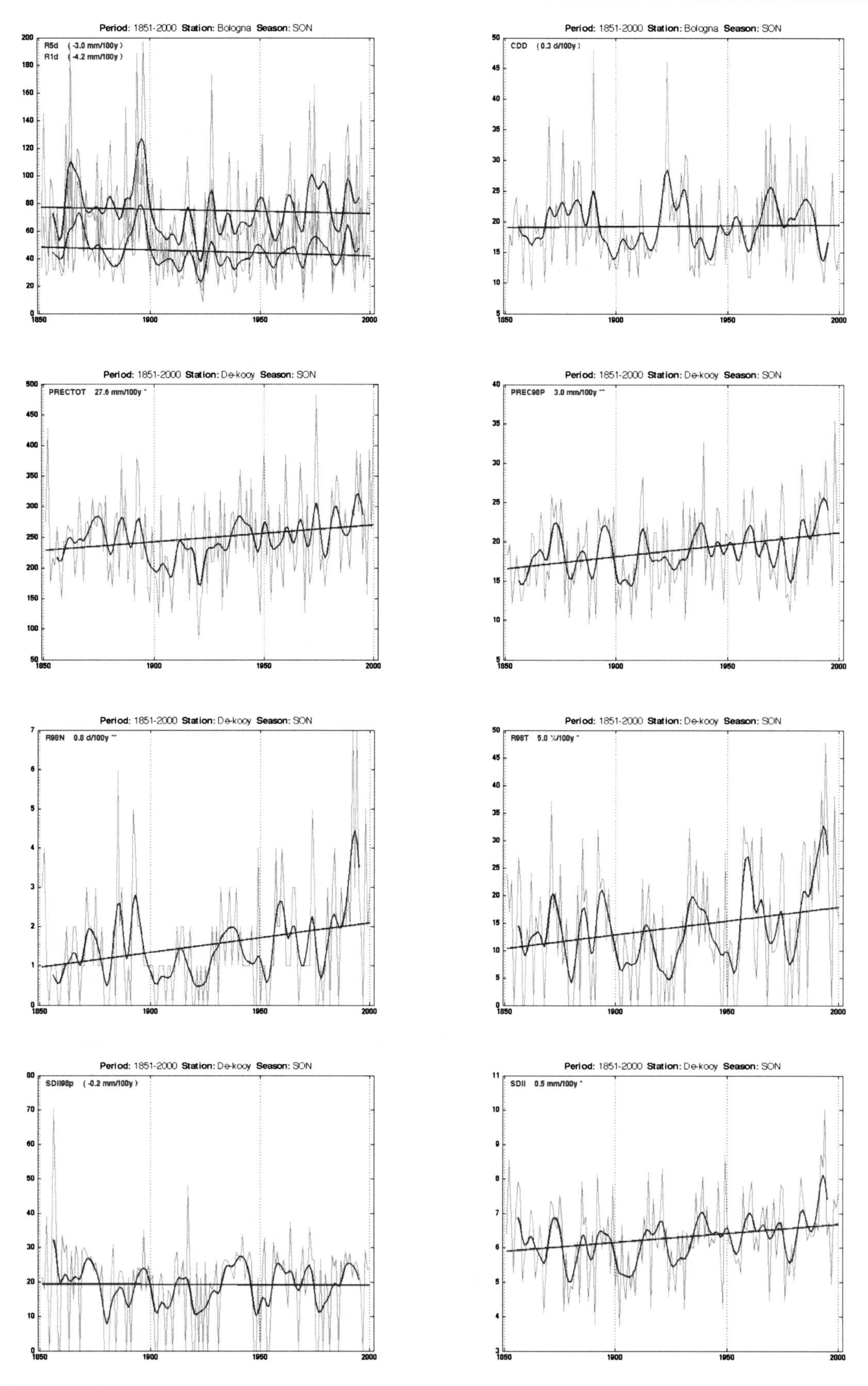

Fig. 3.92 1851–2000 SON Prec Bologna

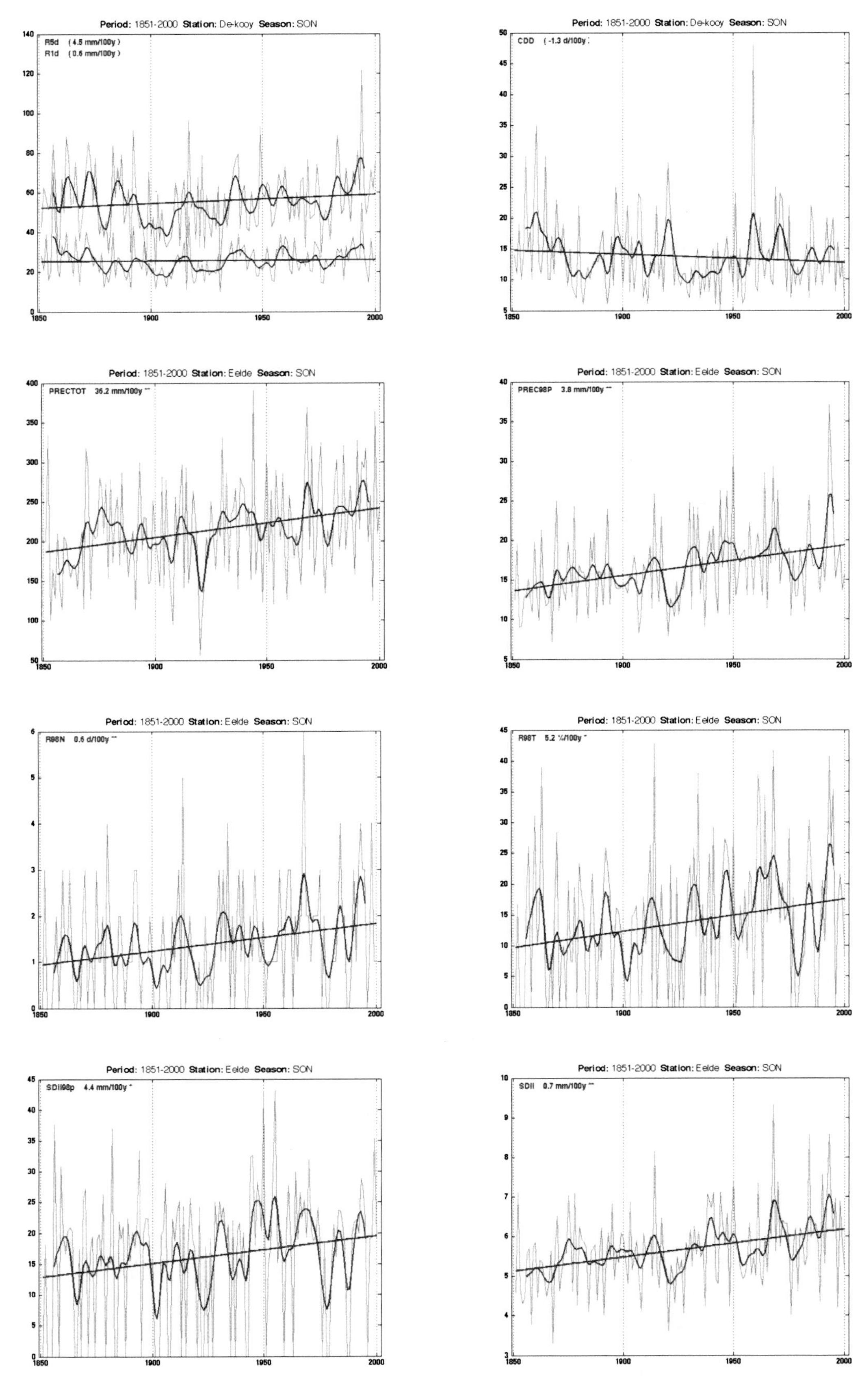

Fig. 3.93 1851–2000 SON Prec De-kooy

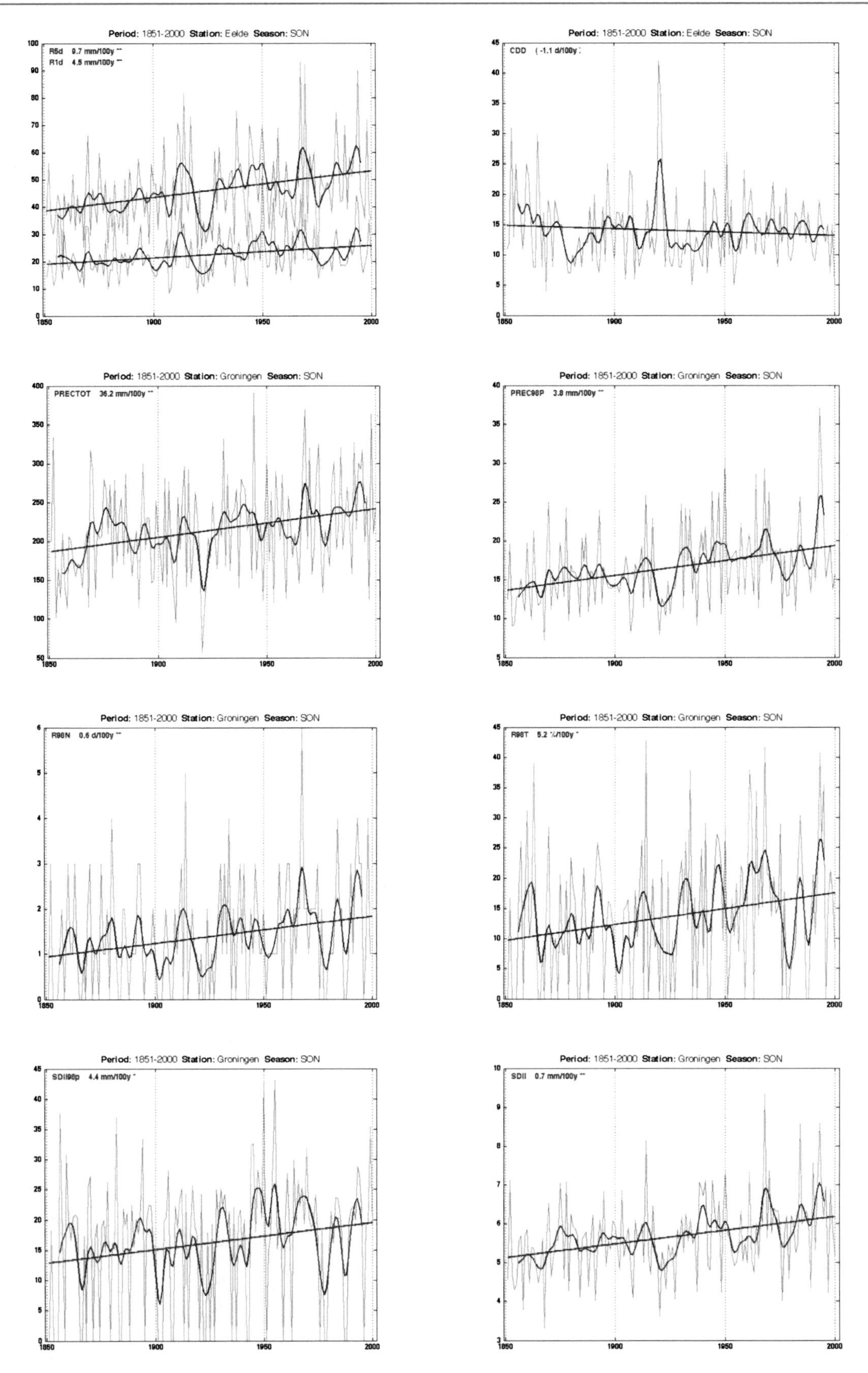

Fig. 3.94 1851–2000 SON Prec Eelde

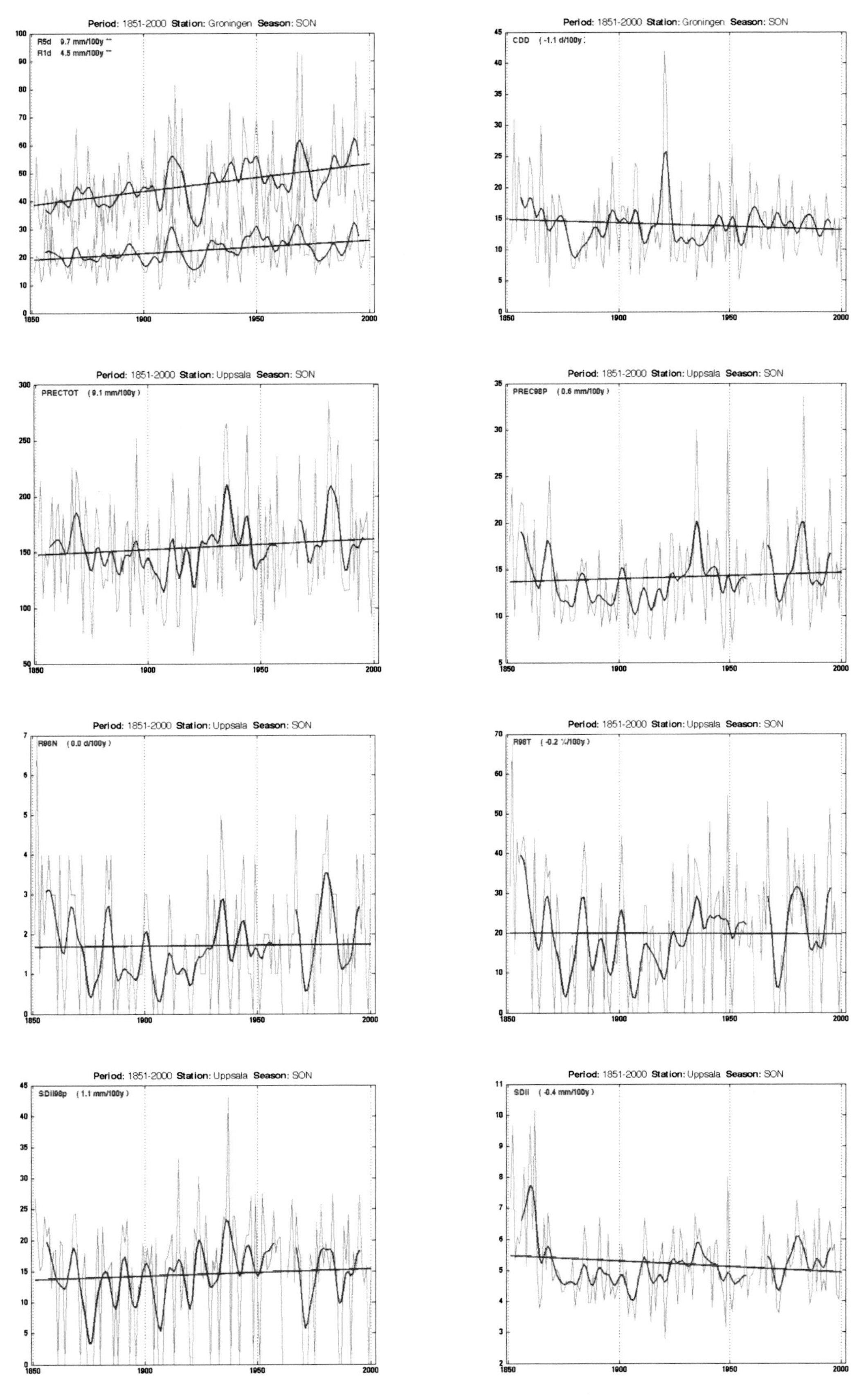

Fig. 3.95 1851–2000 SON Prec Groningen

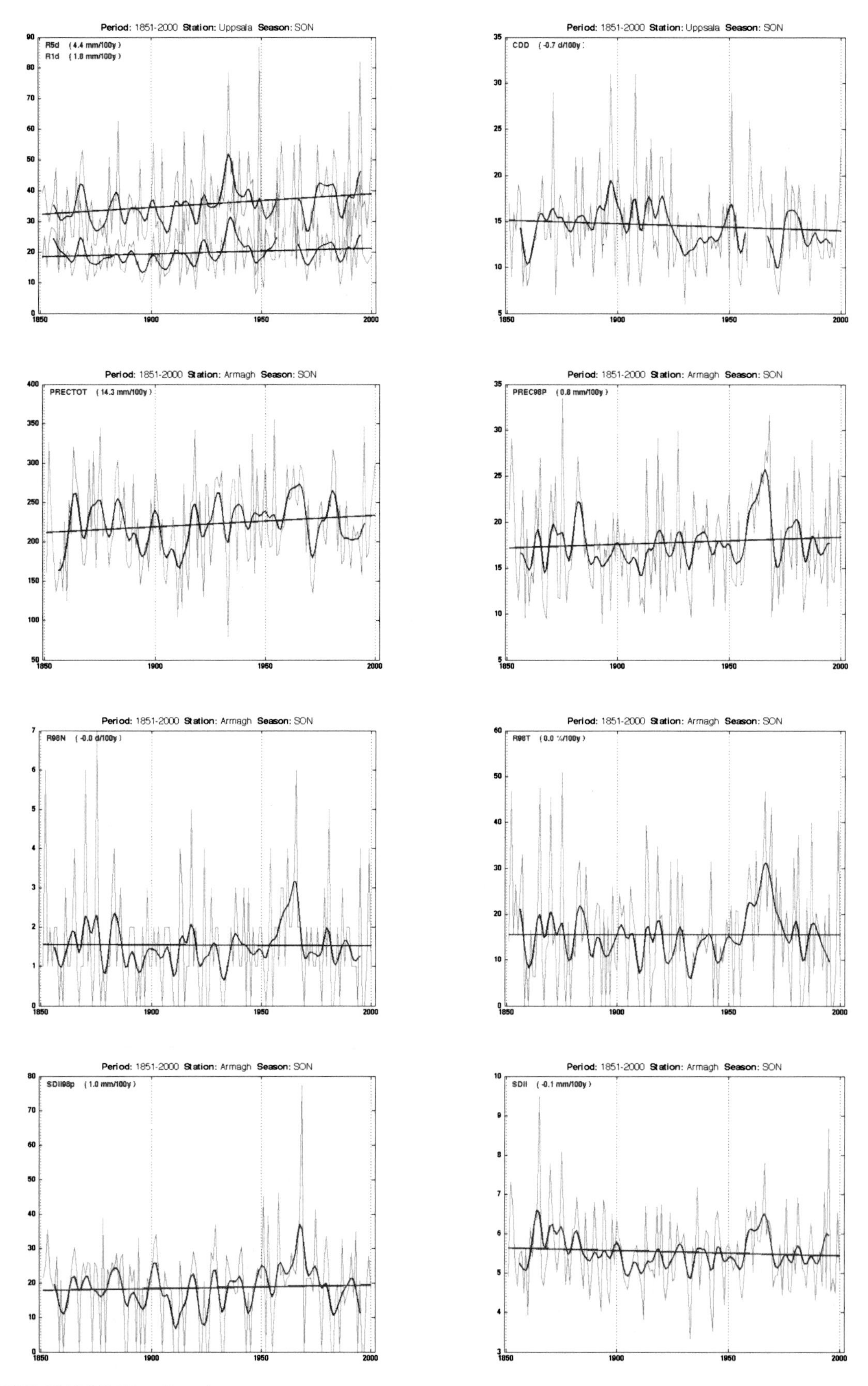

Fig. 3.96 1851–2000 SON Prec Uppsala

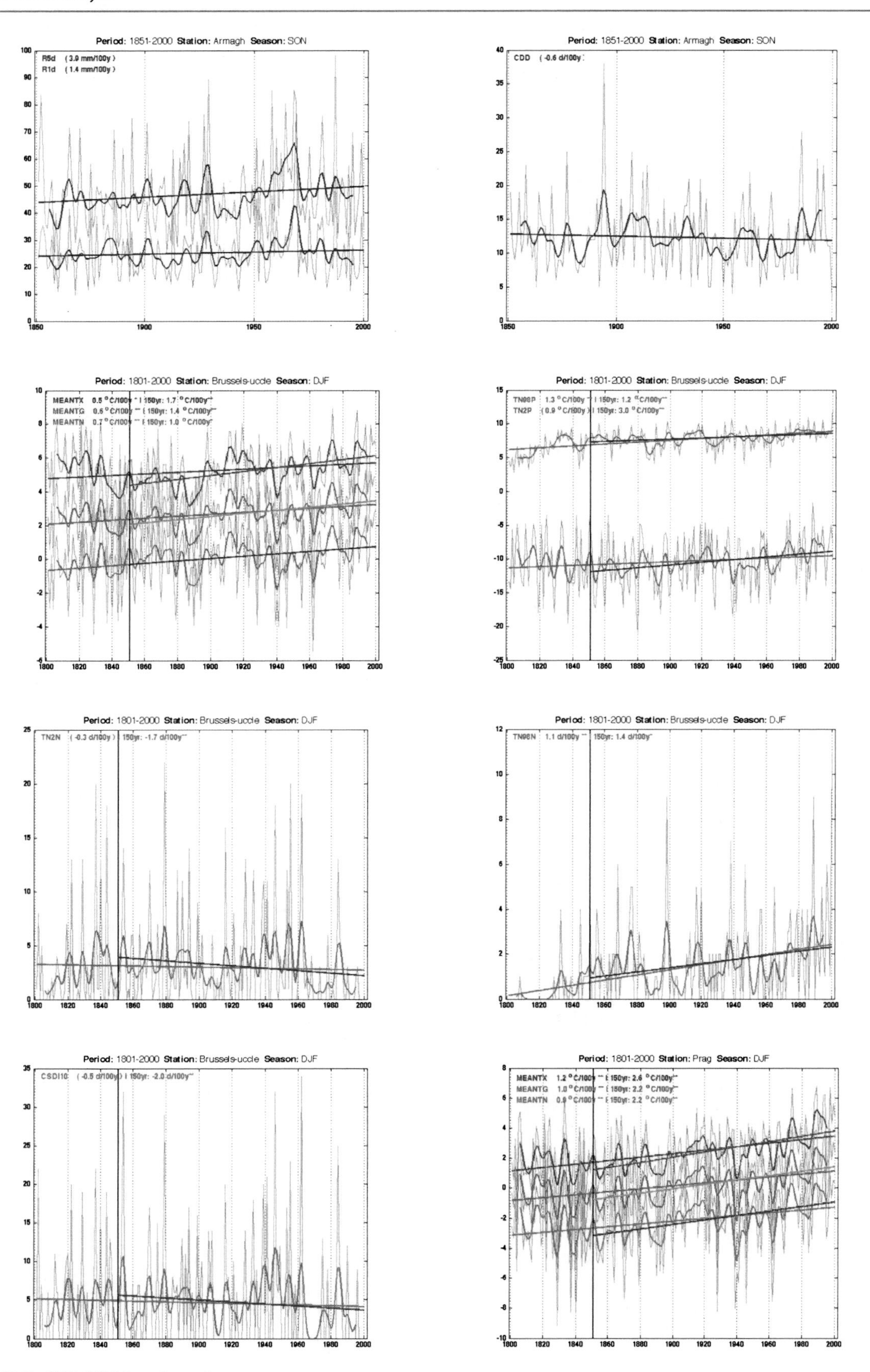

Fig. 3.97 1851–2000 SON Prec Armagh

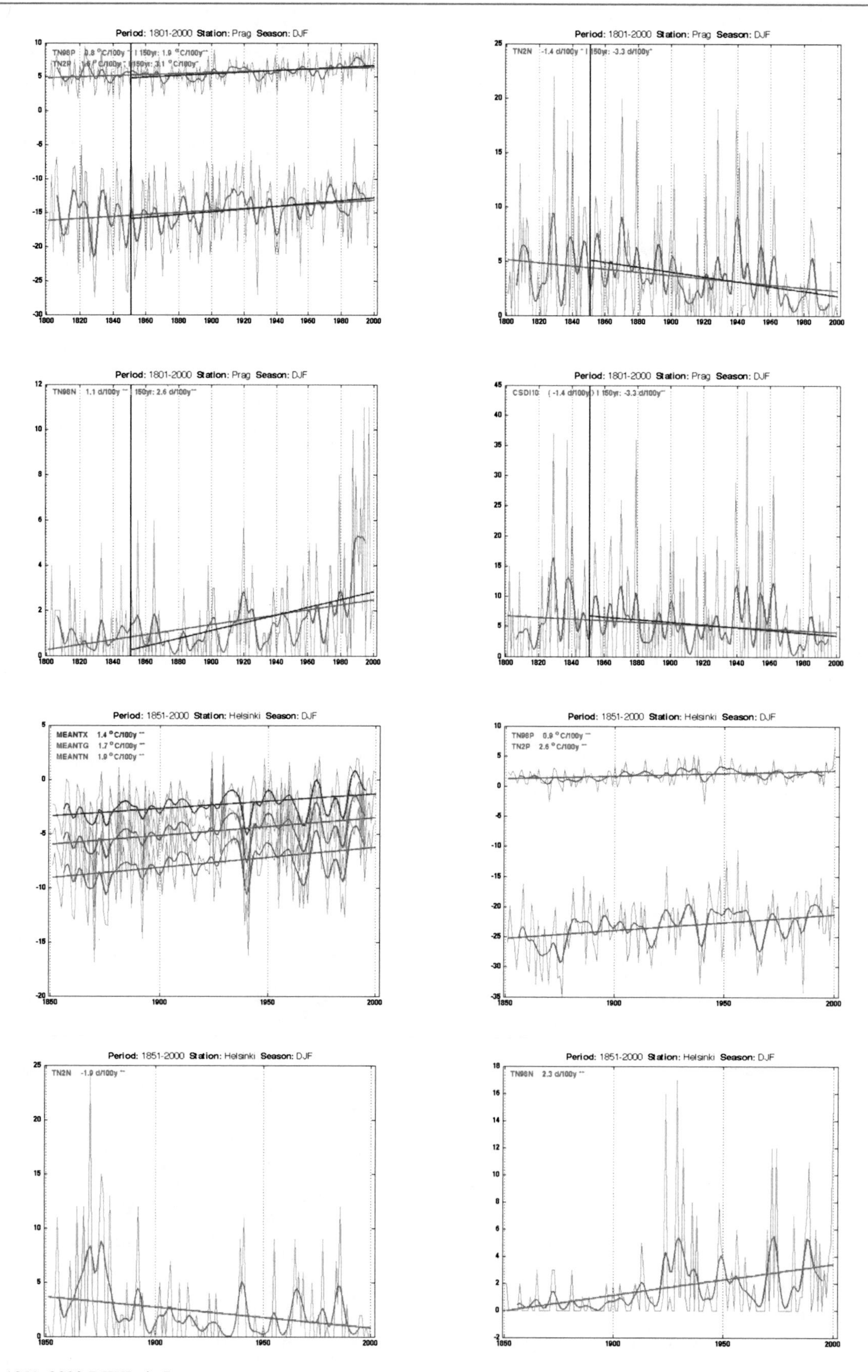

Fig. 3.98 1851–2000 DJF Tmin Prag

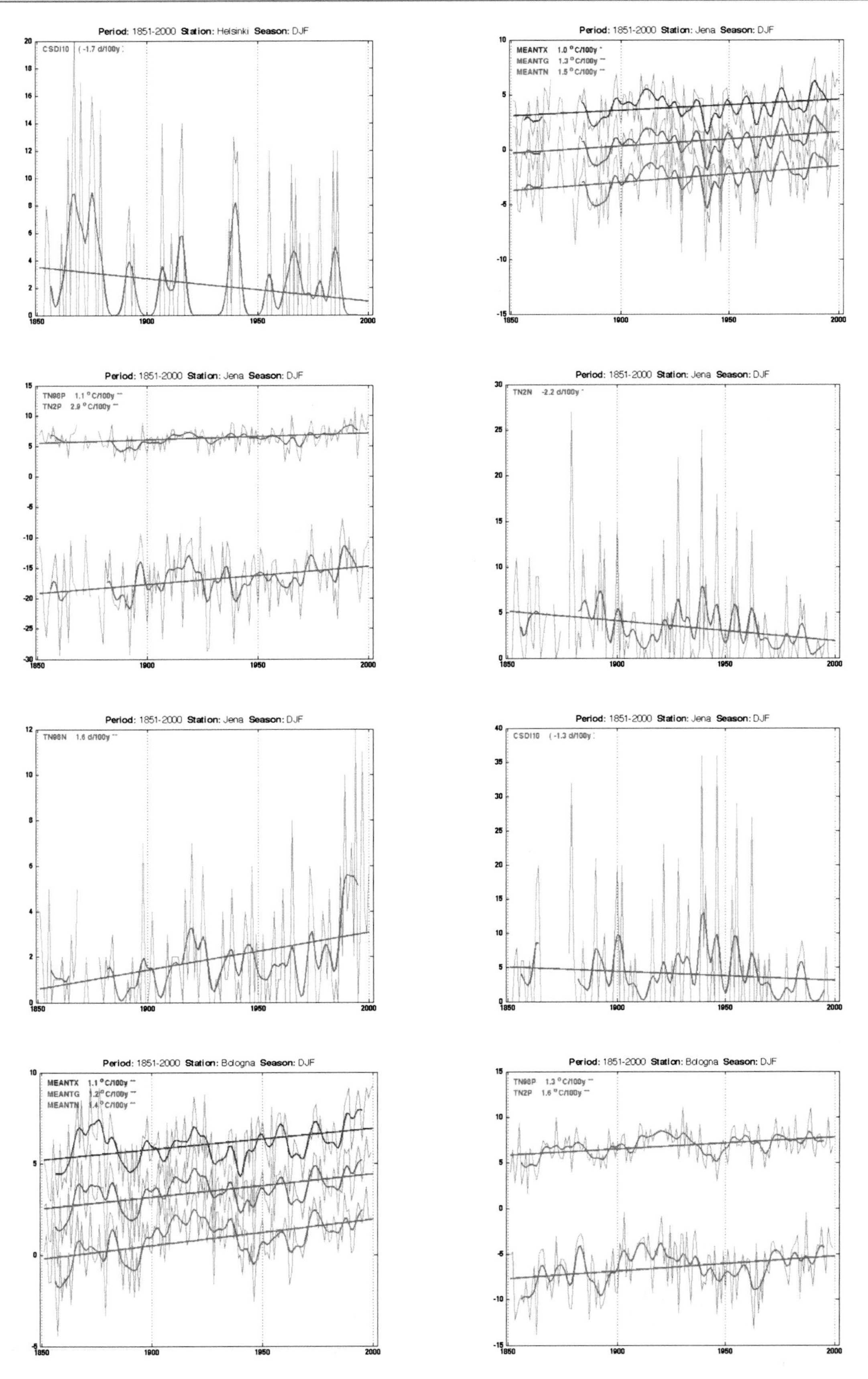

Fig. 3.99 1851–2000 DJF Tmin Helsinki

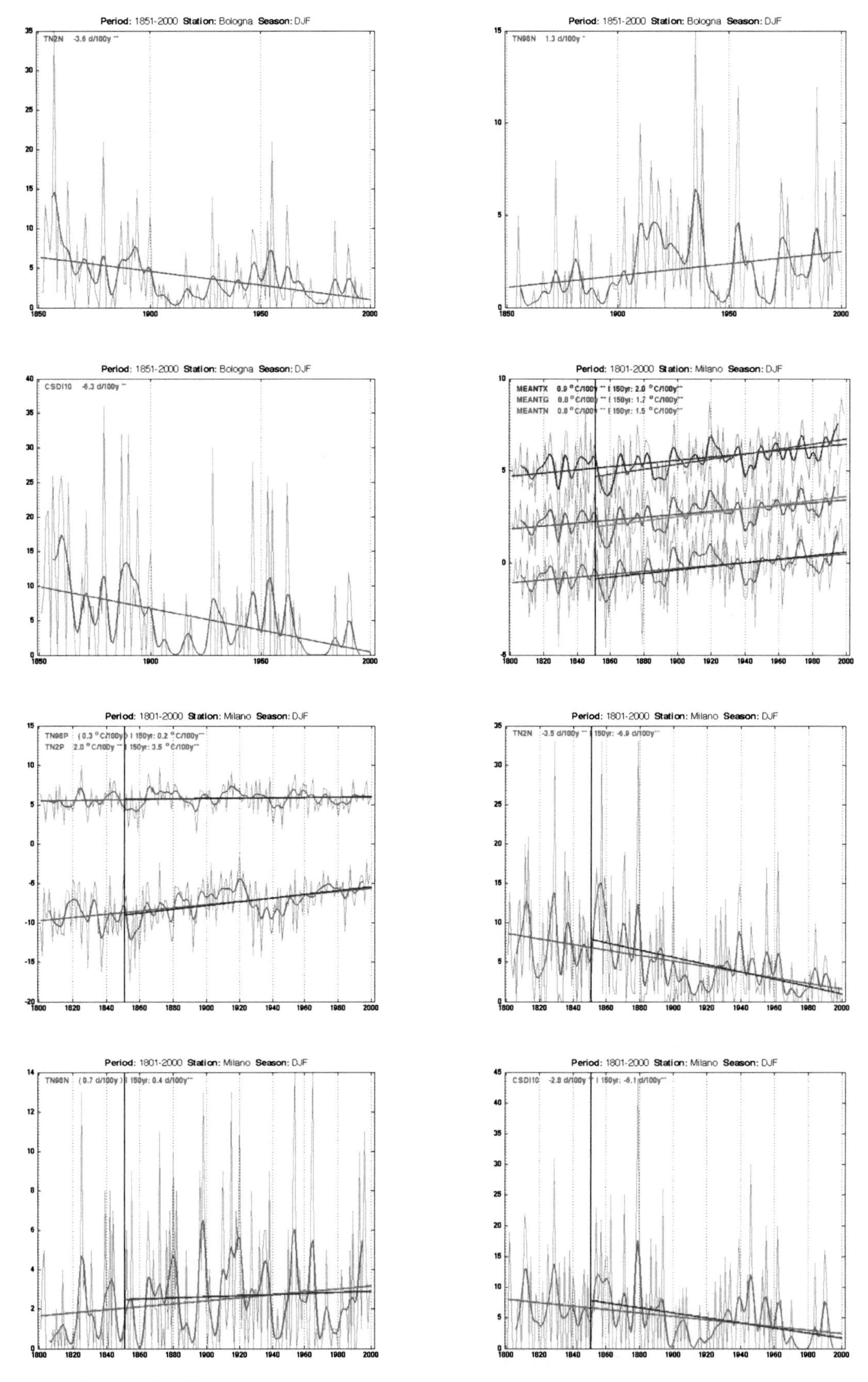

Fig. 3.100 1851–2000 DJF Tmin Bologna

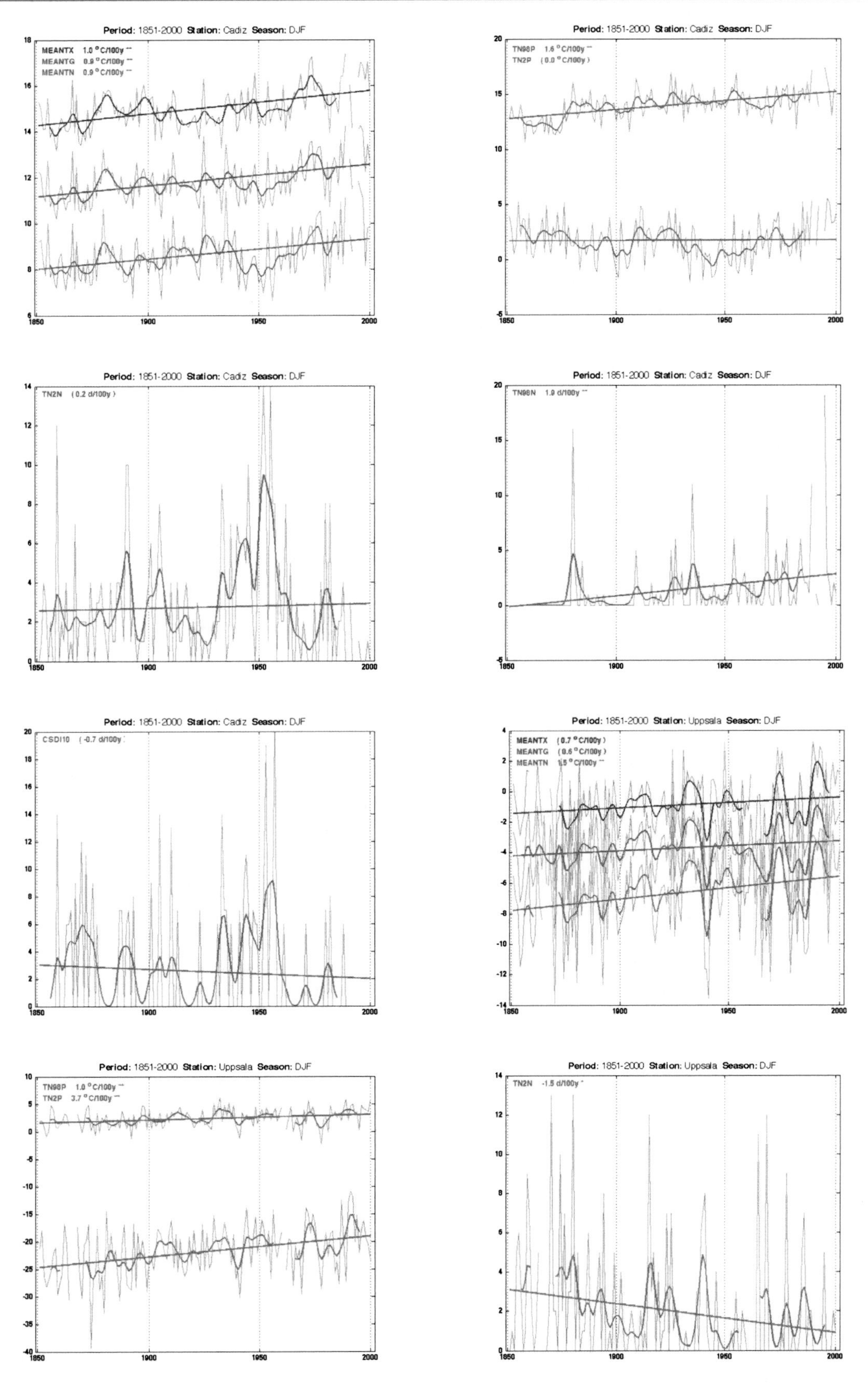

Fig. 3.101 1851–2000 DJF Tmin Cadiz

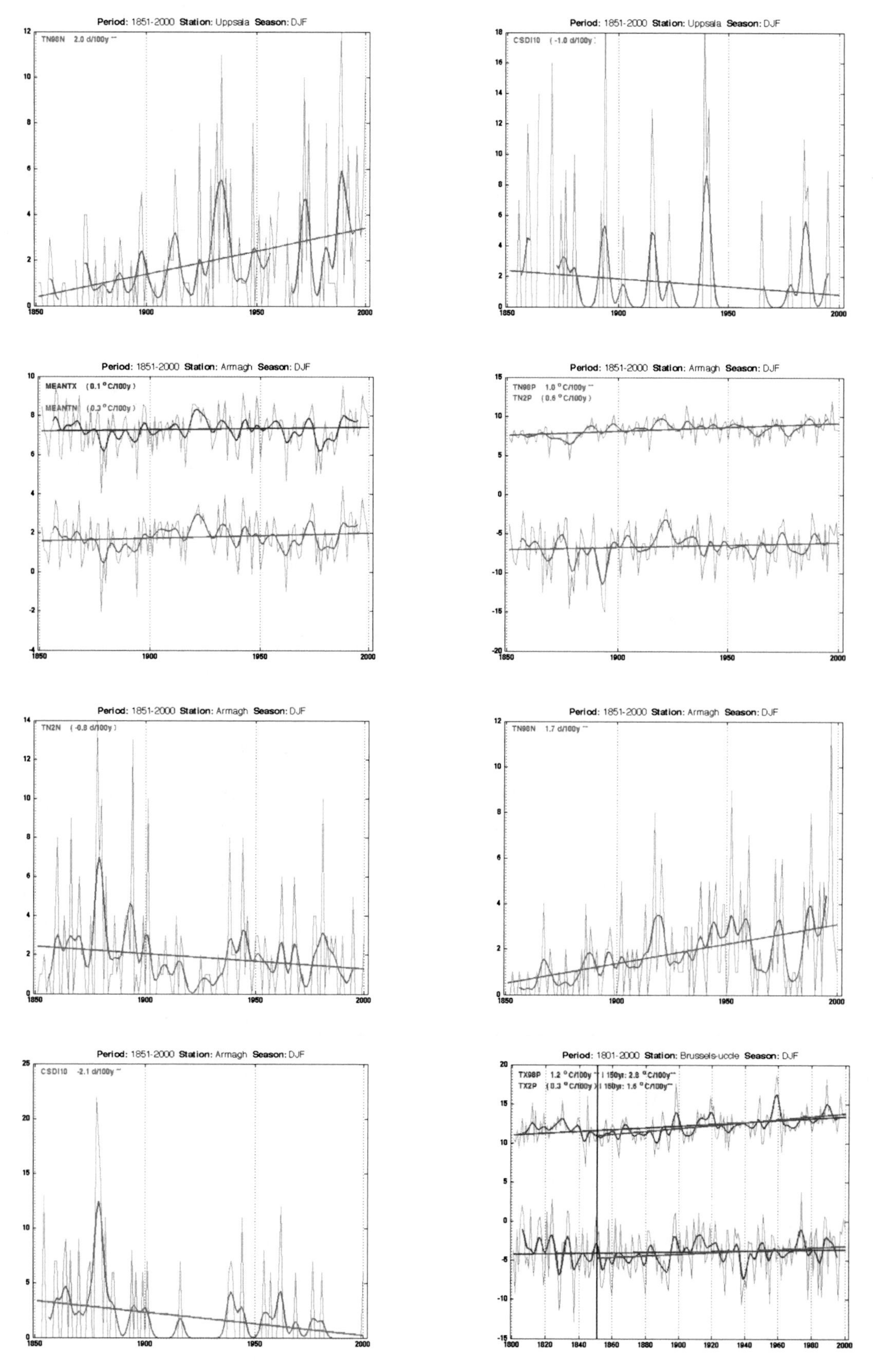

Fig. 3.102 1851–2000 DJF Tmin Uppsala

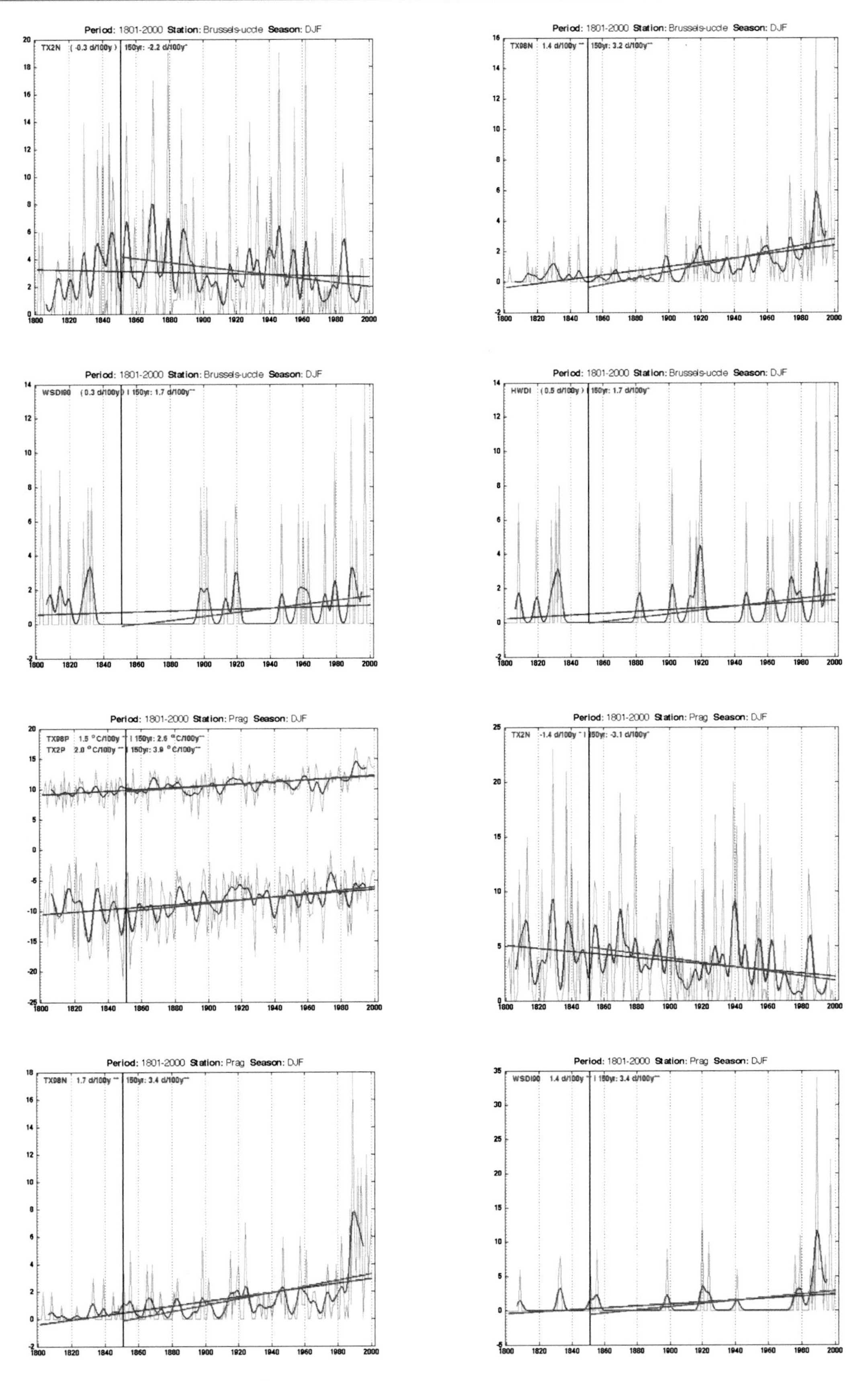

Fig. 3.103 1851–2000 DJF Tmax Brussels-uccle

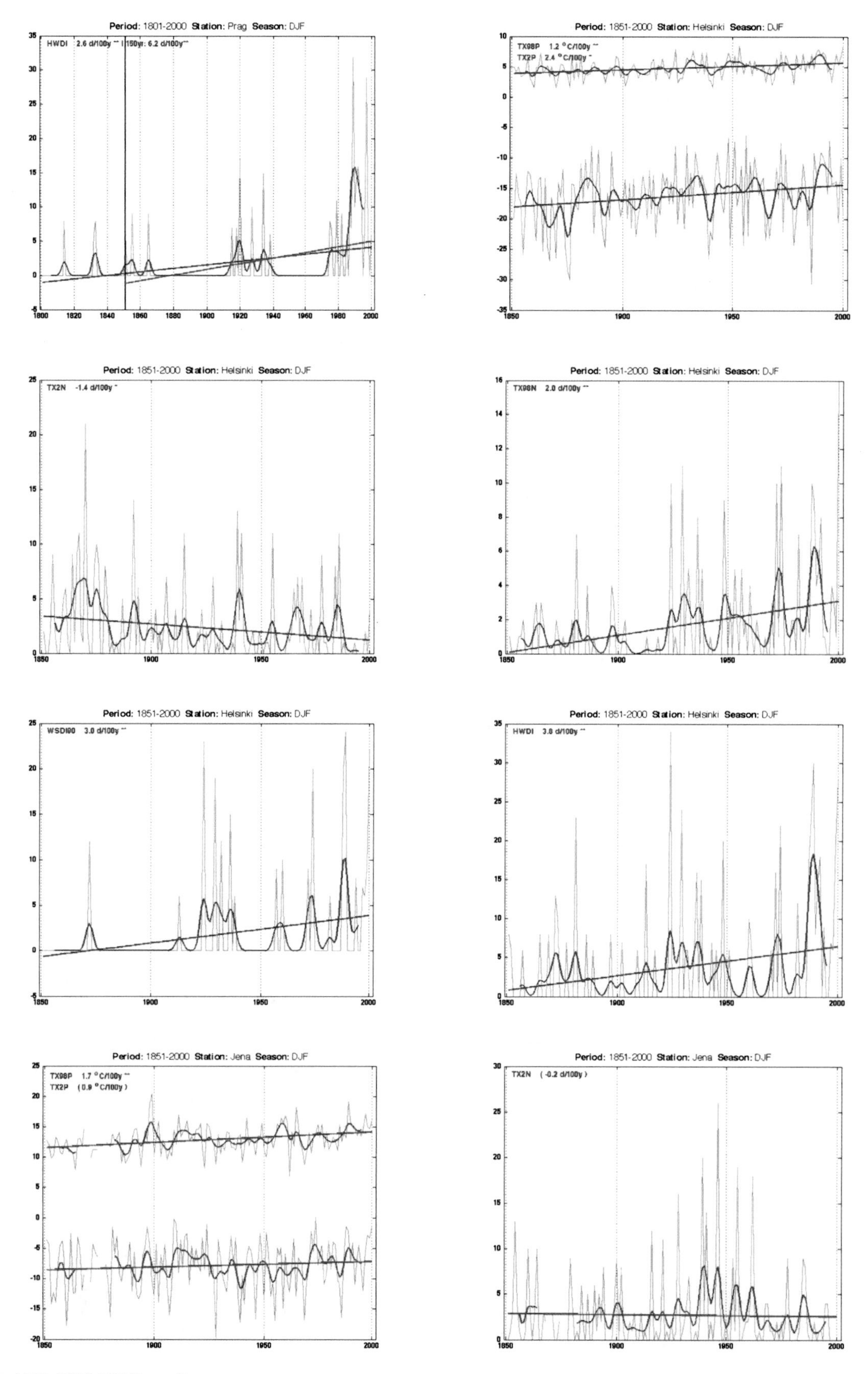

Fig. 3.104 1851–2000 DJF Tmax Prag

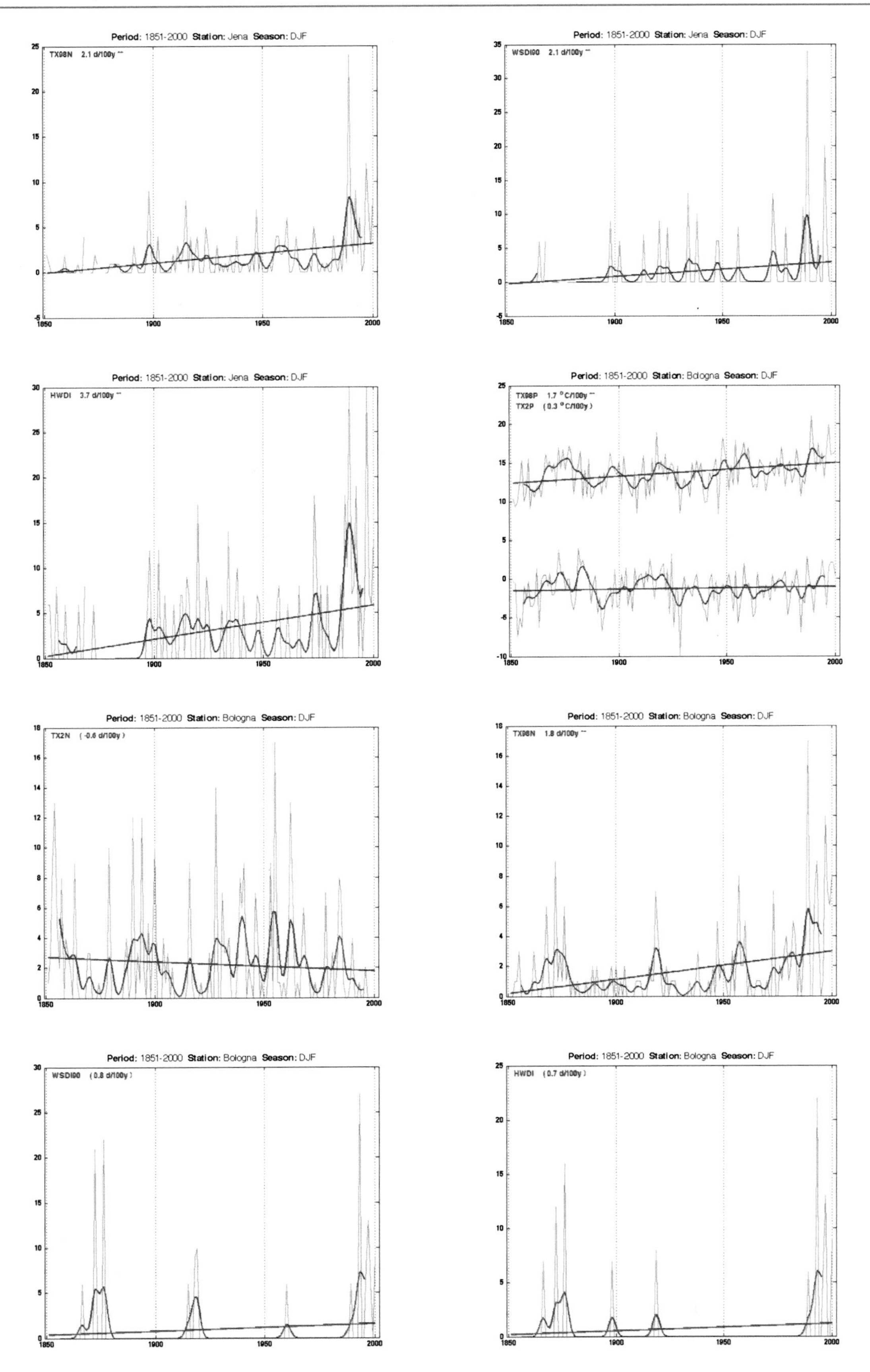

Fig. 3.105 1851–2000 DJF Tmax Jena

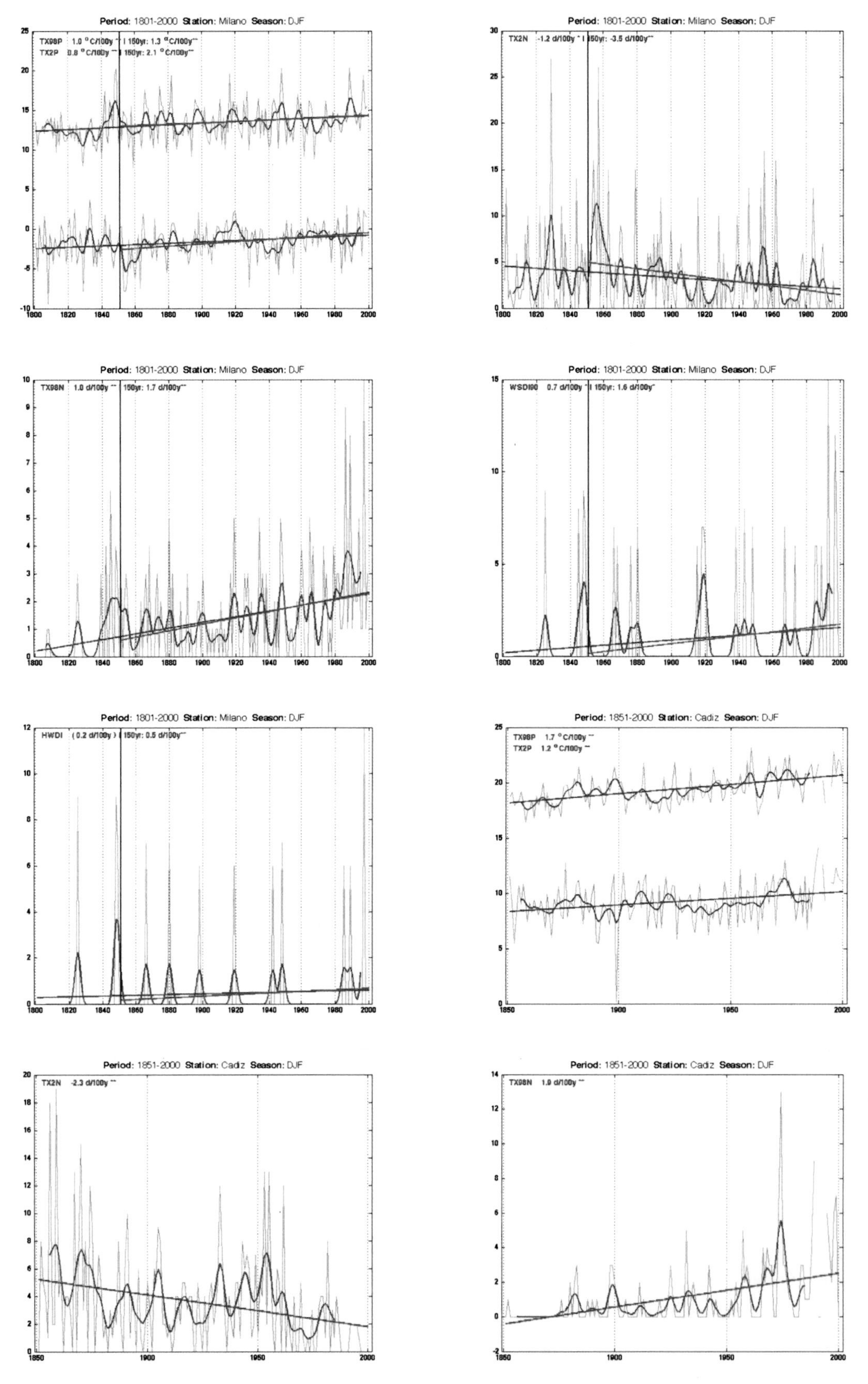

Fig. 3.106 1851–2000 DJF Tmax Milano

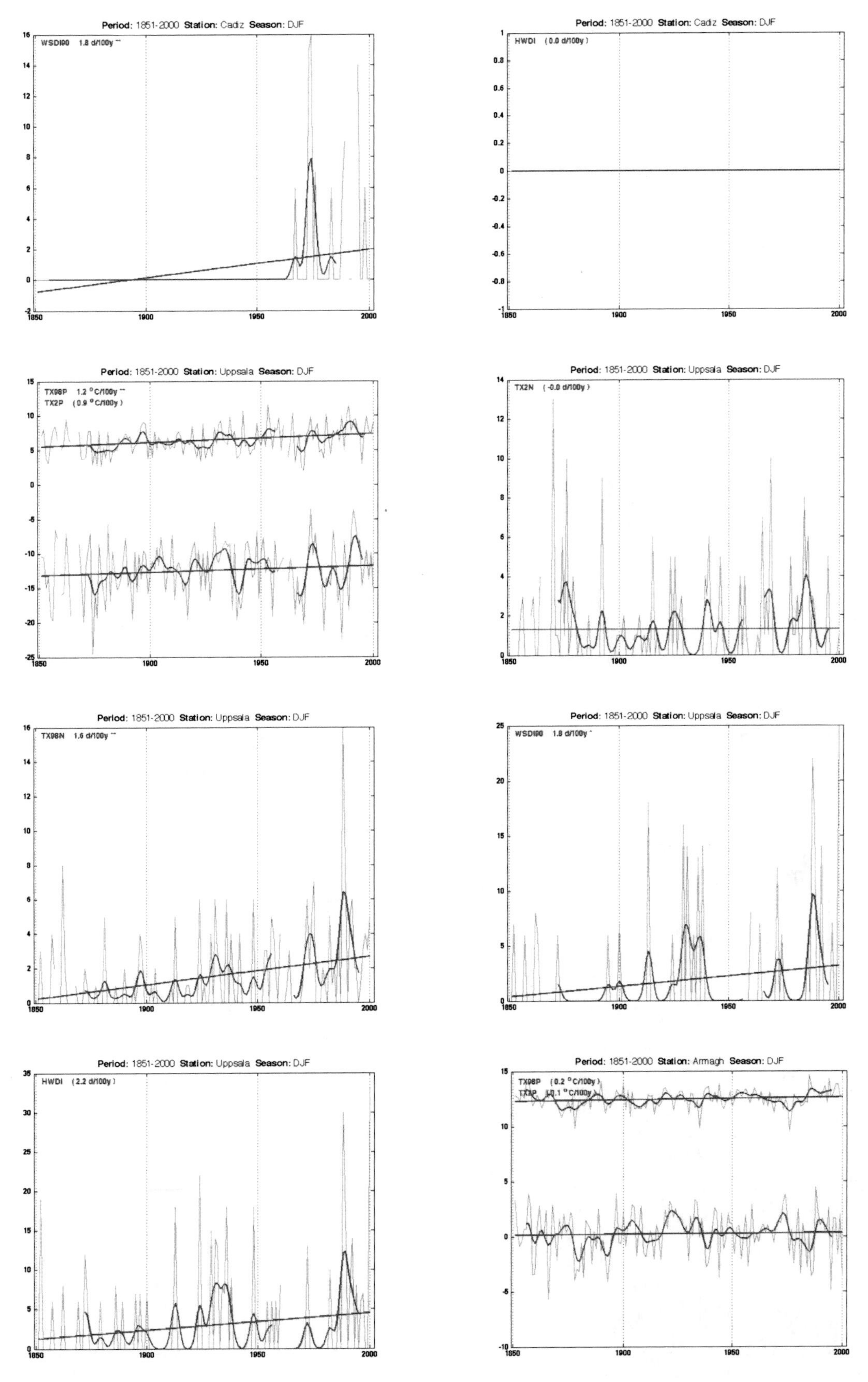

Fig. 3.107 1851–2000 DJF Tmax Cadiz

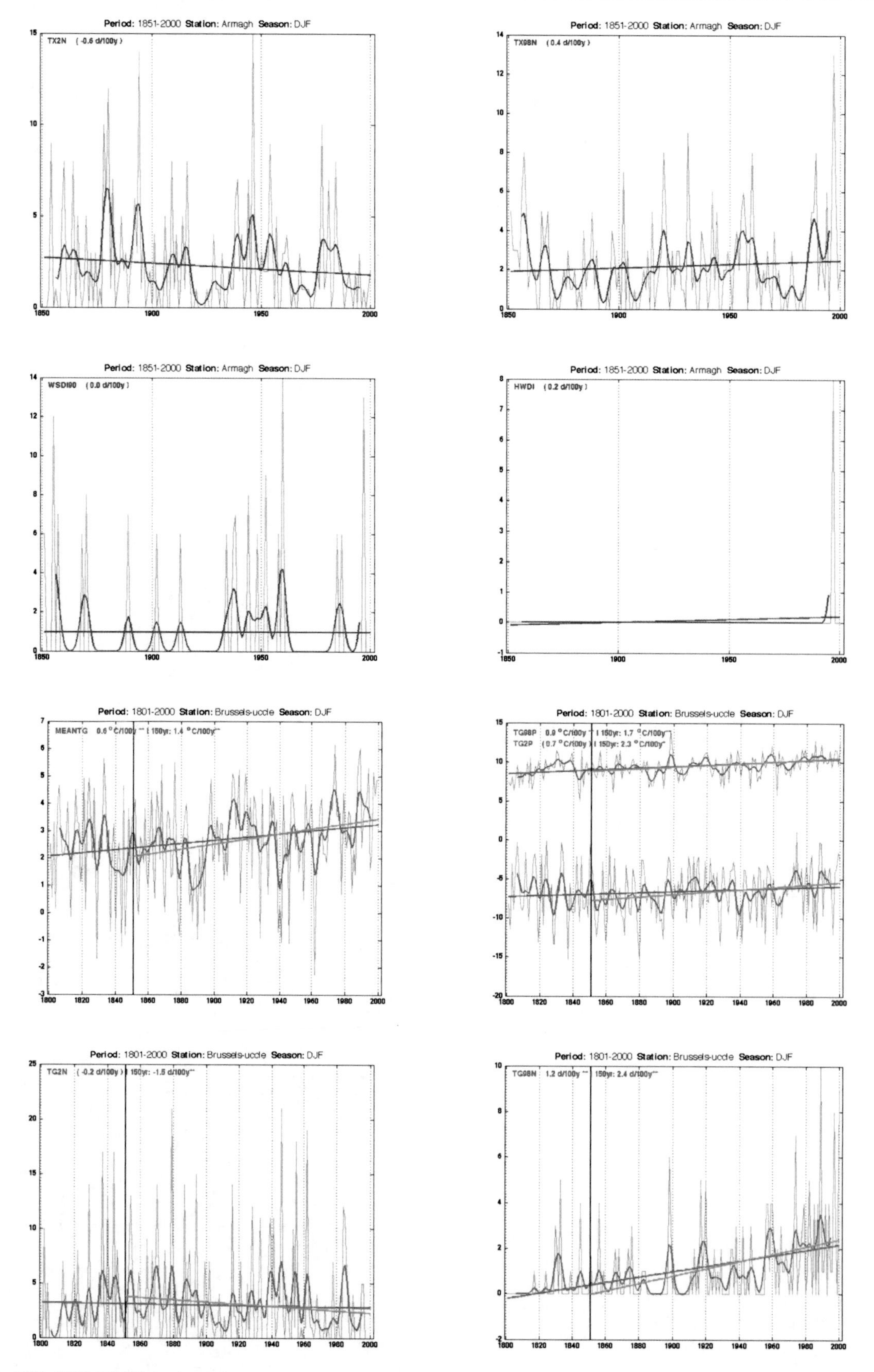

Fig. 3.108 1851–2000 DJF Tmax Armagh

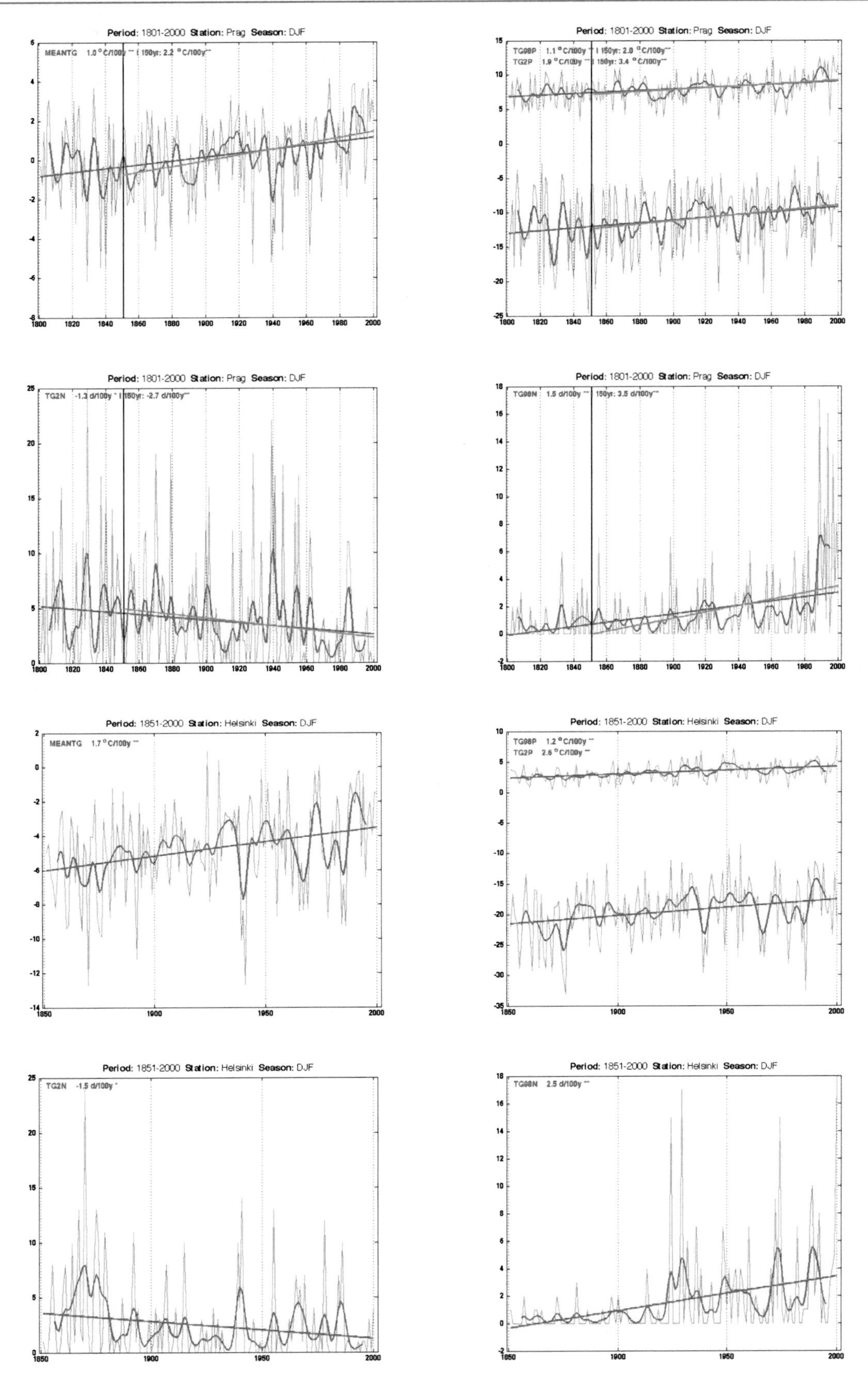

Fig. 3.109 1851–2000 DJF Tmean Prag

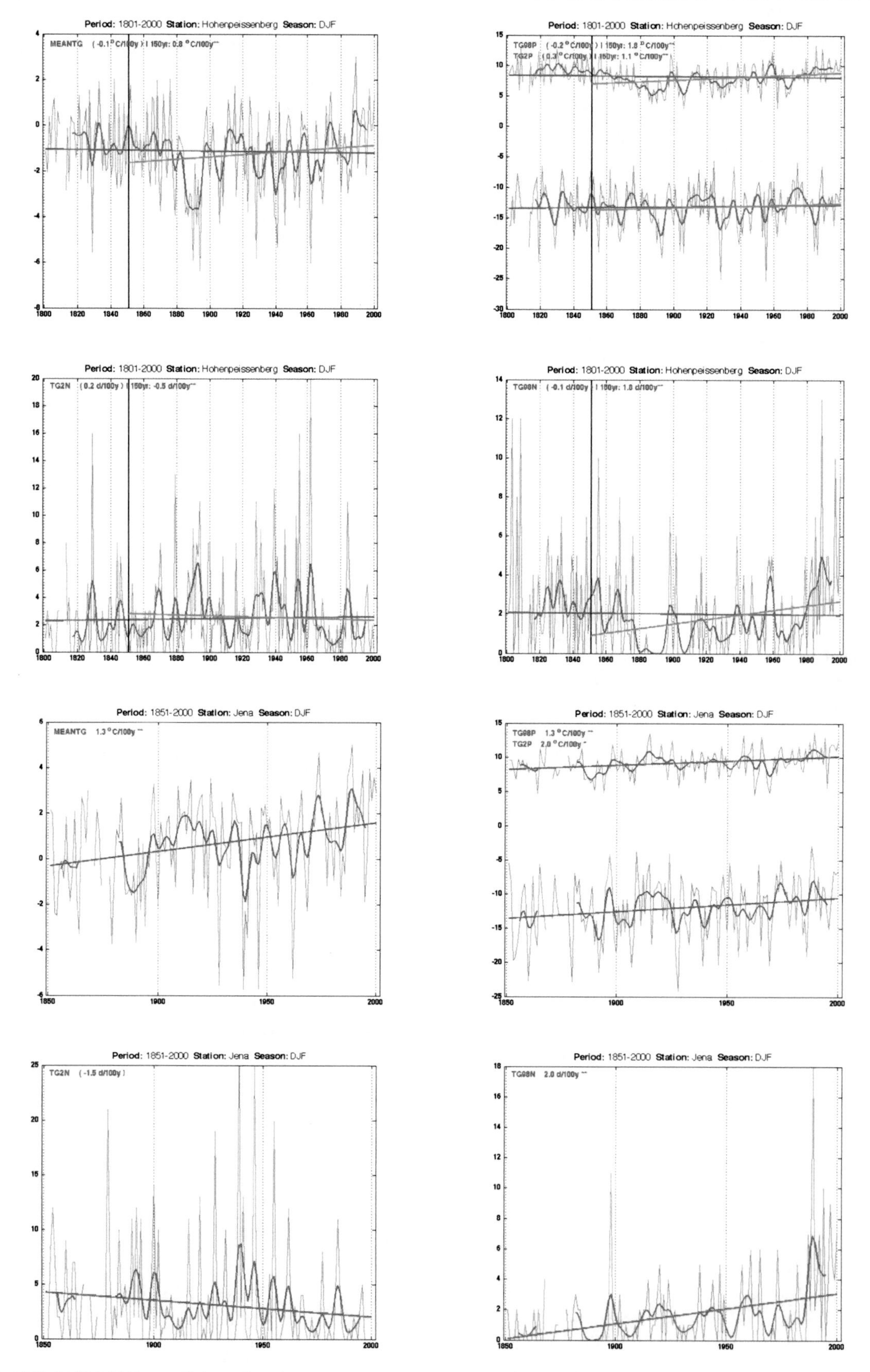

Fig. 3.110 1851–2000 DJF Tmean Hohenpeissenberg

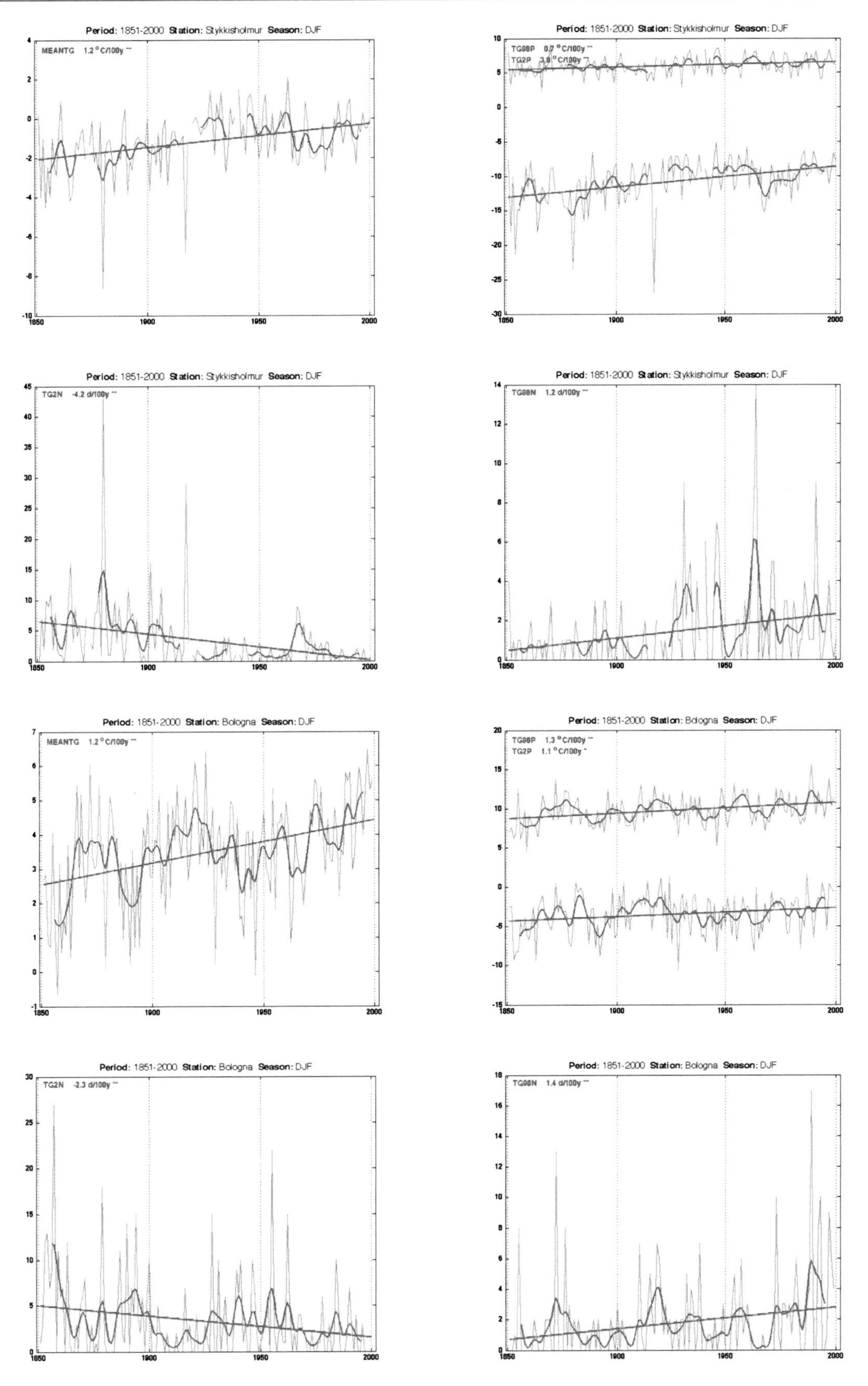

Fig. 3.111 1851–2000 DJF Tmean Stykkisholmur

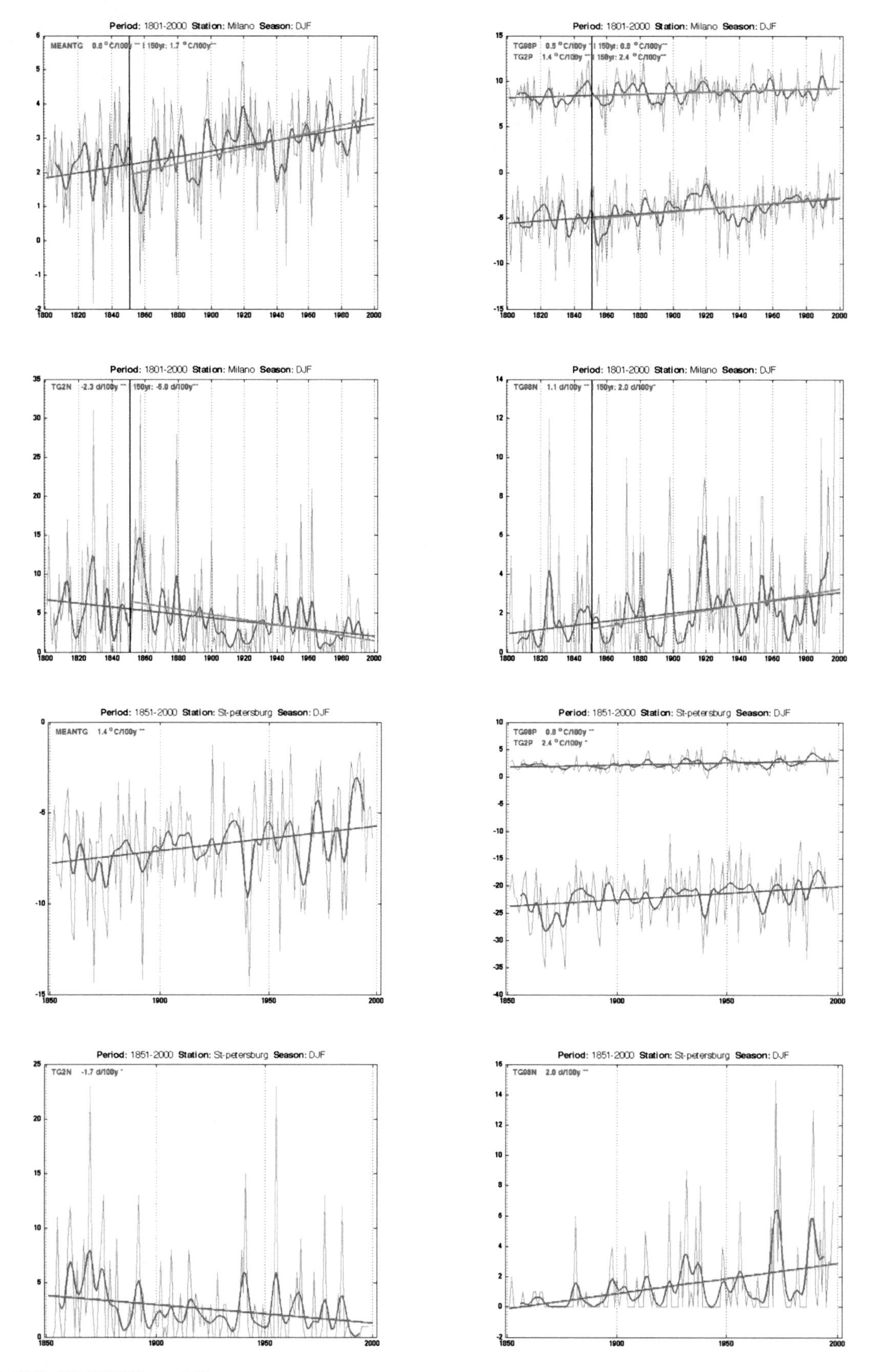

Fig. 3.112 1851–2000 DJF Tmean Milano

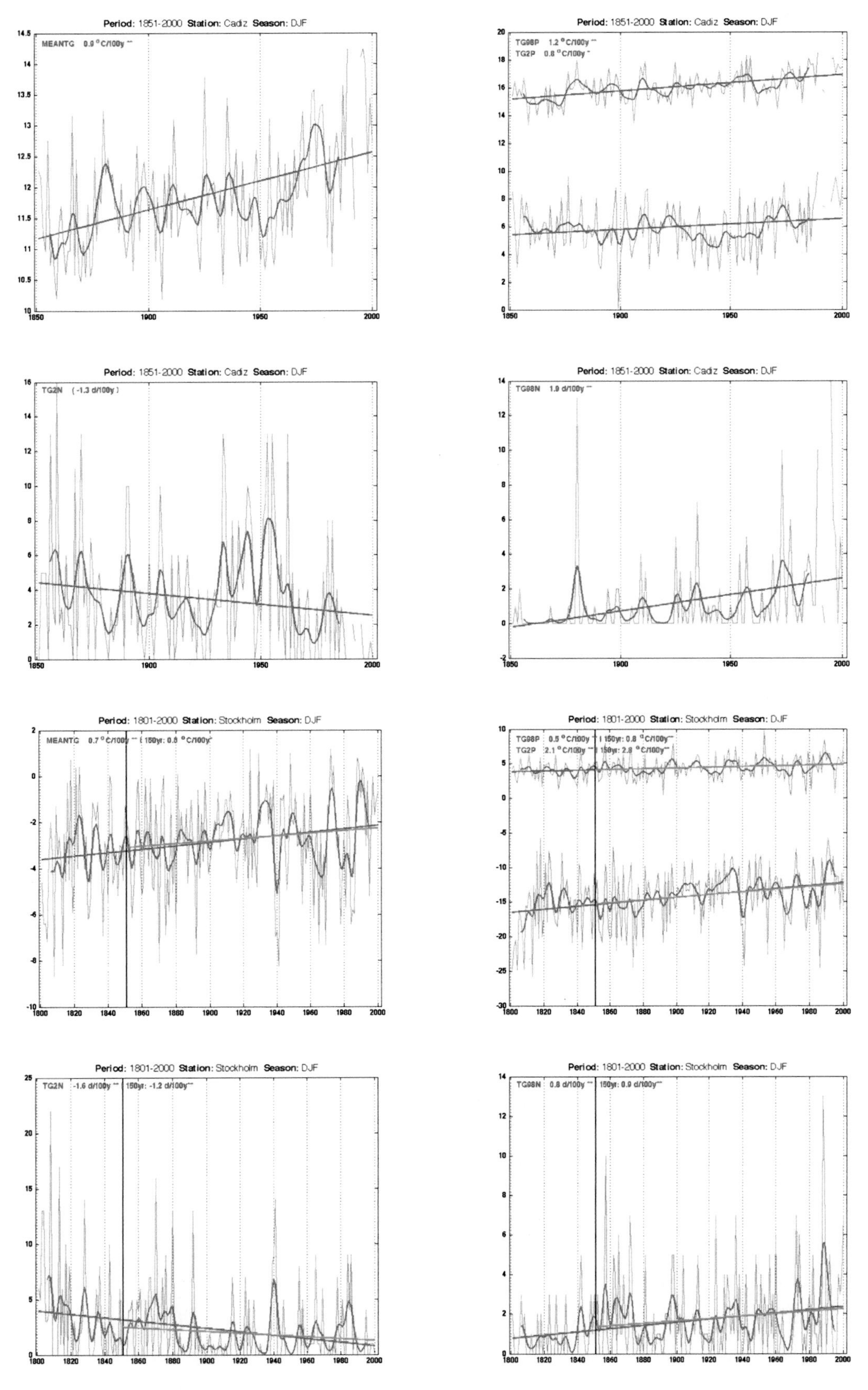

Fig. 3.113 1851–2000 DJF Tmean Cadiz

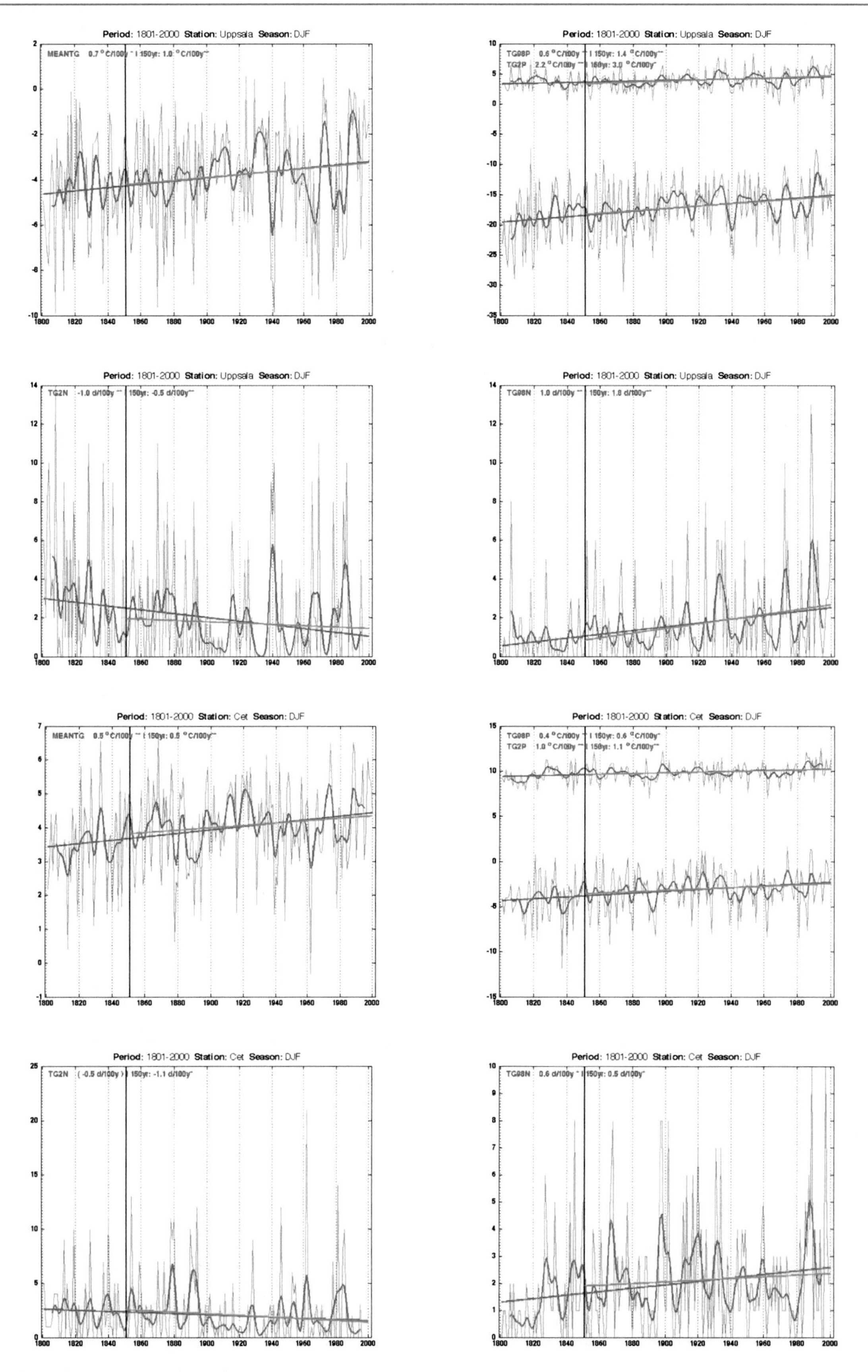

Fig. 3.114 1851–2000 DJF Tmean Uppsala

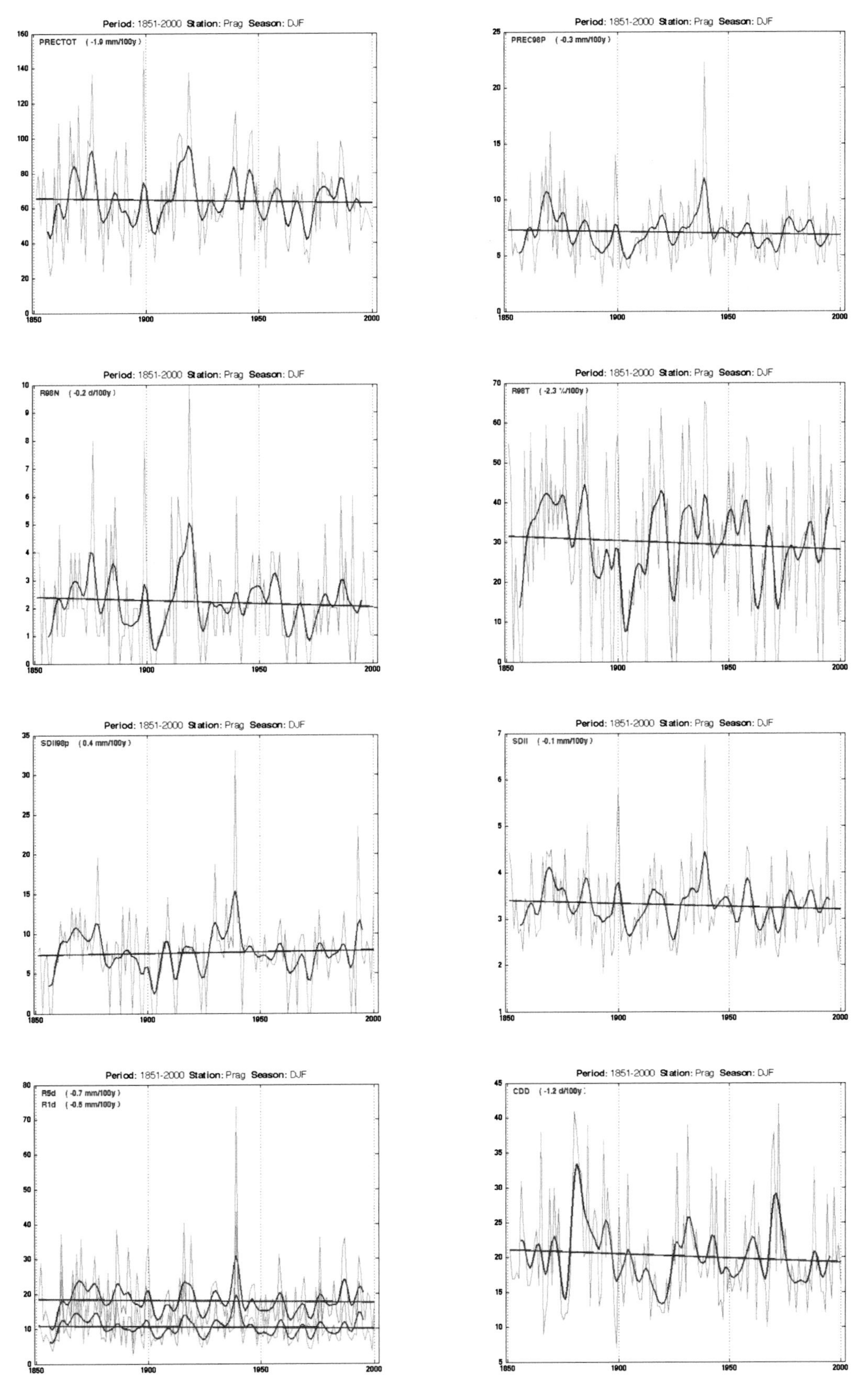

Fig. 3.115 1851–2000 DJF Prec Prag

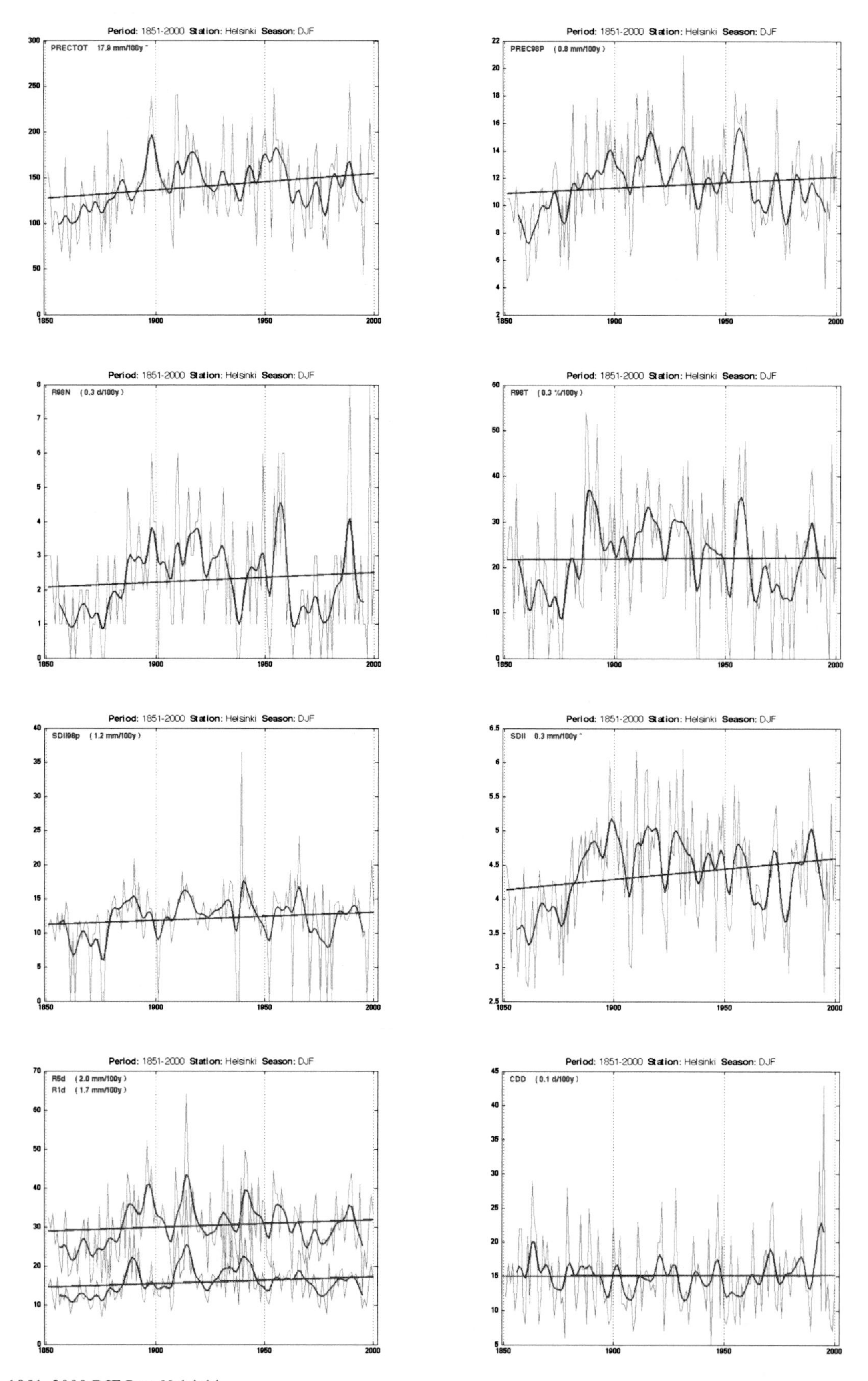

Fig. 3.116 1851–2000 DJF Prec Helsinki

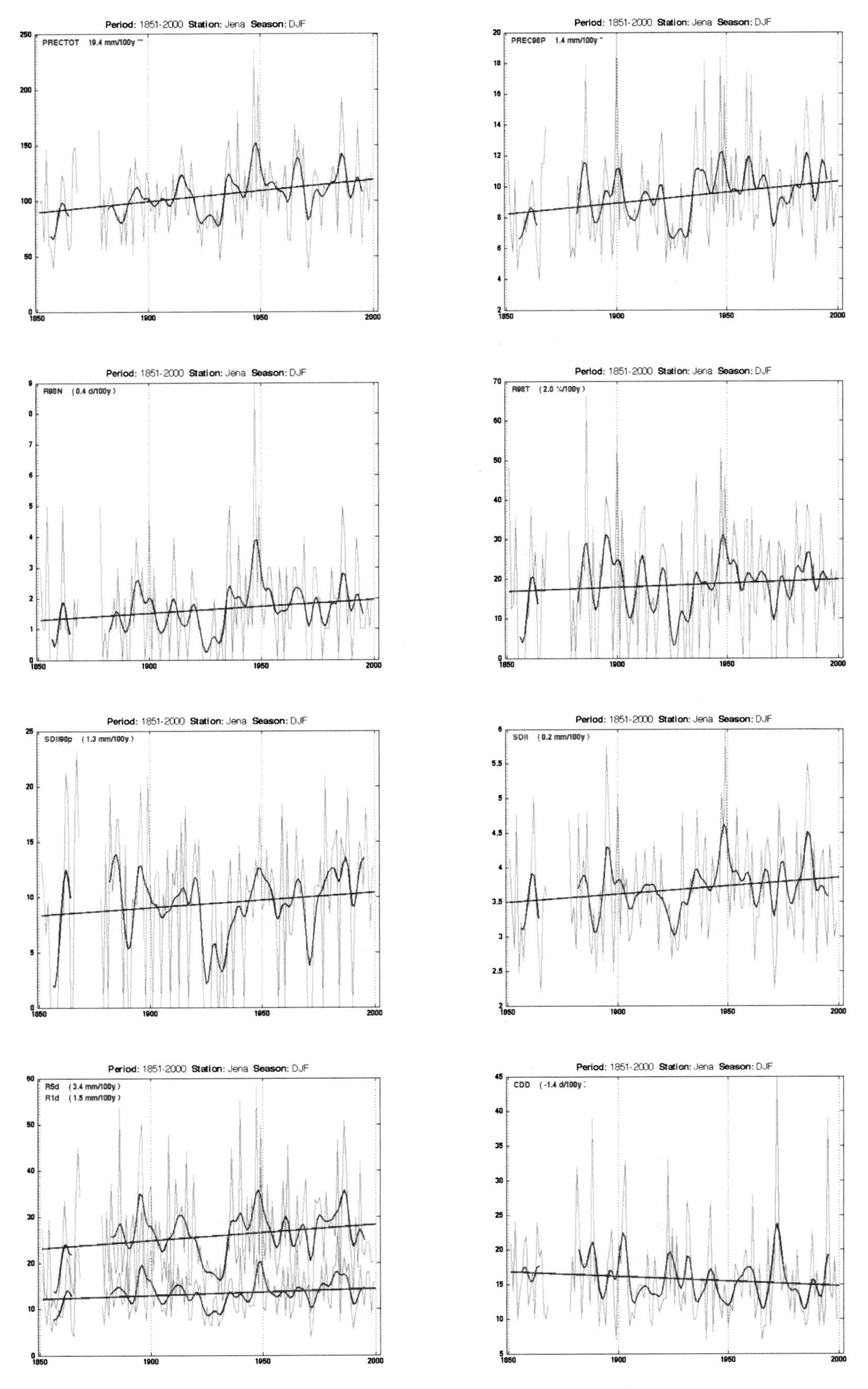

Fig. 3.117 1851–2000 DJF Prec Jena

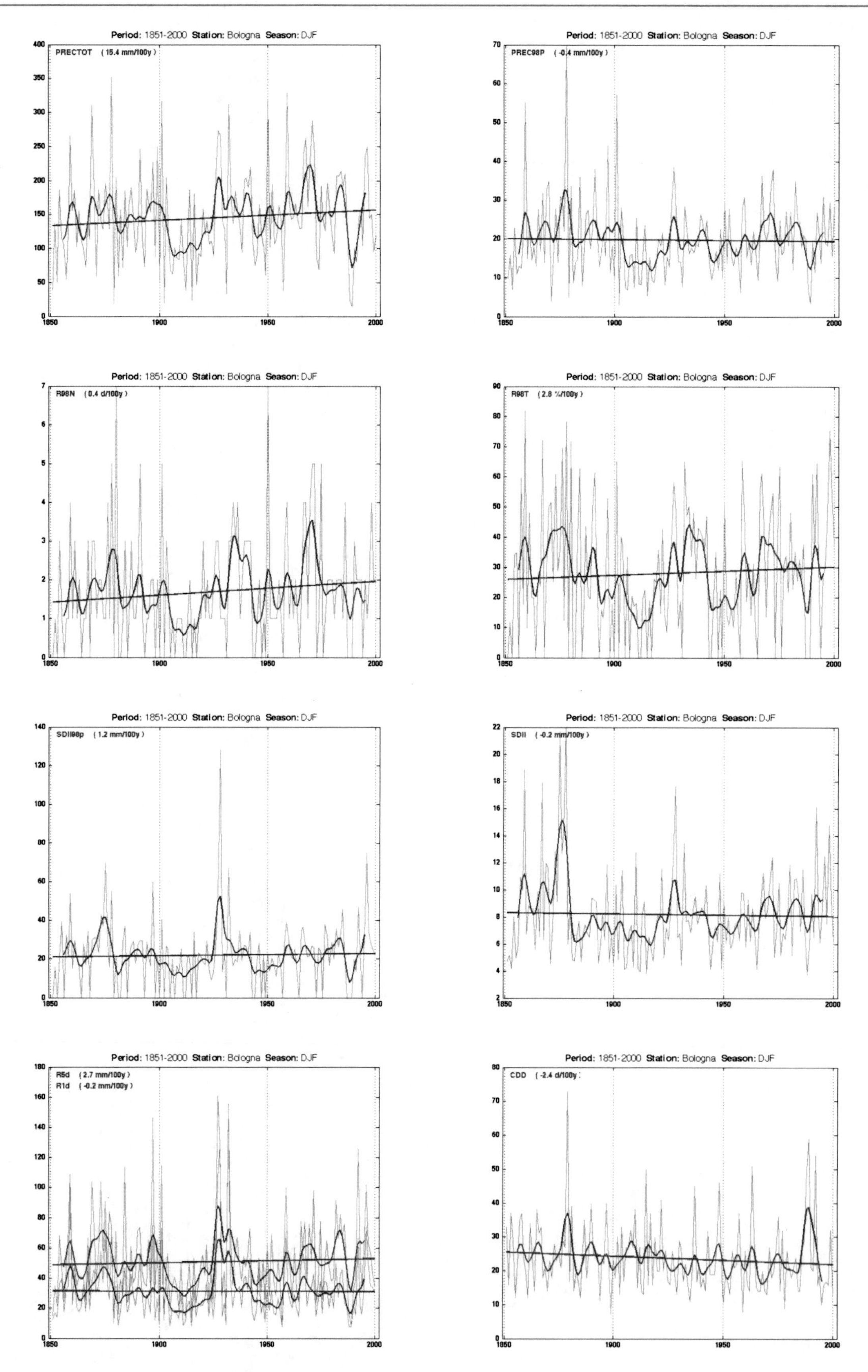

Fig. 3.118 1851–2000 DJF Prec Bologna

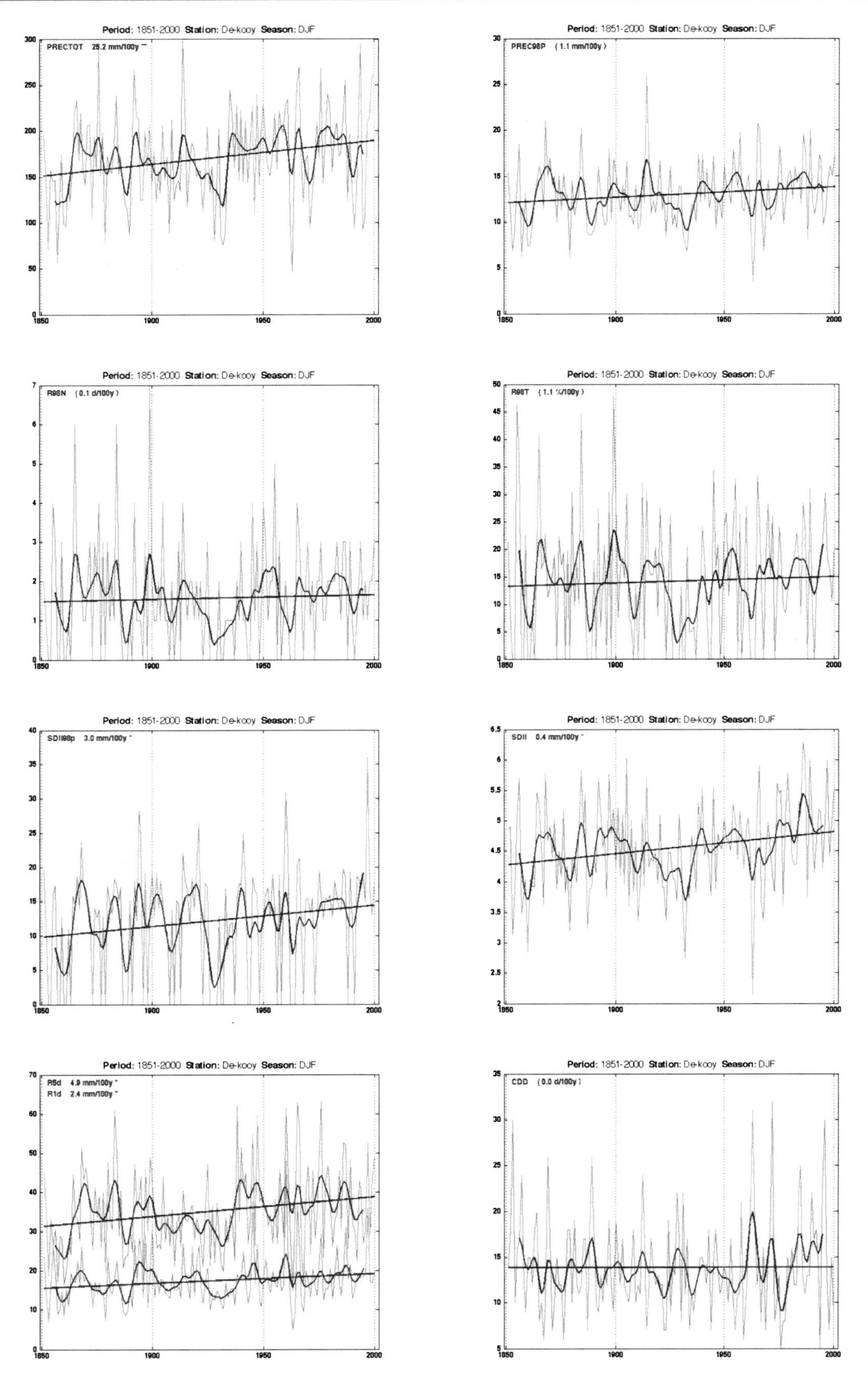

Fig. 3.119 1851–2000 DJF Prec De-kooy

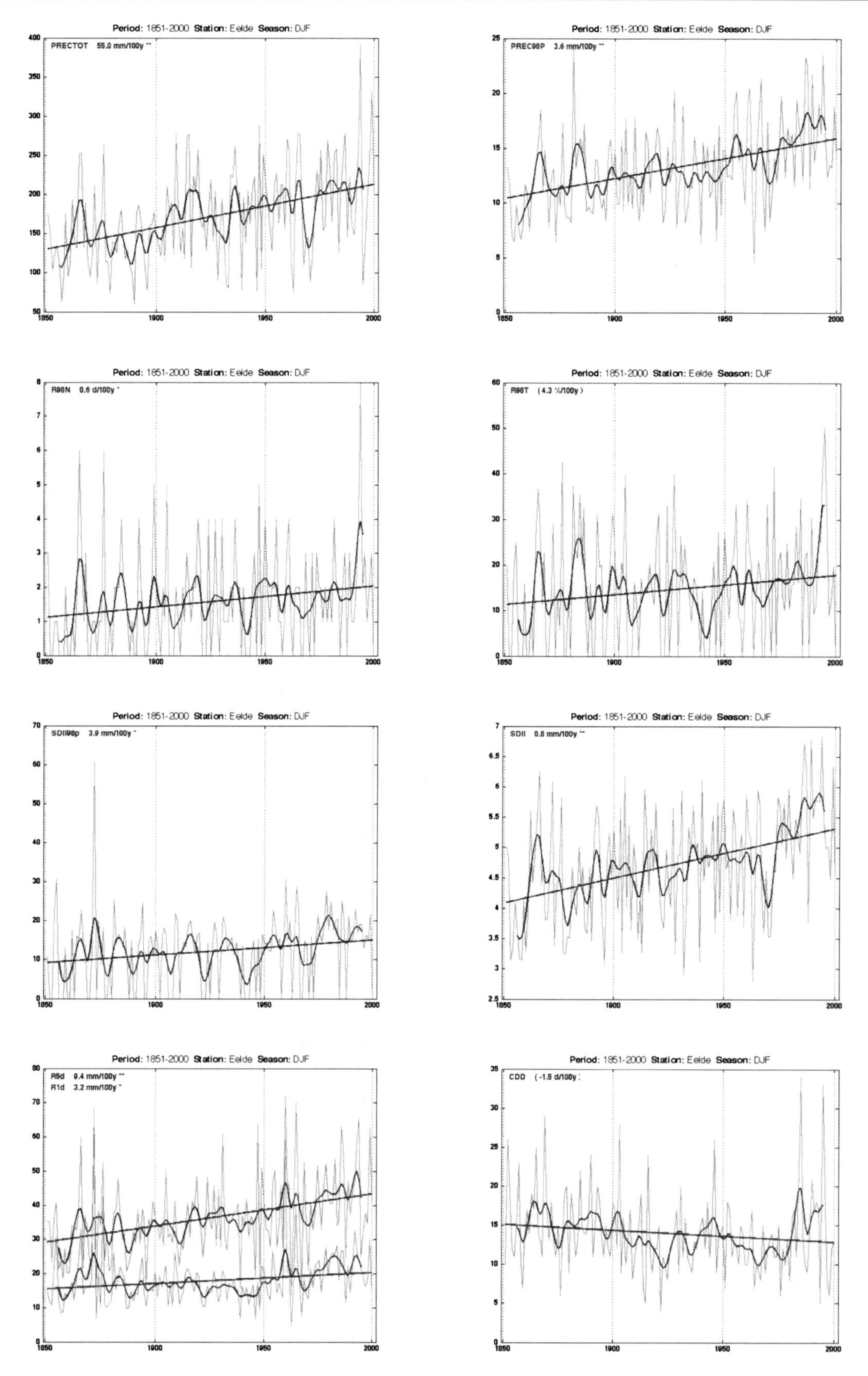

Fig. 3.120 1851–2000 DJF Prec Eelde

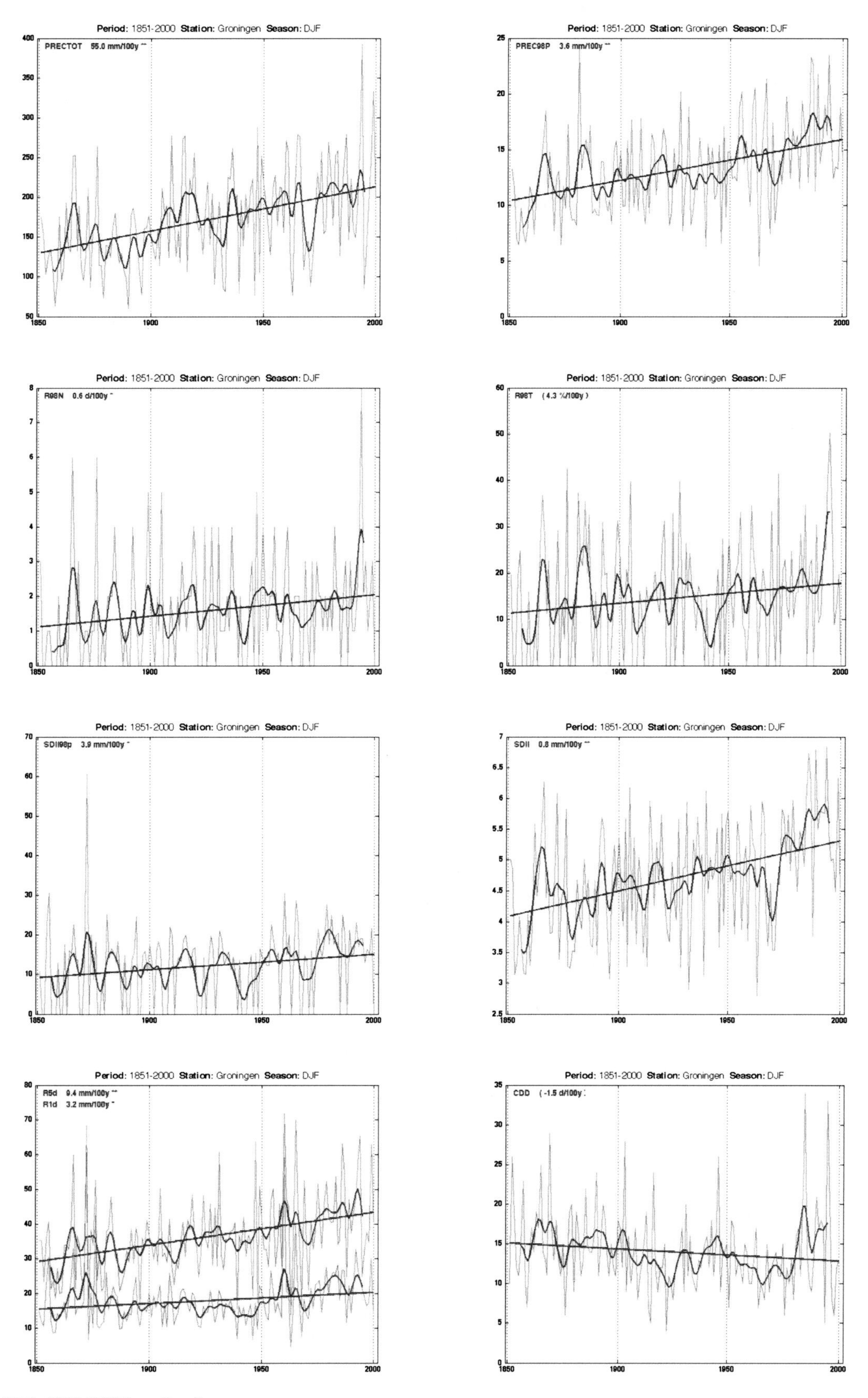

Fig. 3.121 1851–2000 DJF Prec Groningen

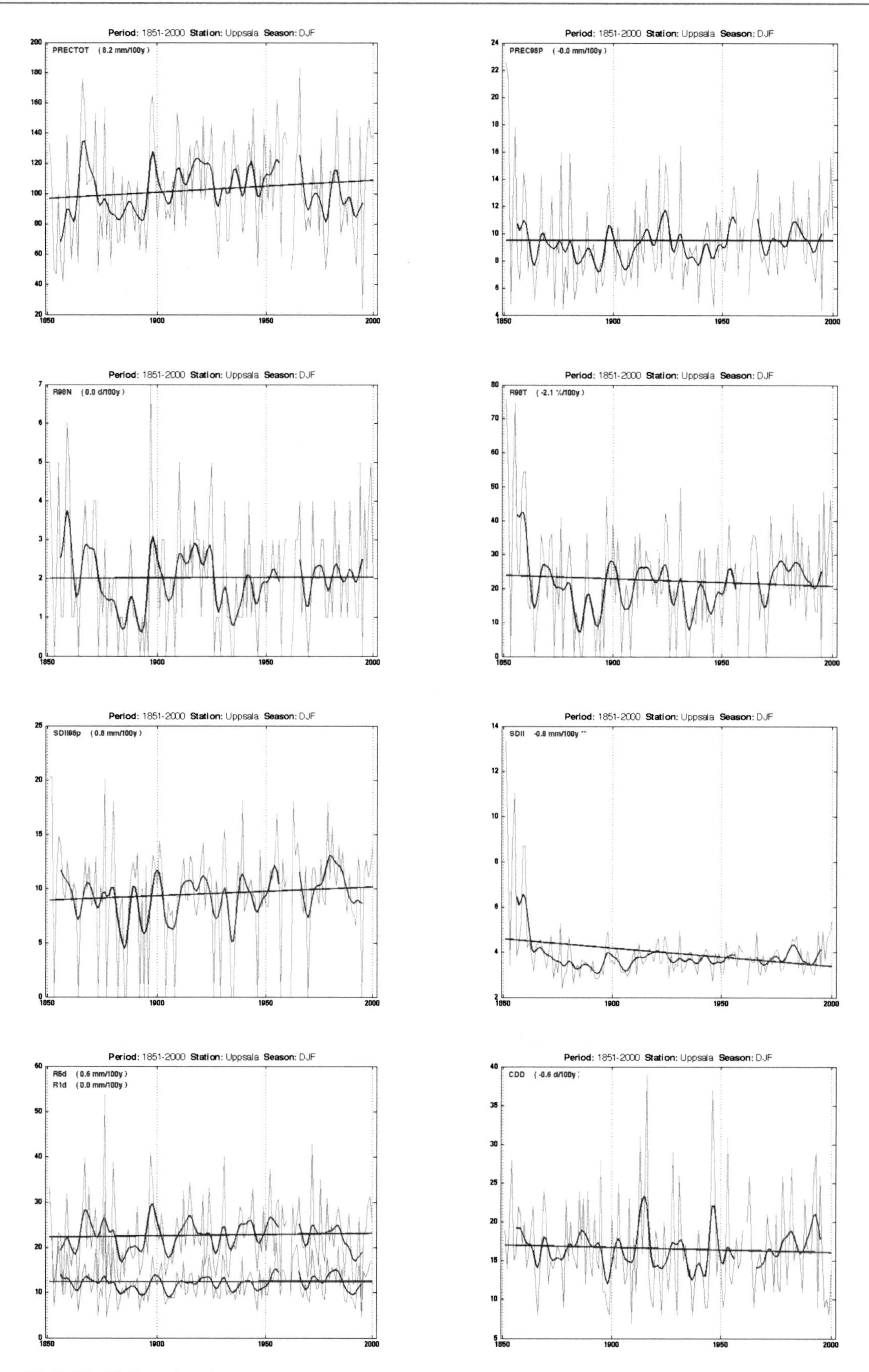

Fig. 3.122 1851–2000 DJF Prec Uppsala

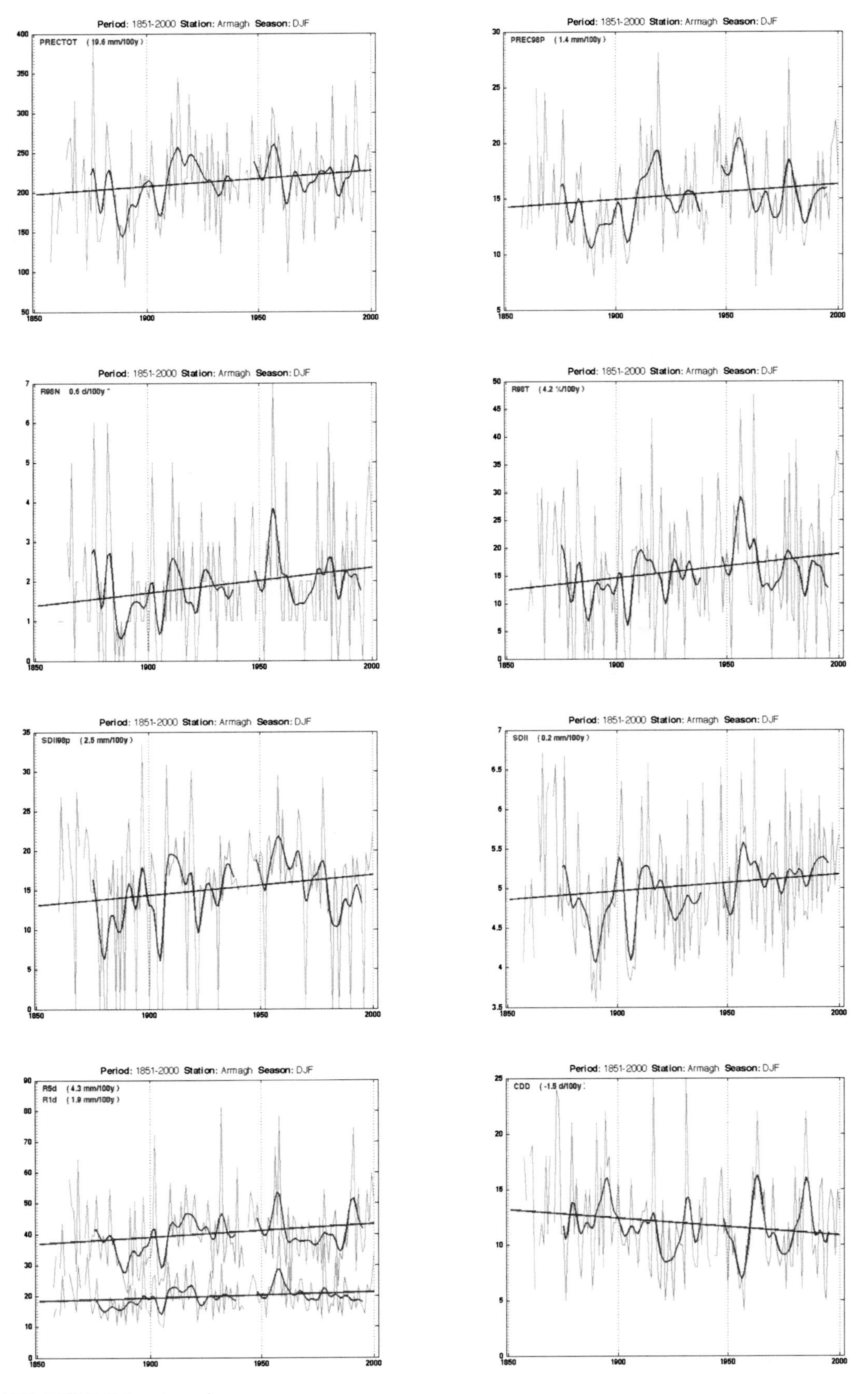

Fig. 3.123 1851–2000 DJF Prec Armagh

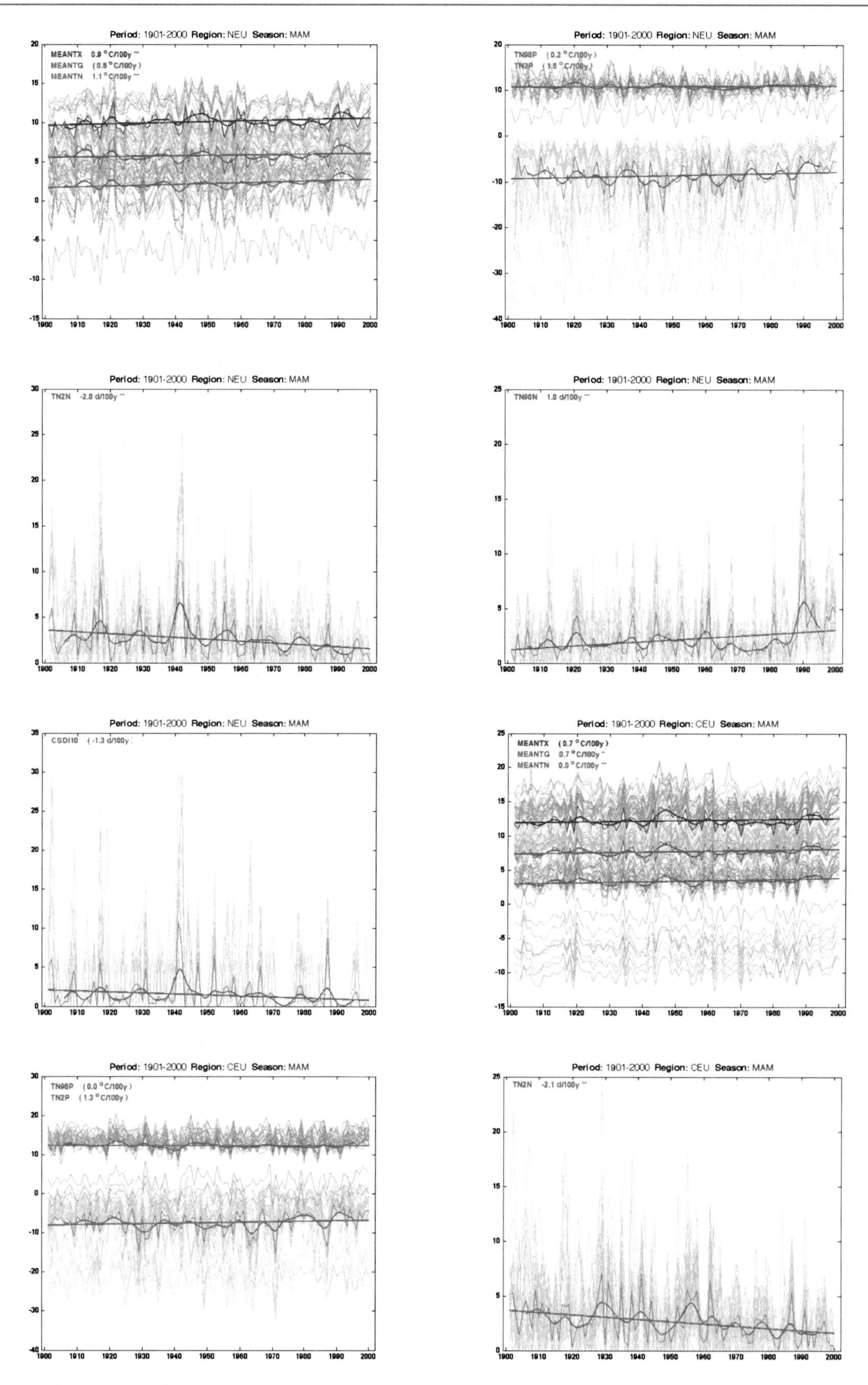

Fig. 3.124 1901–2000 MAM Tmin NEU

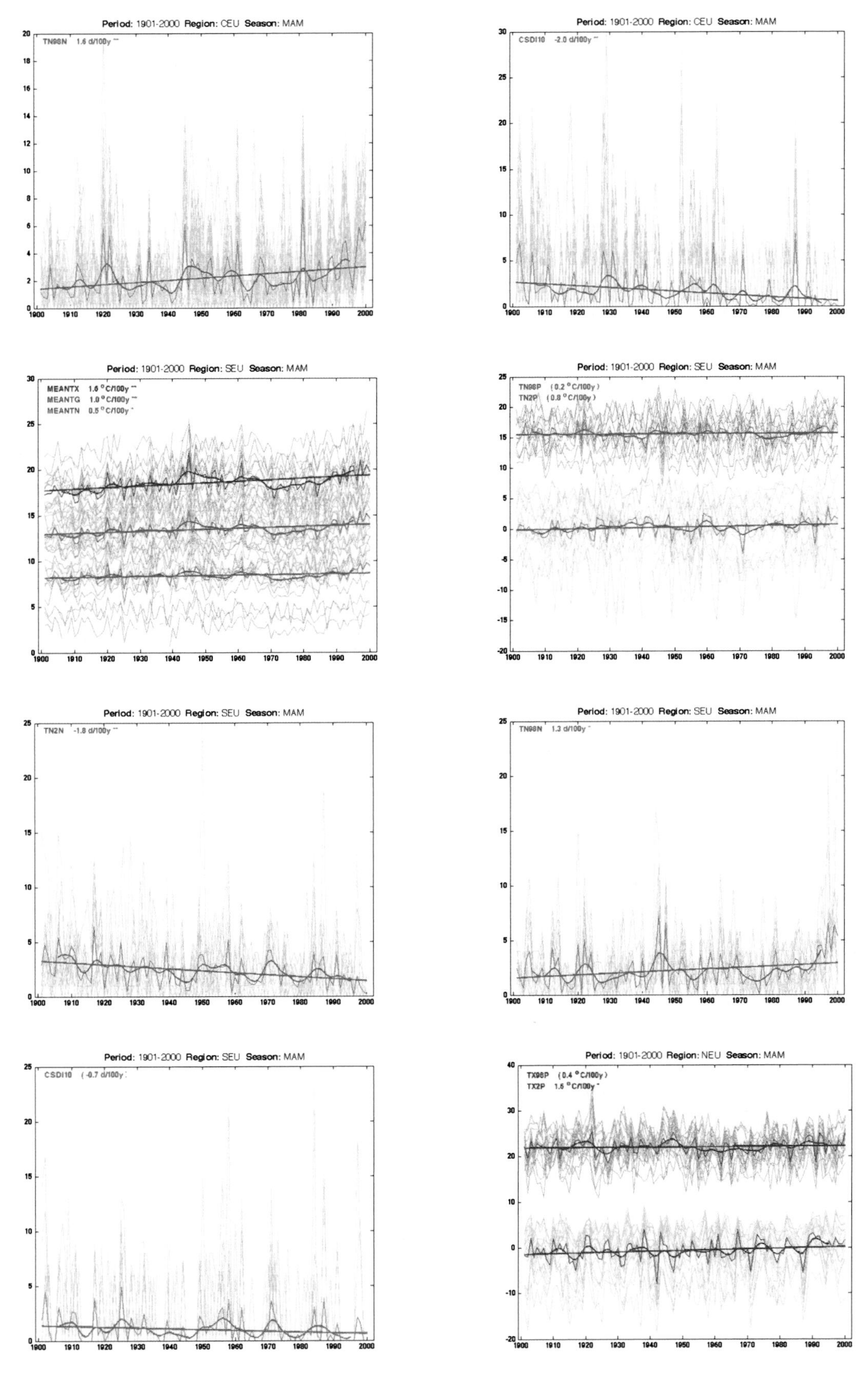

Fig. 3.125 1901–2000 MAM Tmin CEU

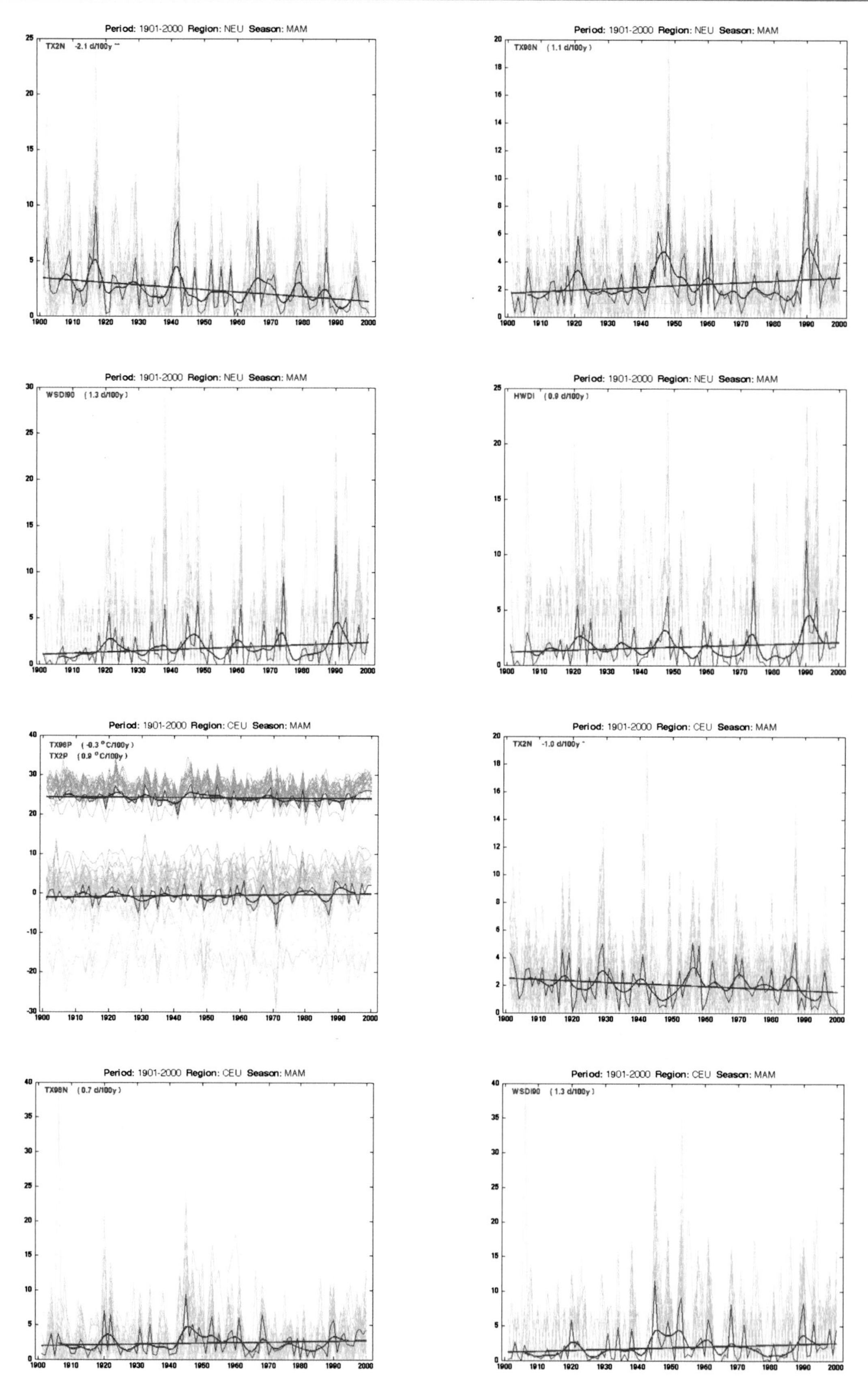

Fig. 3.126 1901–2000 MAM Tmax NEU

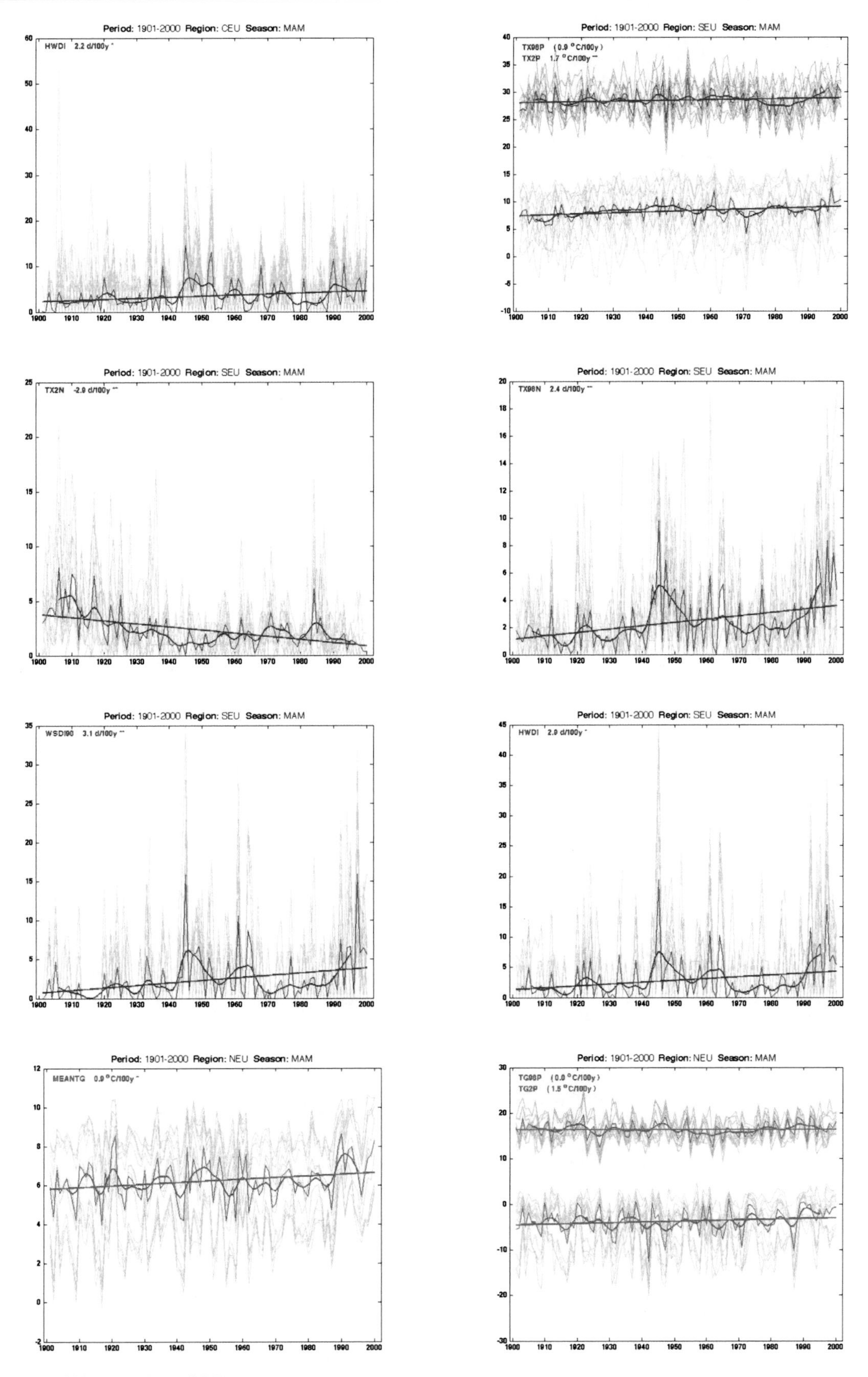

Fig. 3.127 1901–2000 MAM Tmax CEU

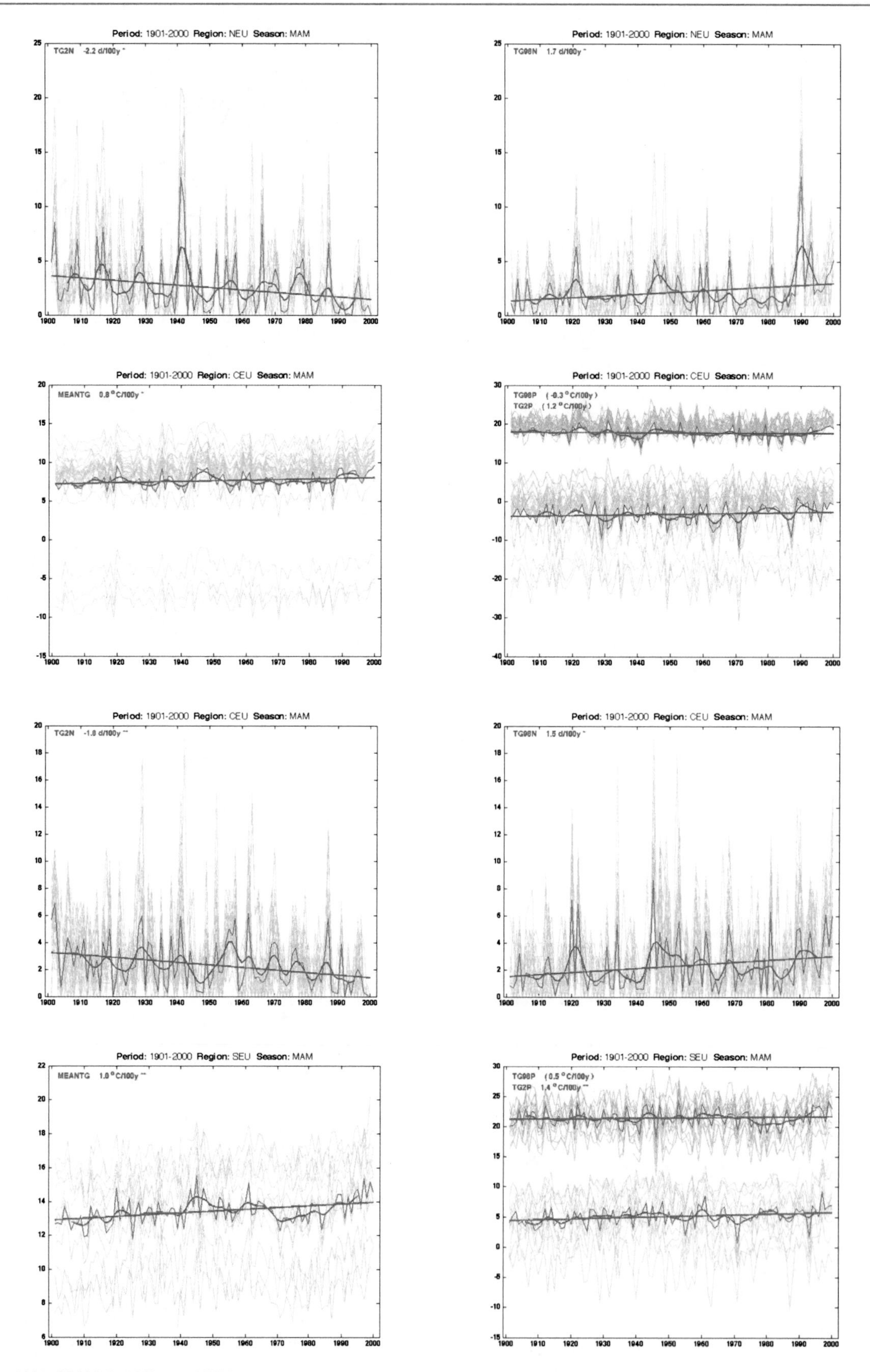

Fig. 3.128 1901–2000 MAM Tmean NEU

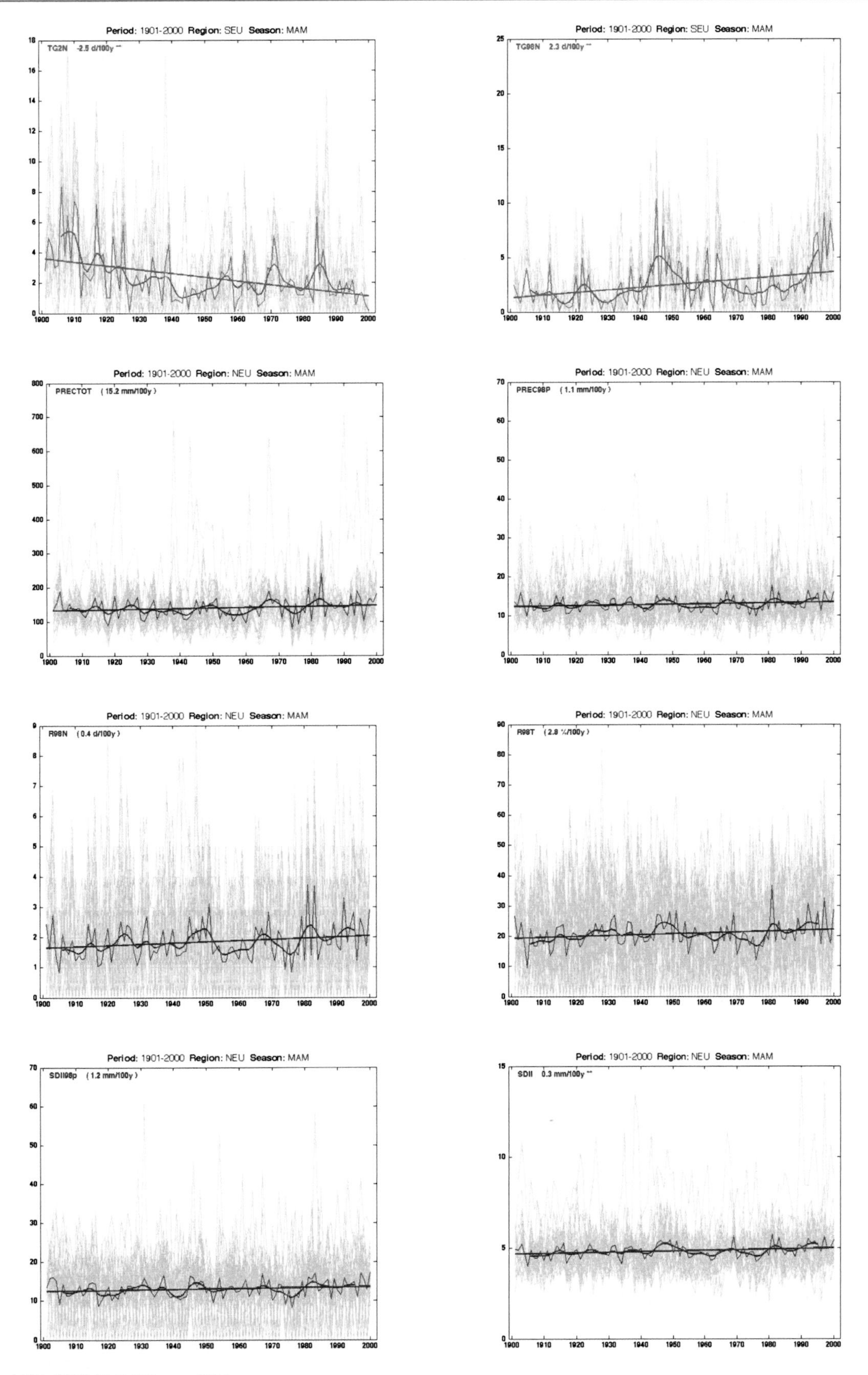

Fig. 3.129 1901–2000 MAM Tmean SEU

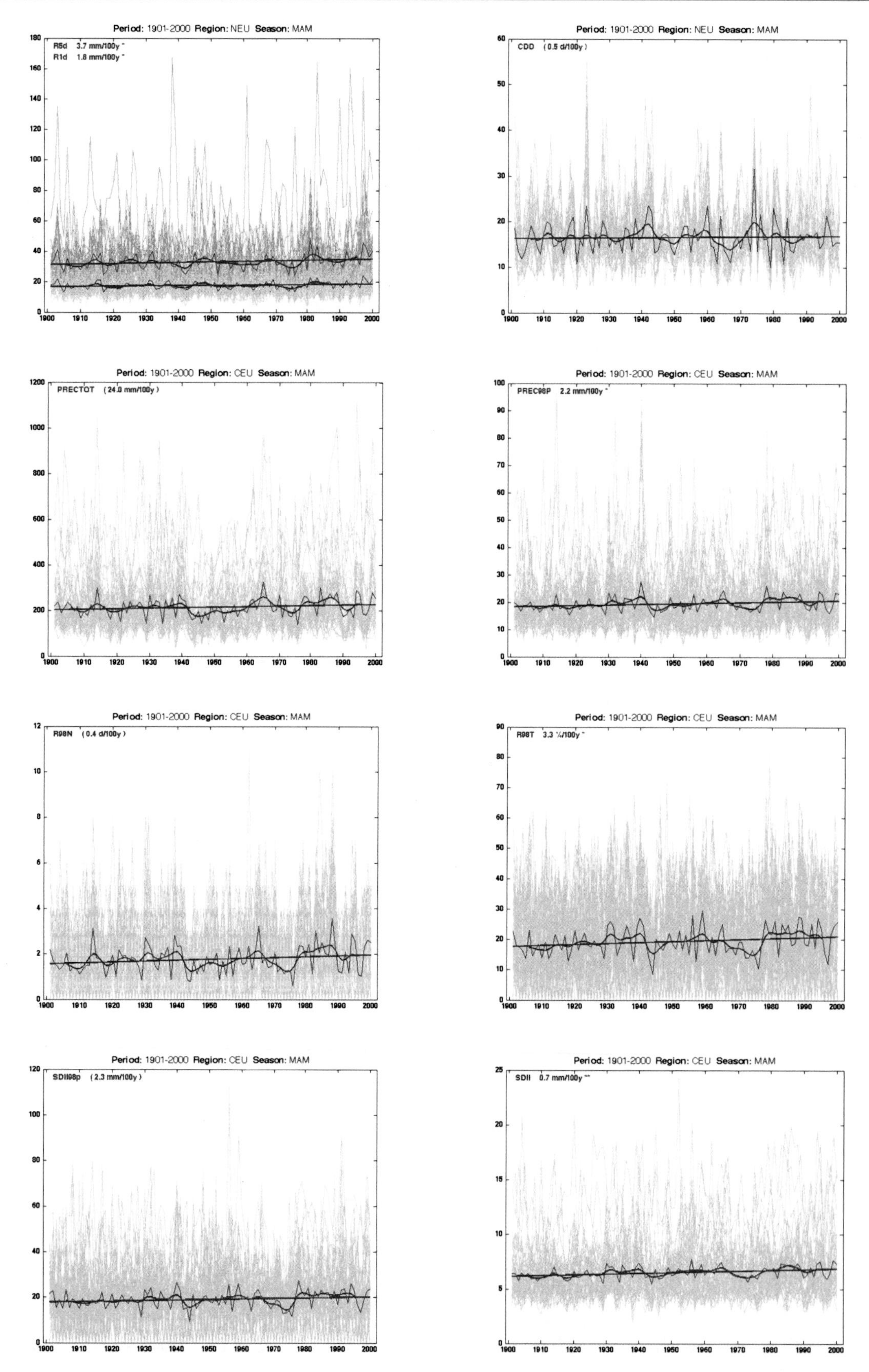

Fig. 3.130 1901–2000 MAM Prec NEU

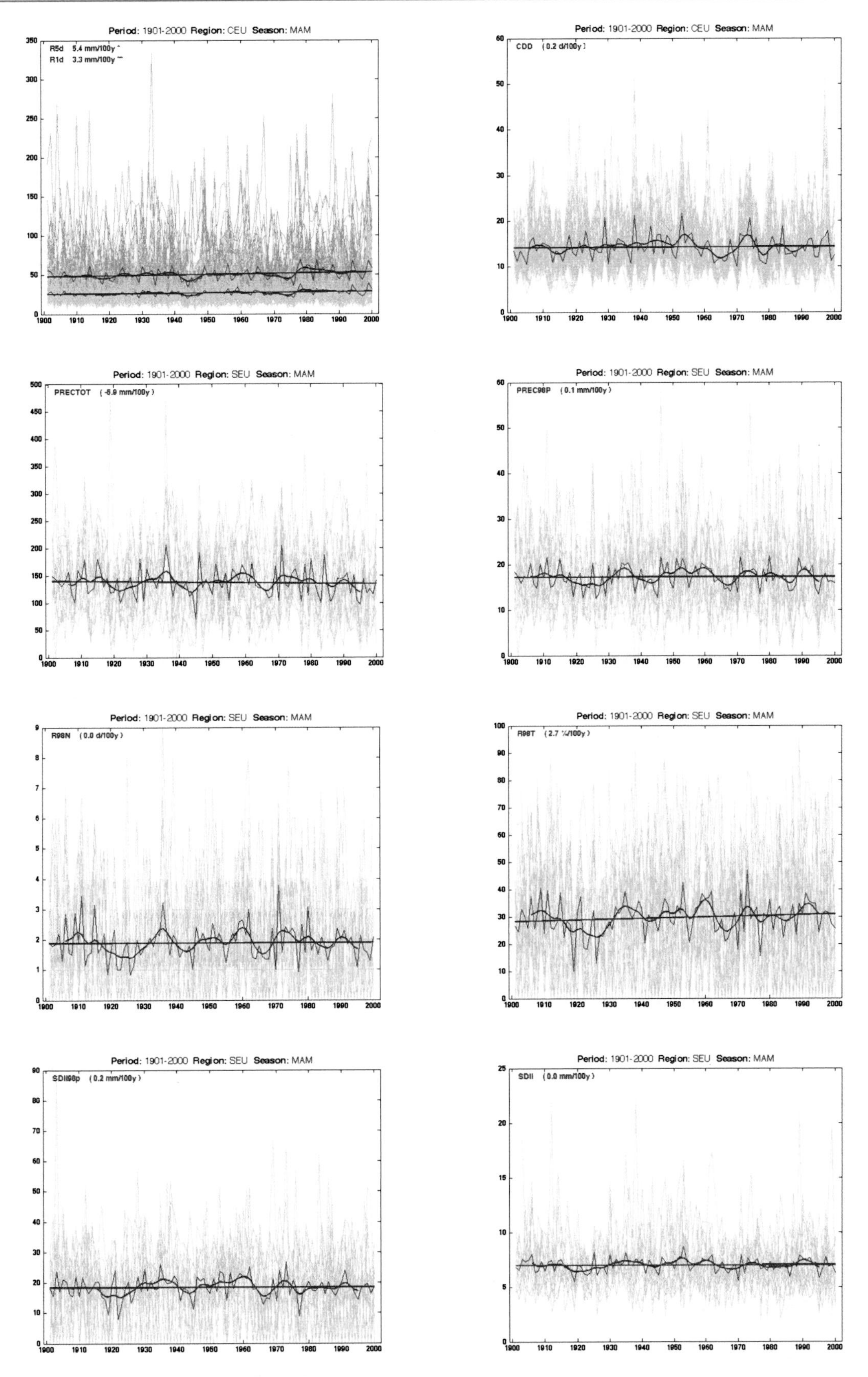

Fig. 3.131 1901–2000 MAM Prec CEU

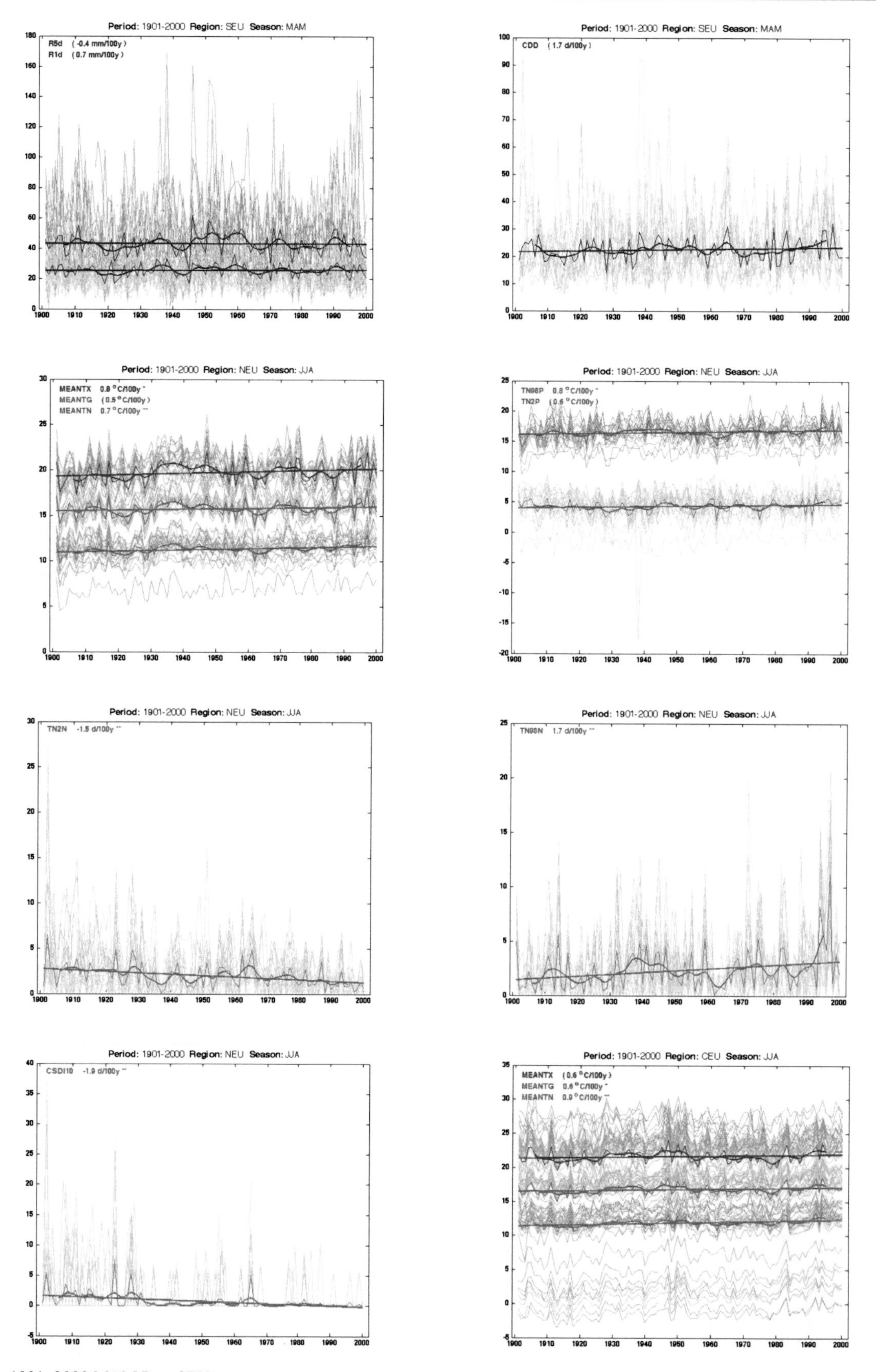

Fig. 3.132 1901–2000 MAM Prec SEU

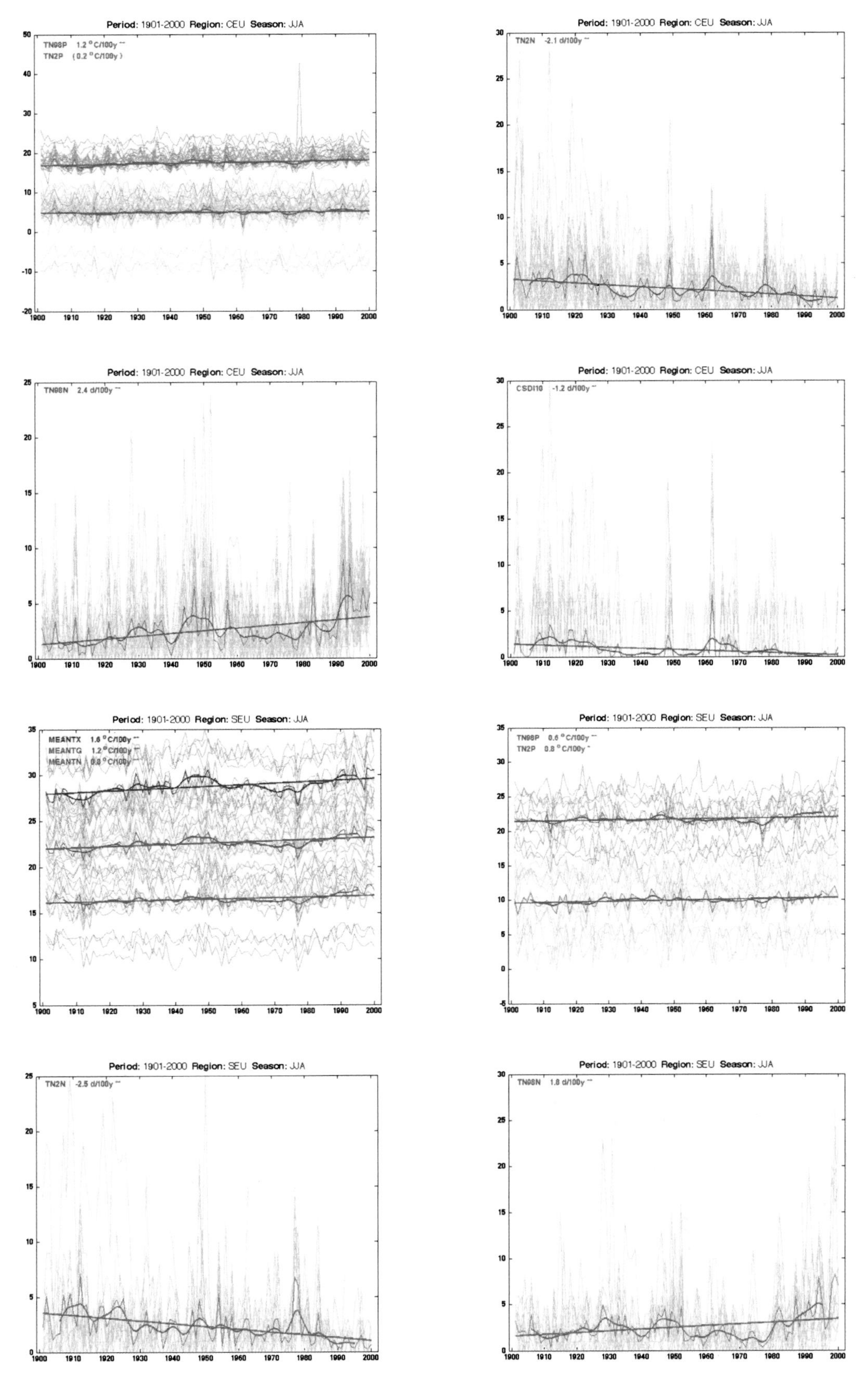

Fig. 3.133 1901–2000 JJA Tmin CEU

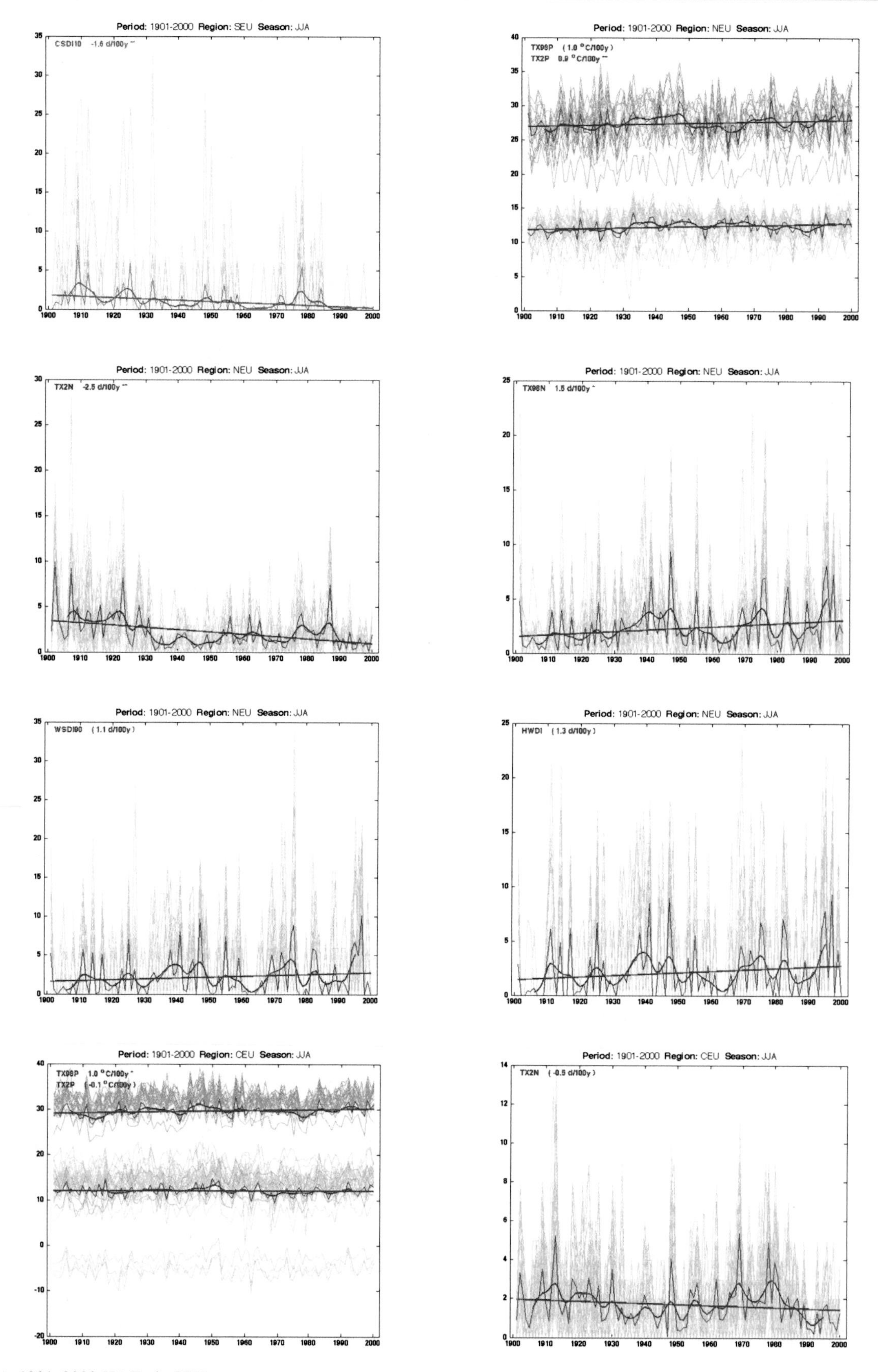

Fig. 3.134 1901–2000 JJA Tmin SEU

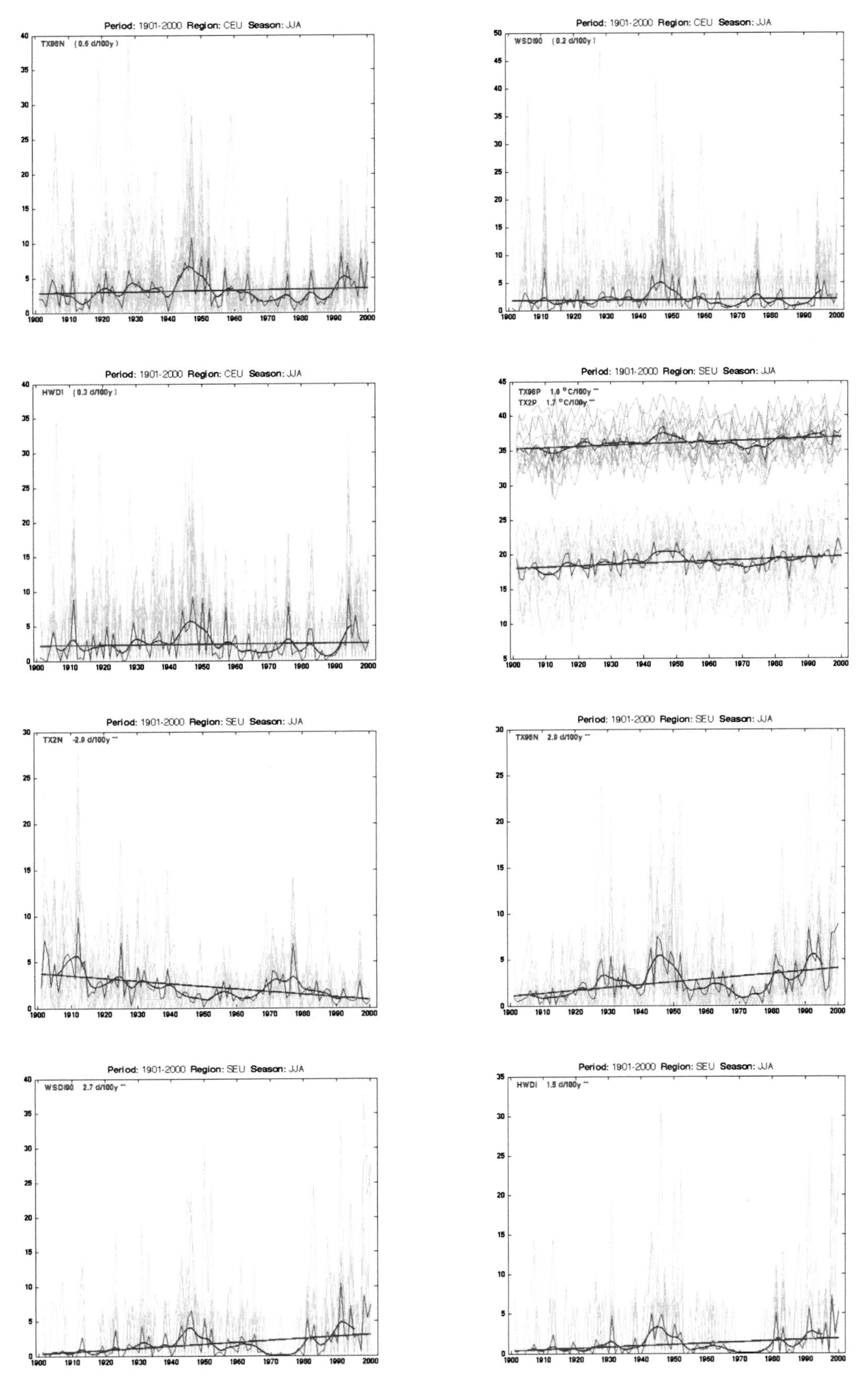

Fig. 3.135 1901–2000 JJA Tmax CEU

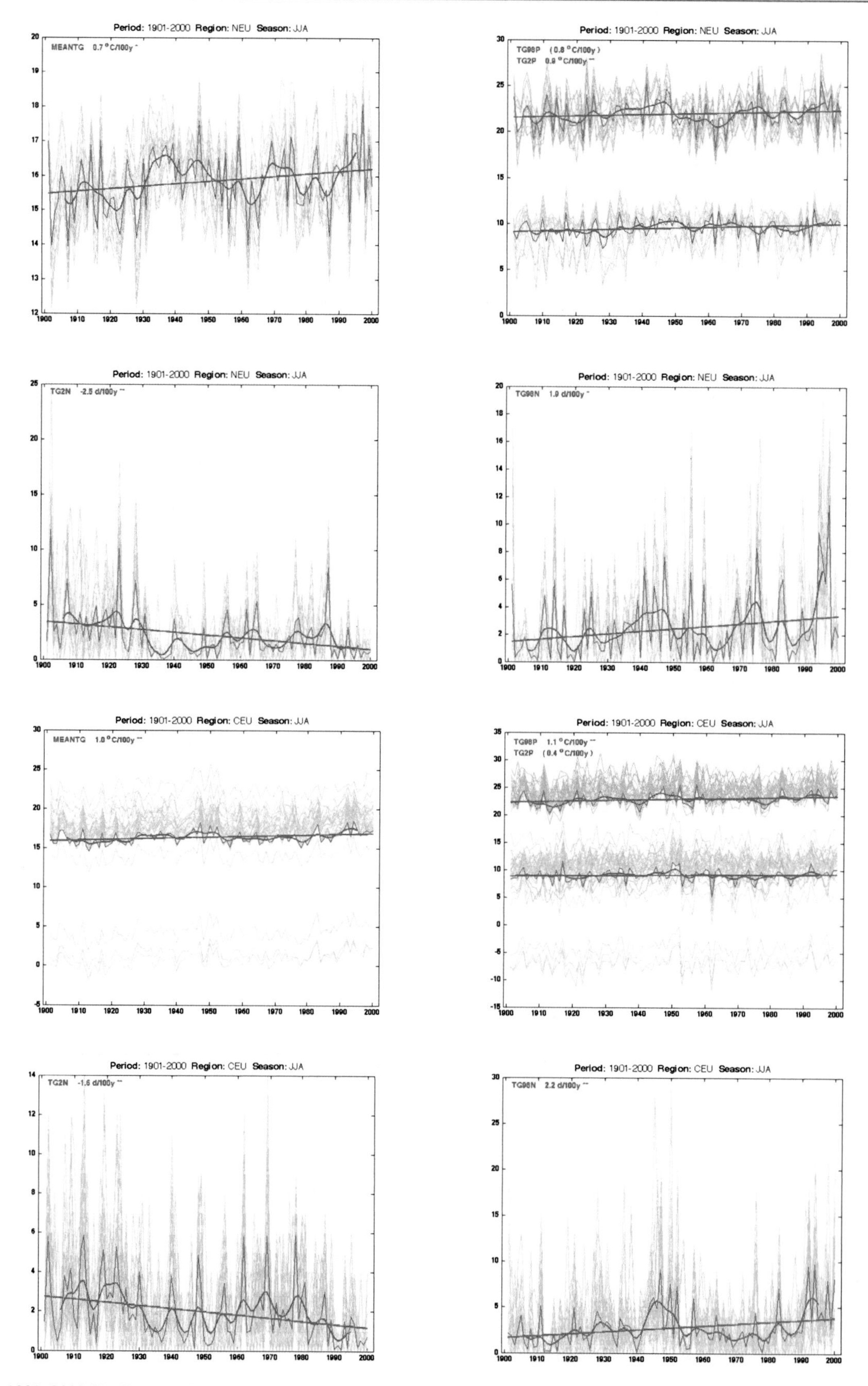

Fig. 3.136 1901–2000 JJA Tmean NEU

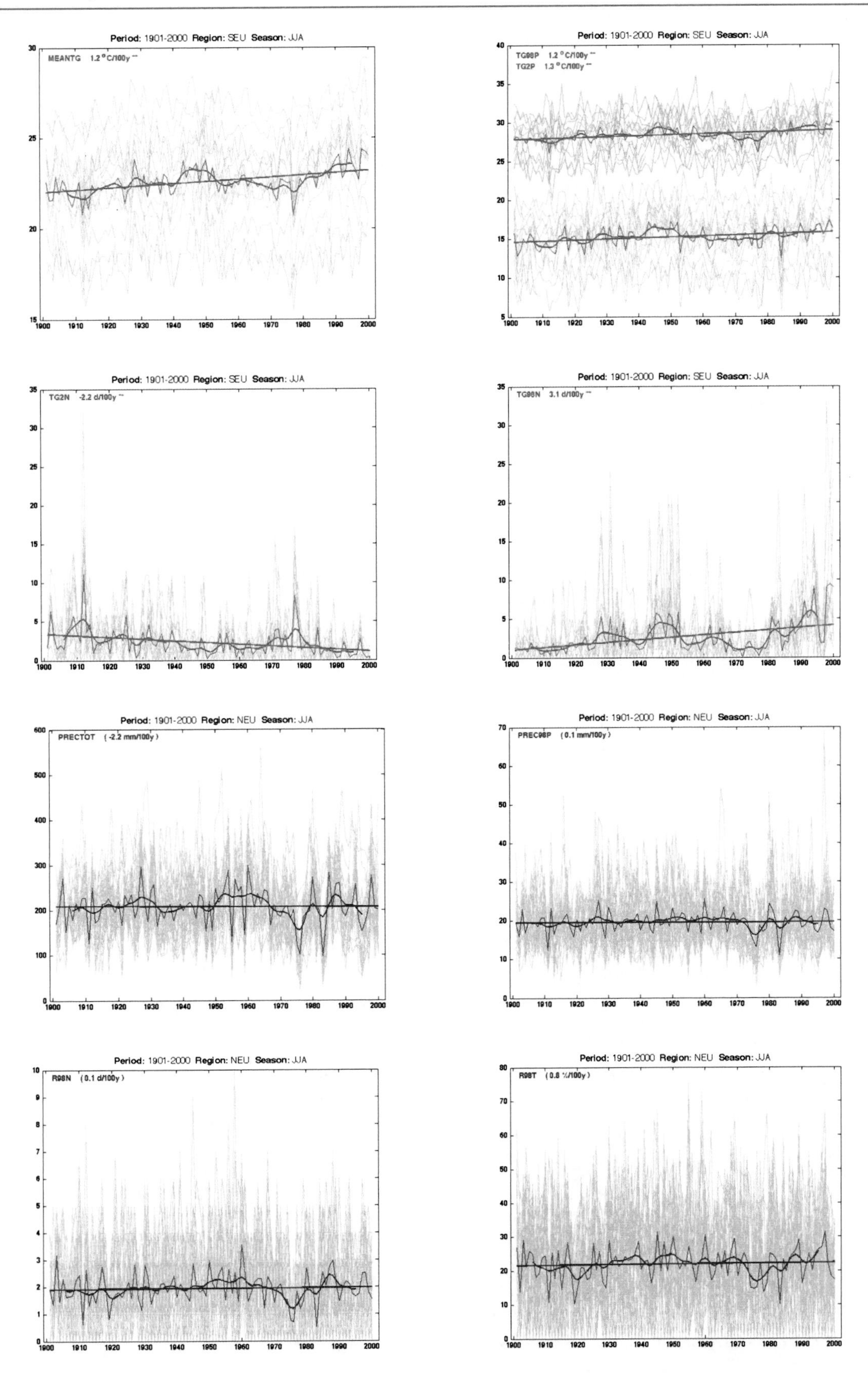

Fig. 3.137 1901–2000 JJA Tmean SEU

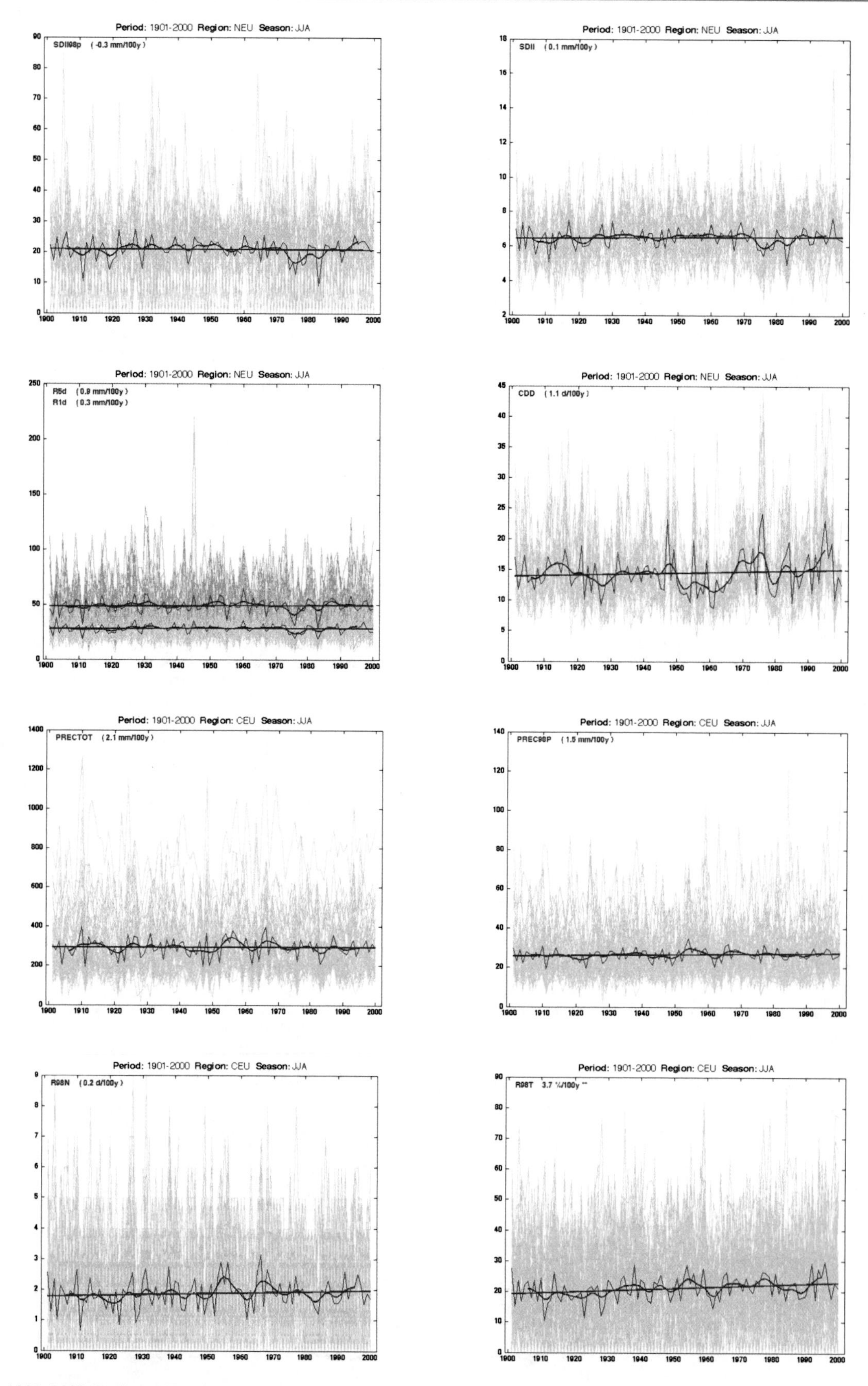

Fig. 3.138 1901–2000 JJA Prec NEU

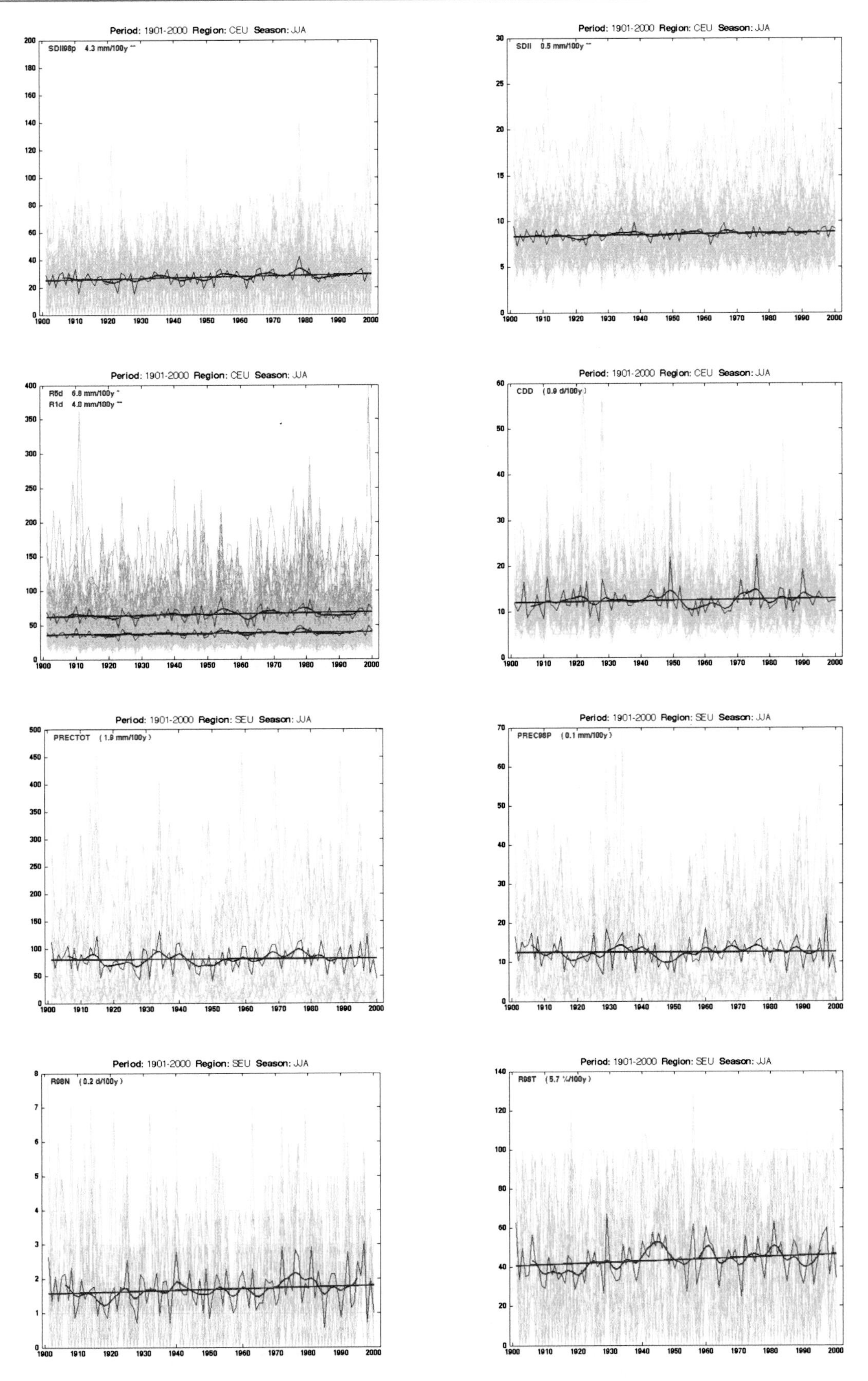

Fig. 3.139 1901–2000 JJA Prec CEU

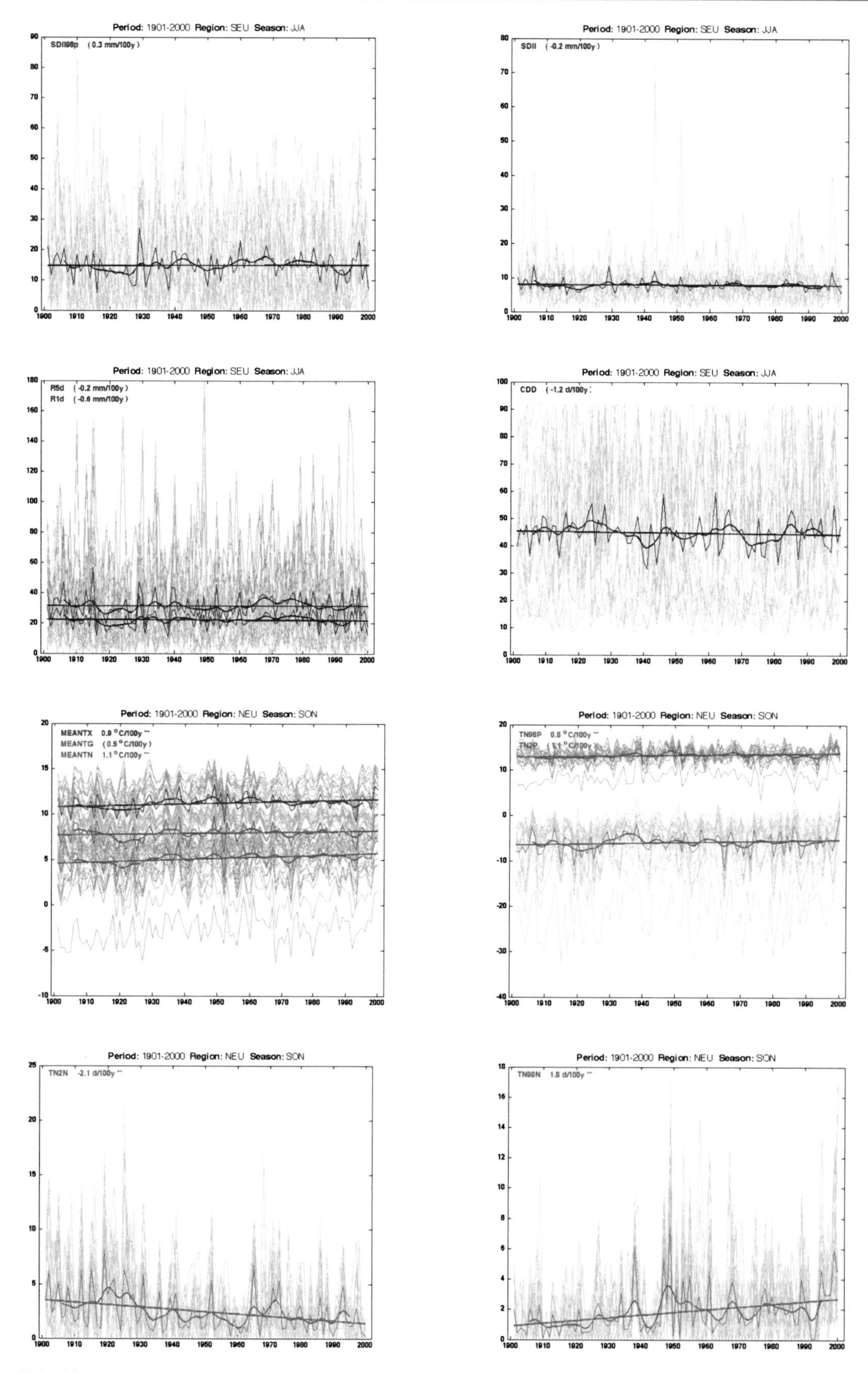

Fig. 3.140 1901–2000 JJA Prec SEU

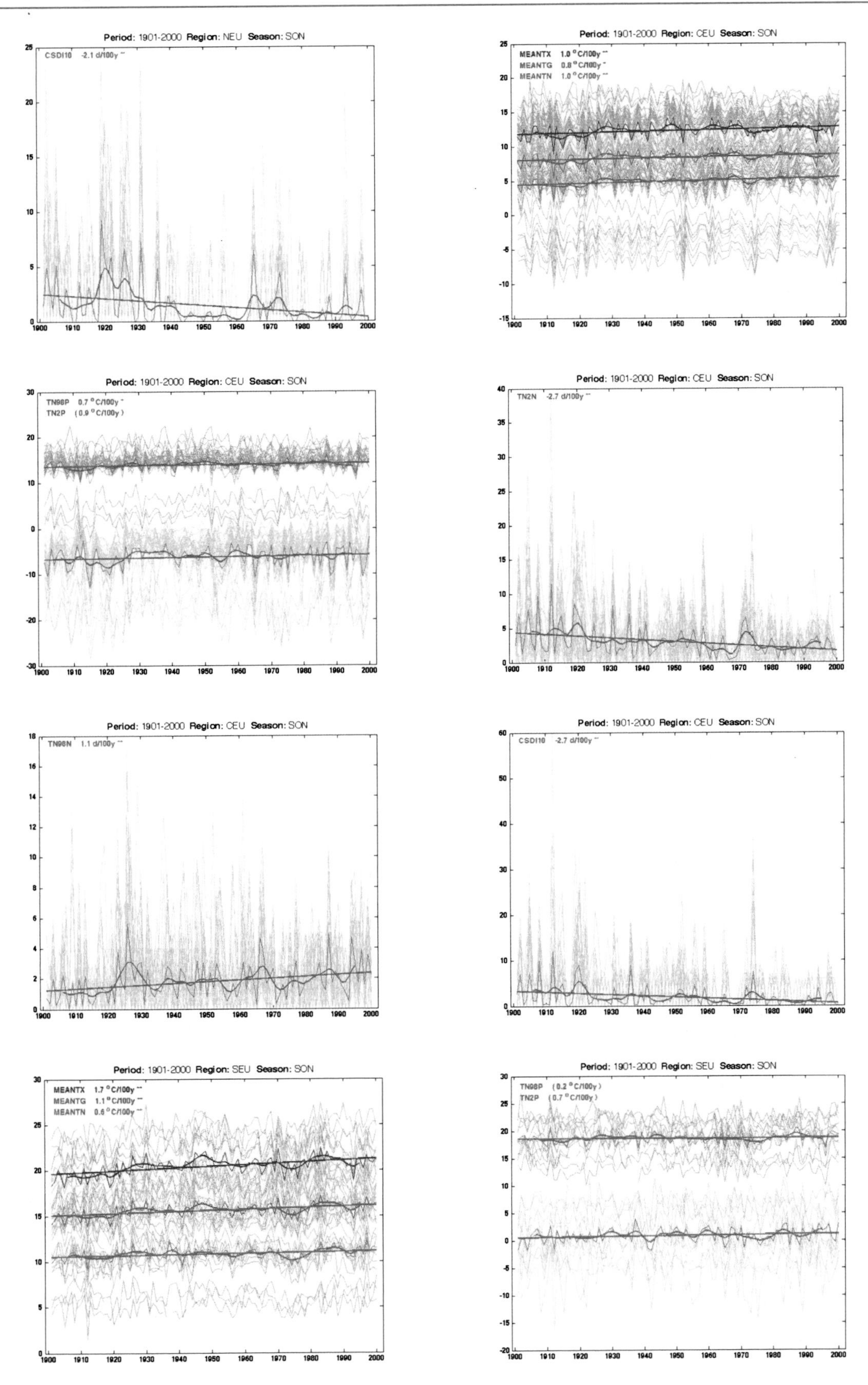

Fig. 3.141 1901–2000 SON Tmin NEU

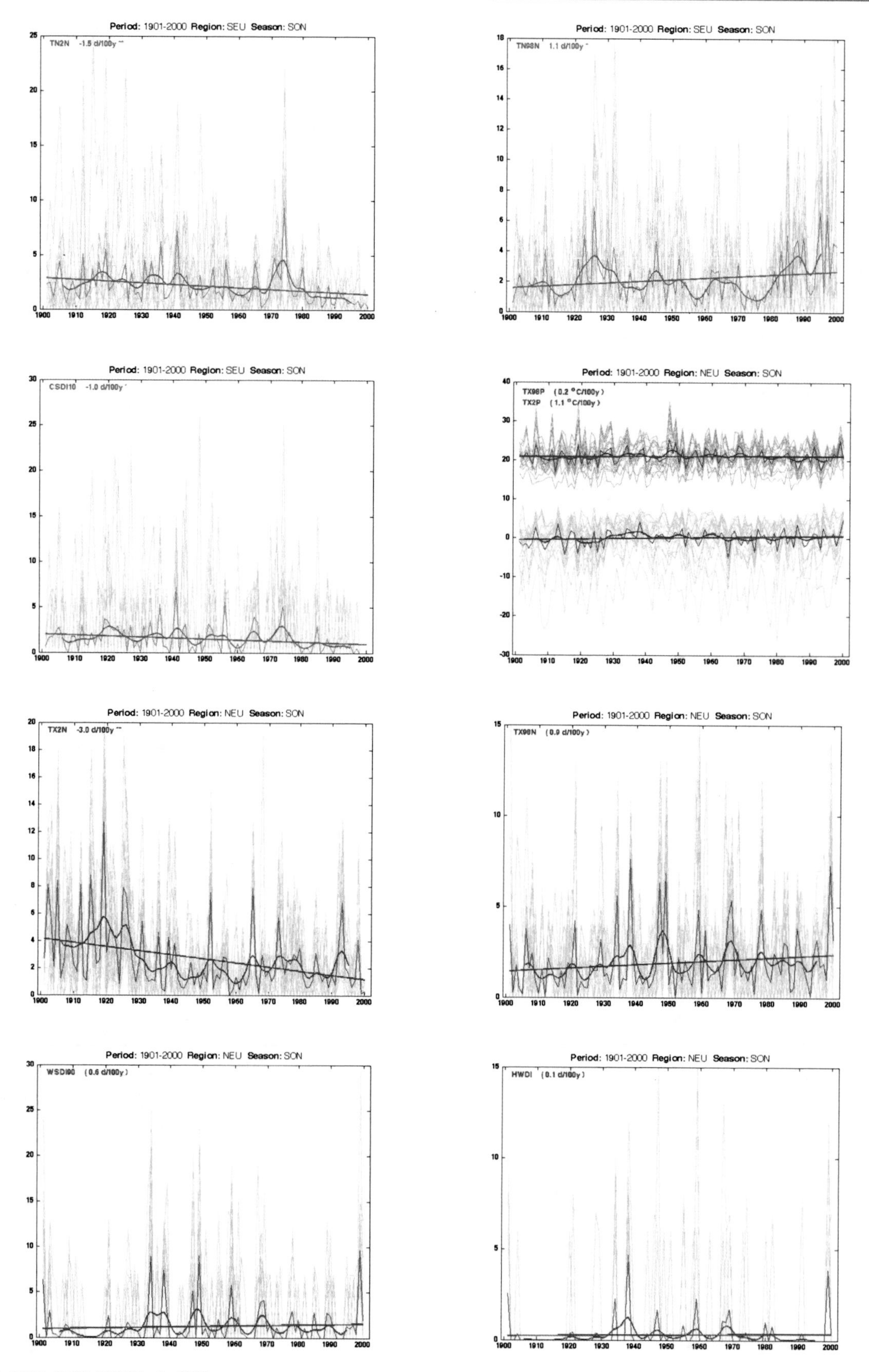

Fig. 3.142 1901–2000 SON Tmin SEU

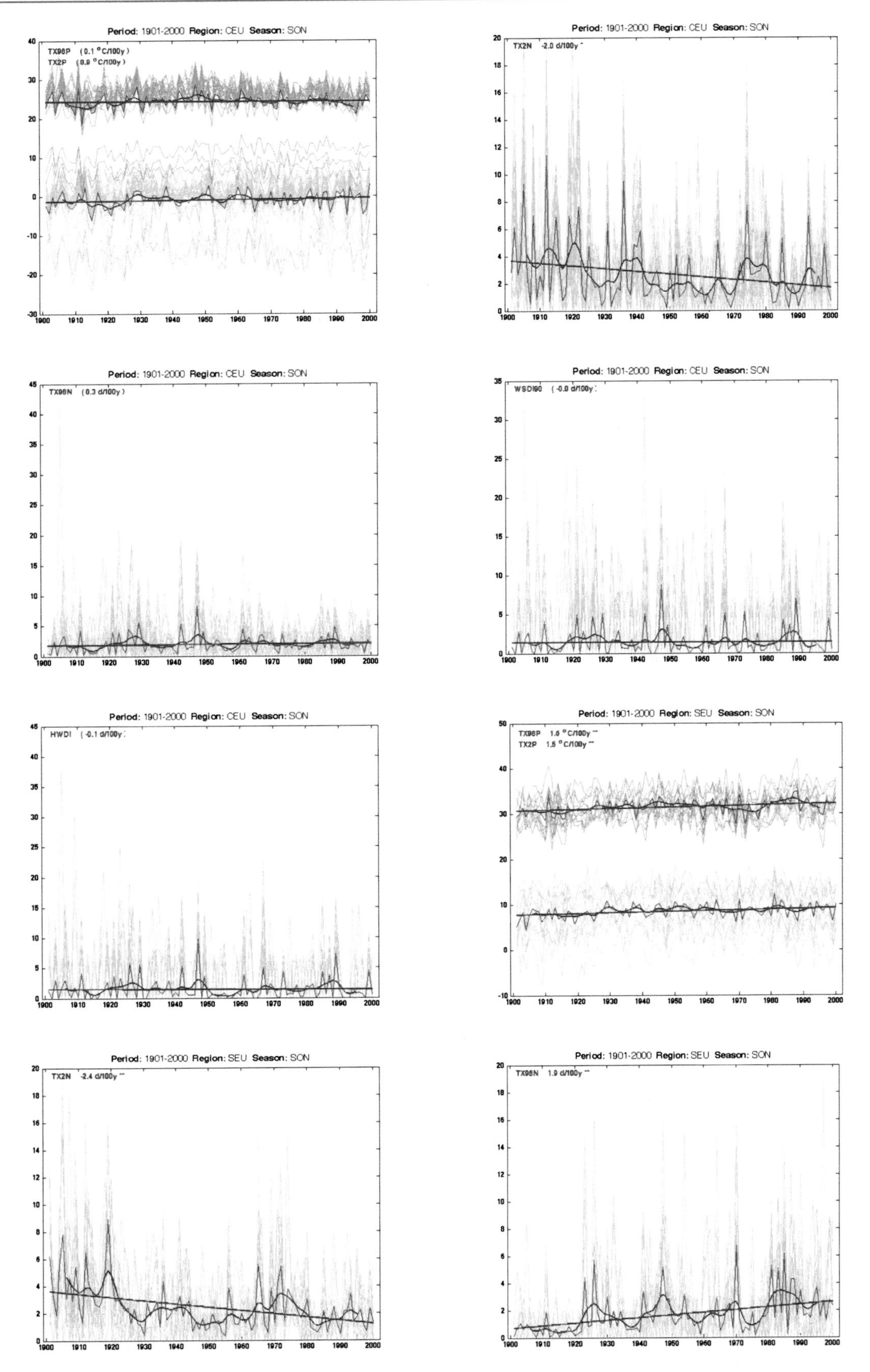

Fig. 3.143 1901–2000 SON Tmax CEU

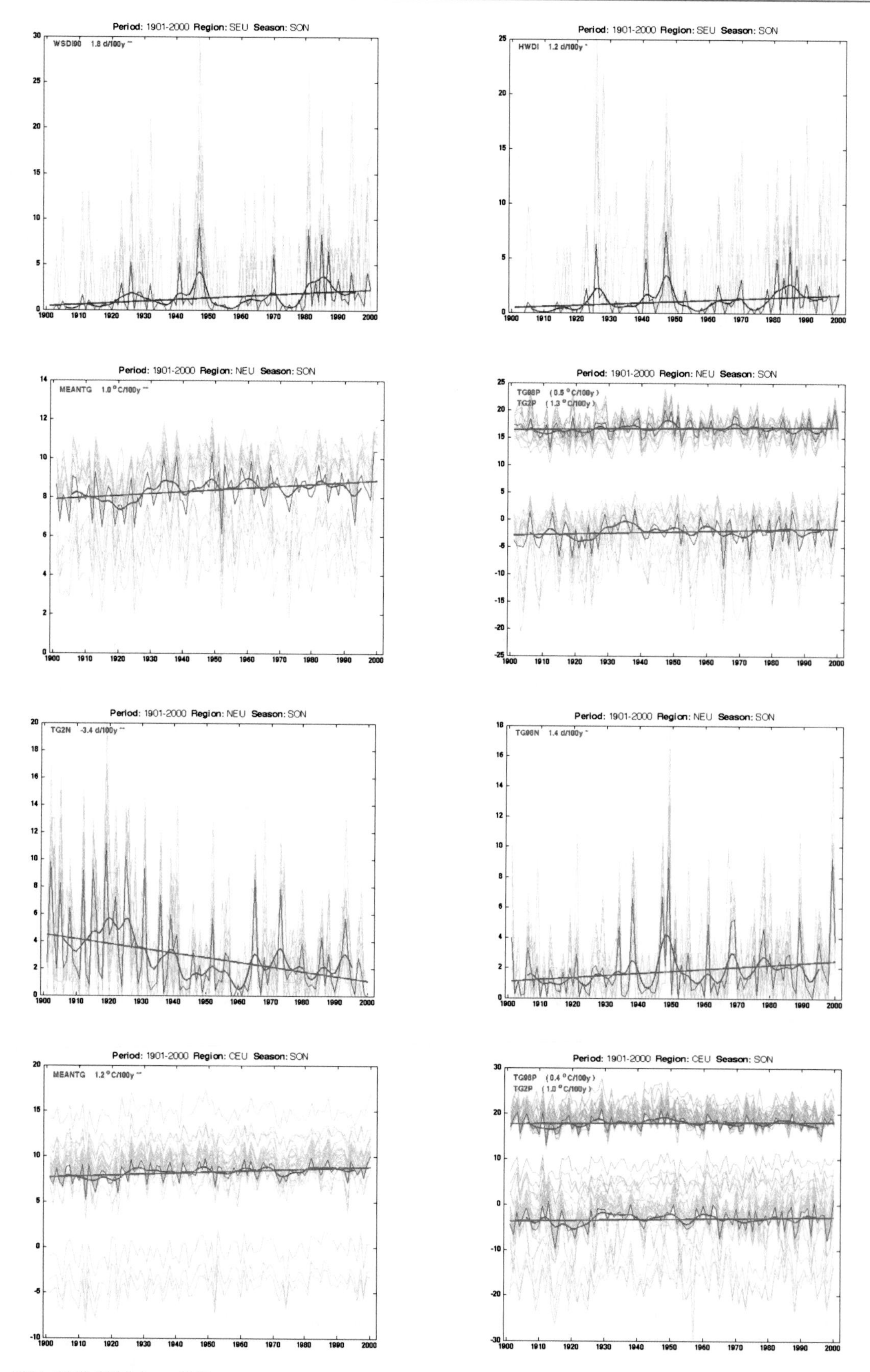

Fig. 3.144 1901–2000 SON Tmax SEU

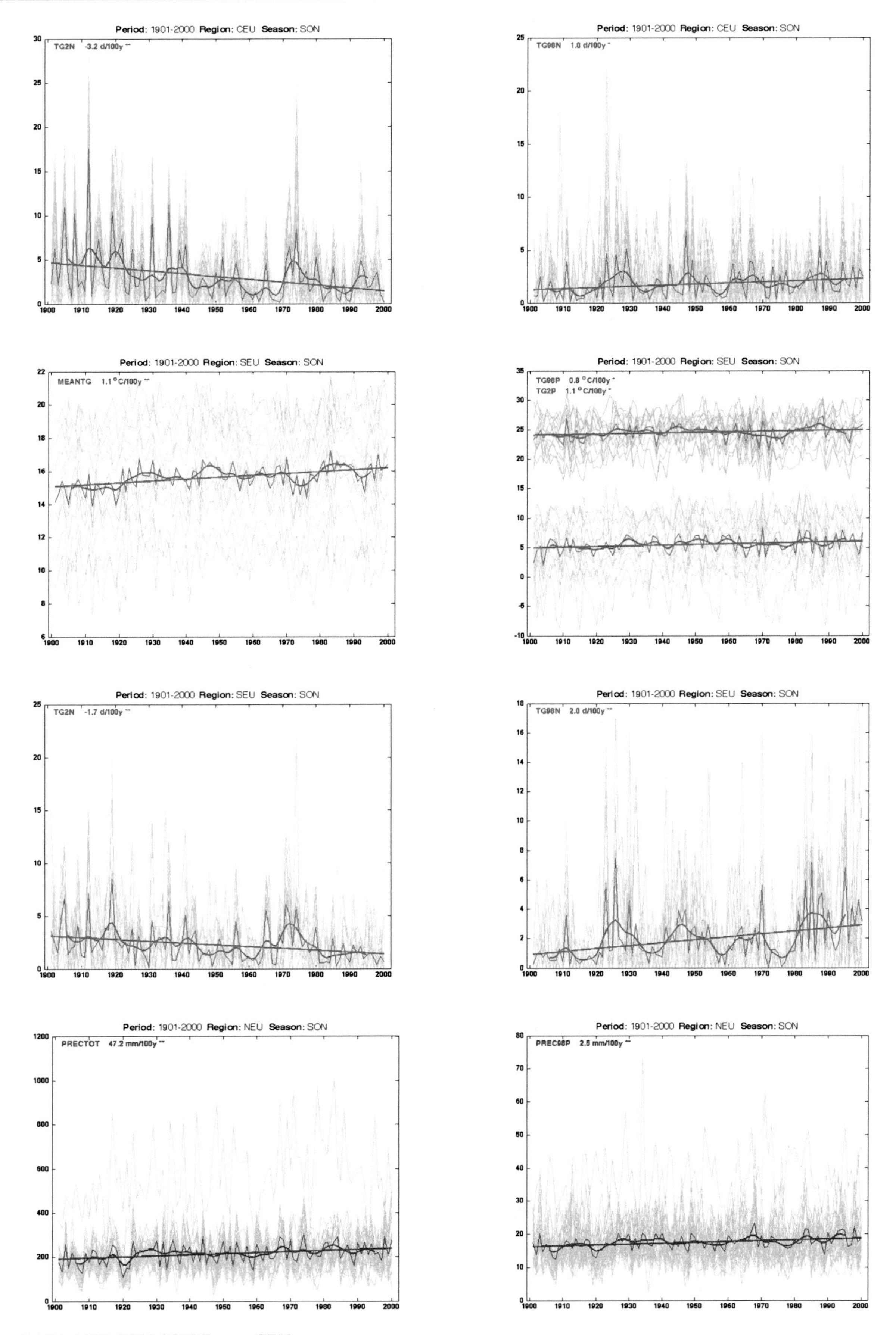

Fig. 3.145 1901–2000 SON Tmean CEU

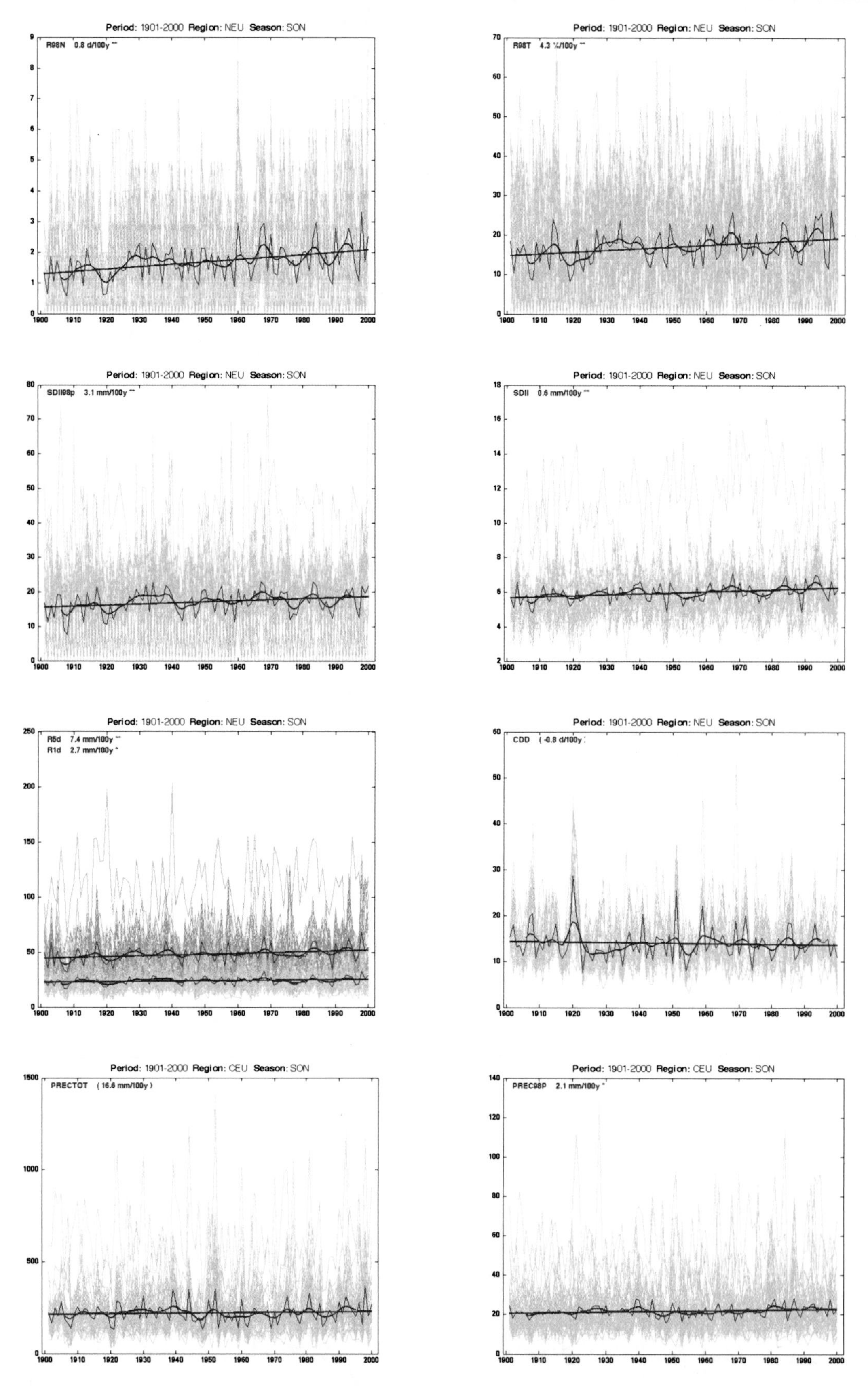

Fig. 3.146 1901–2000 SON Prec NEU

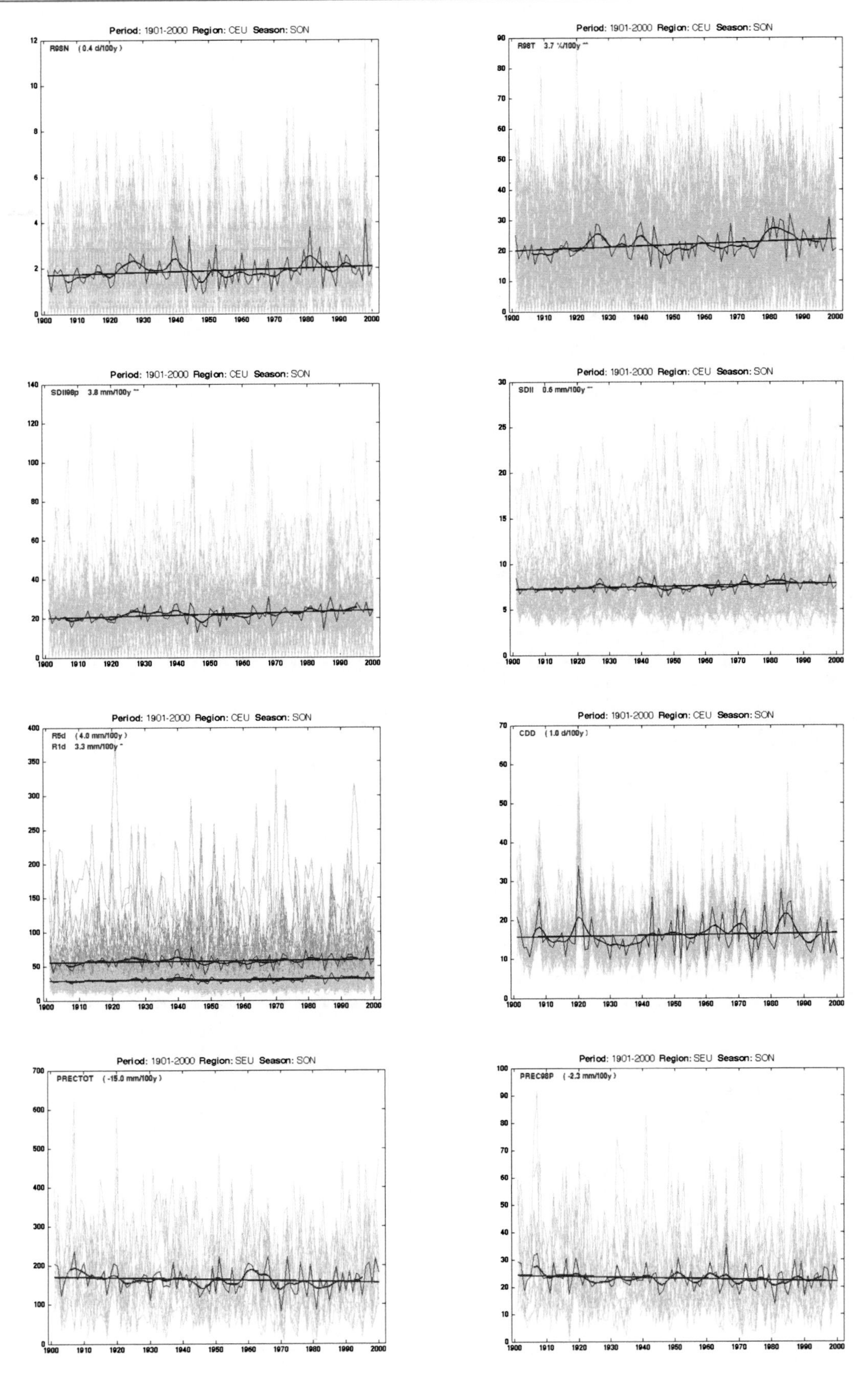

Fig. 3.147 1901–2000 SON Prec CEU

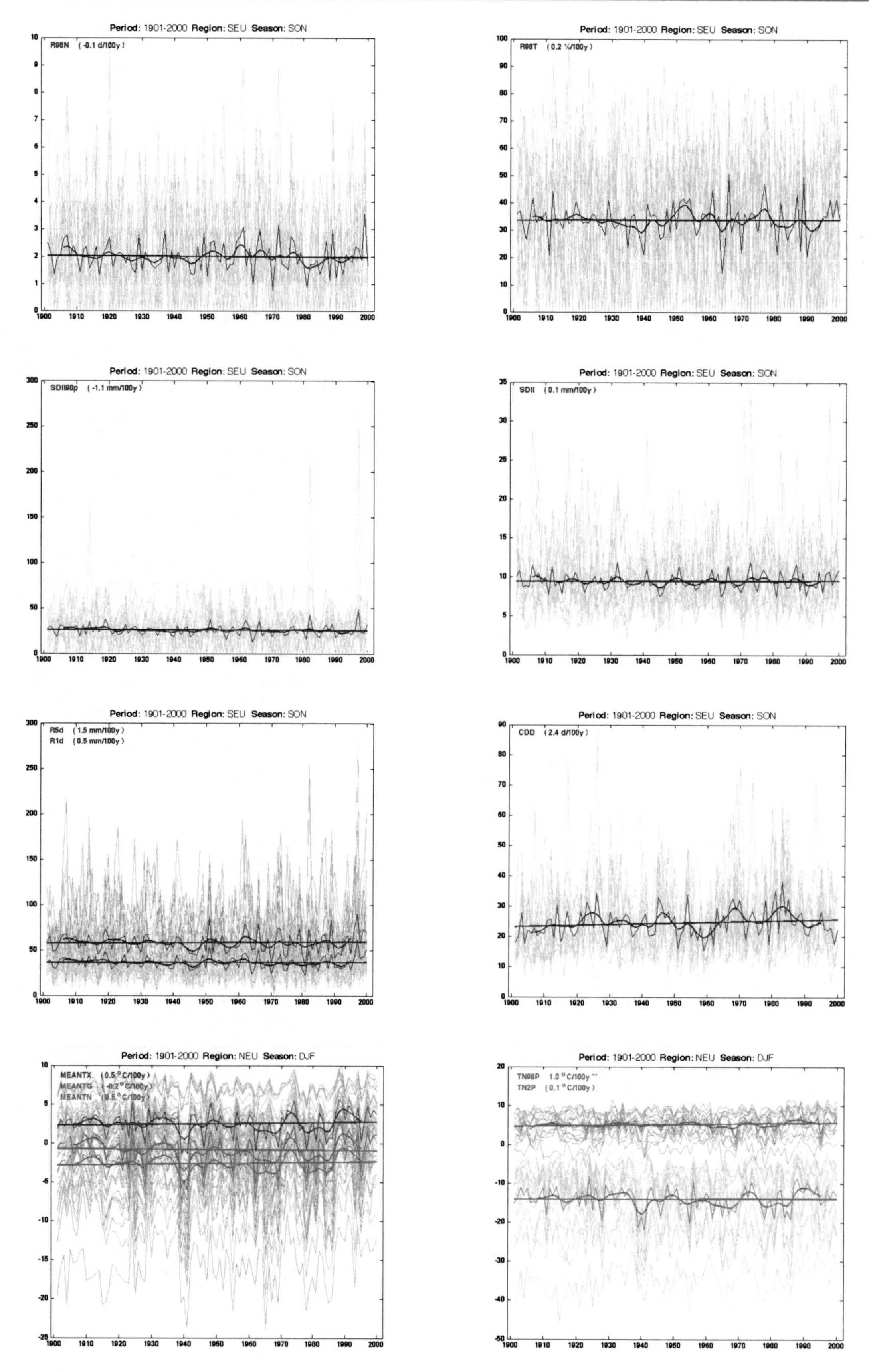

Fig. 3.148 1901–2000 SON Prec SEU

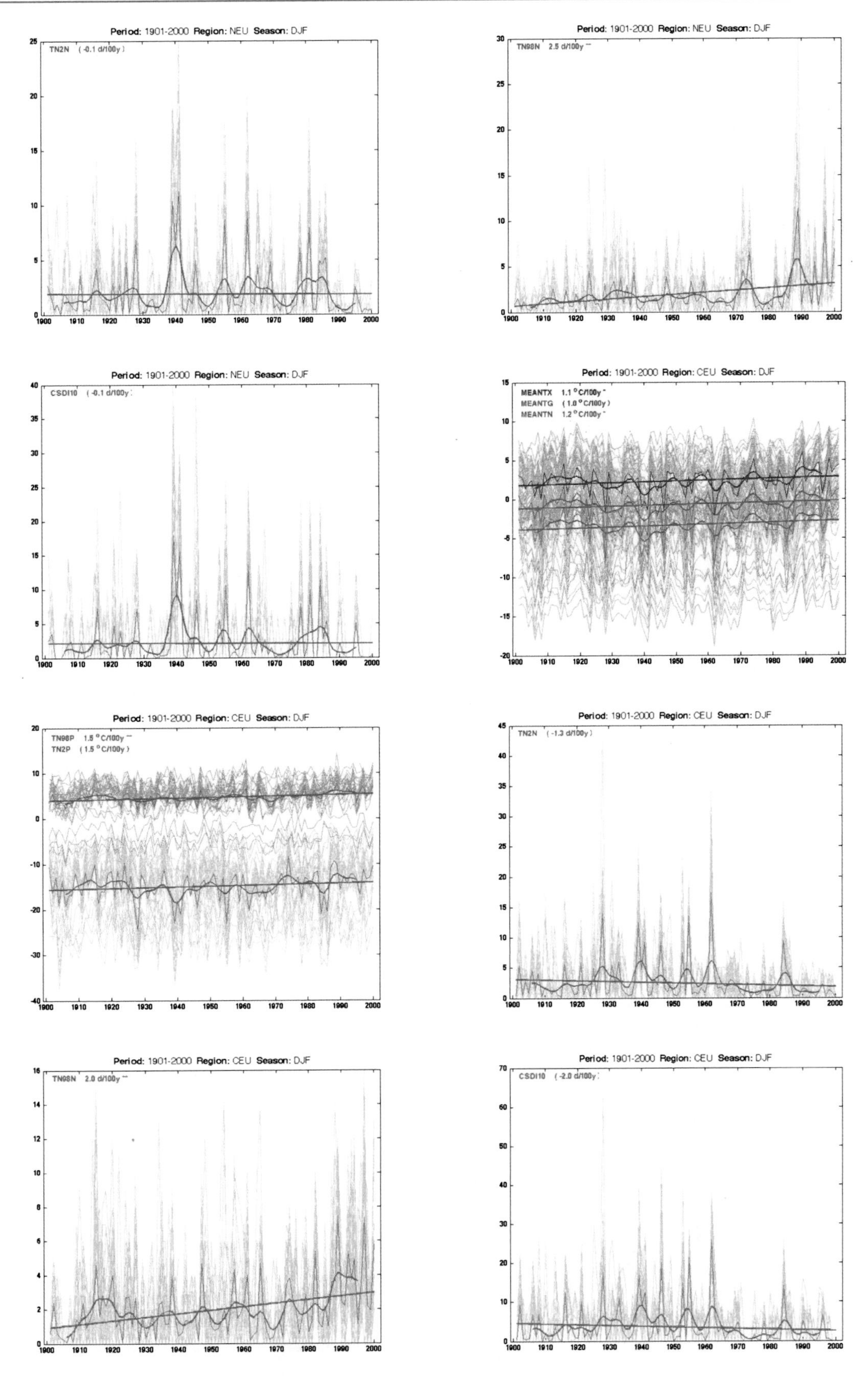

Fig. 3.149 1901–2000 DJF Tmin NEU

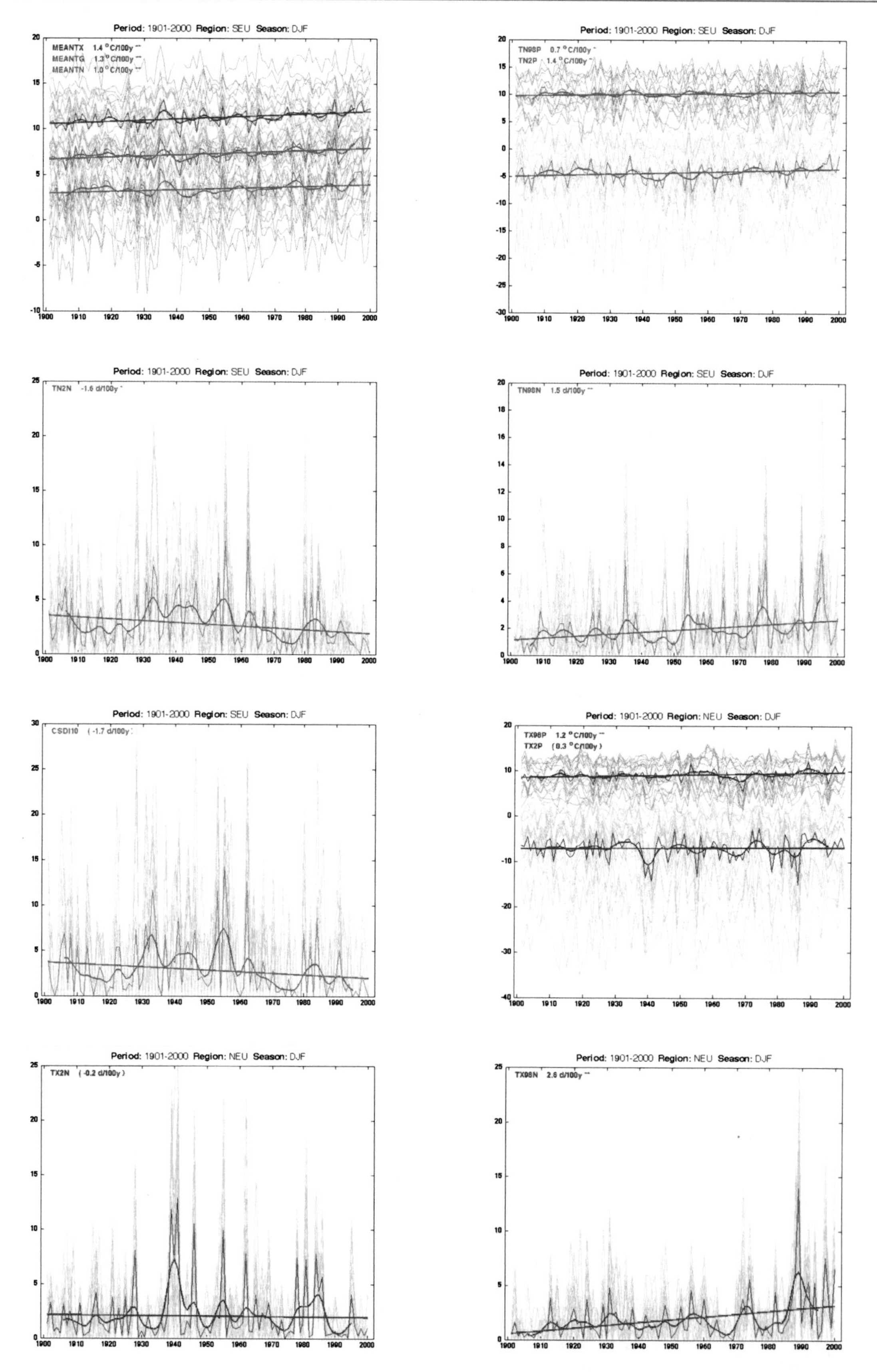

Fig. 3.150 1901–2000 DJF Tmin SEU

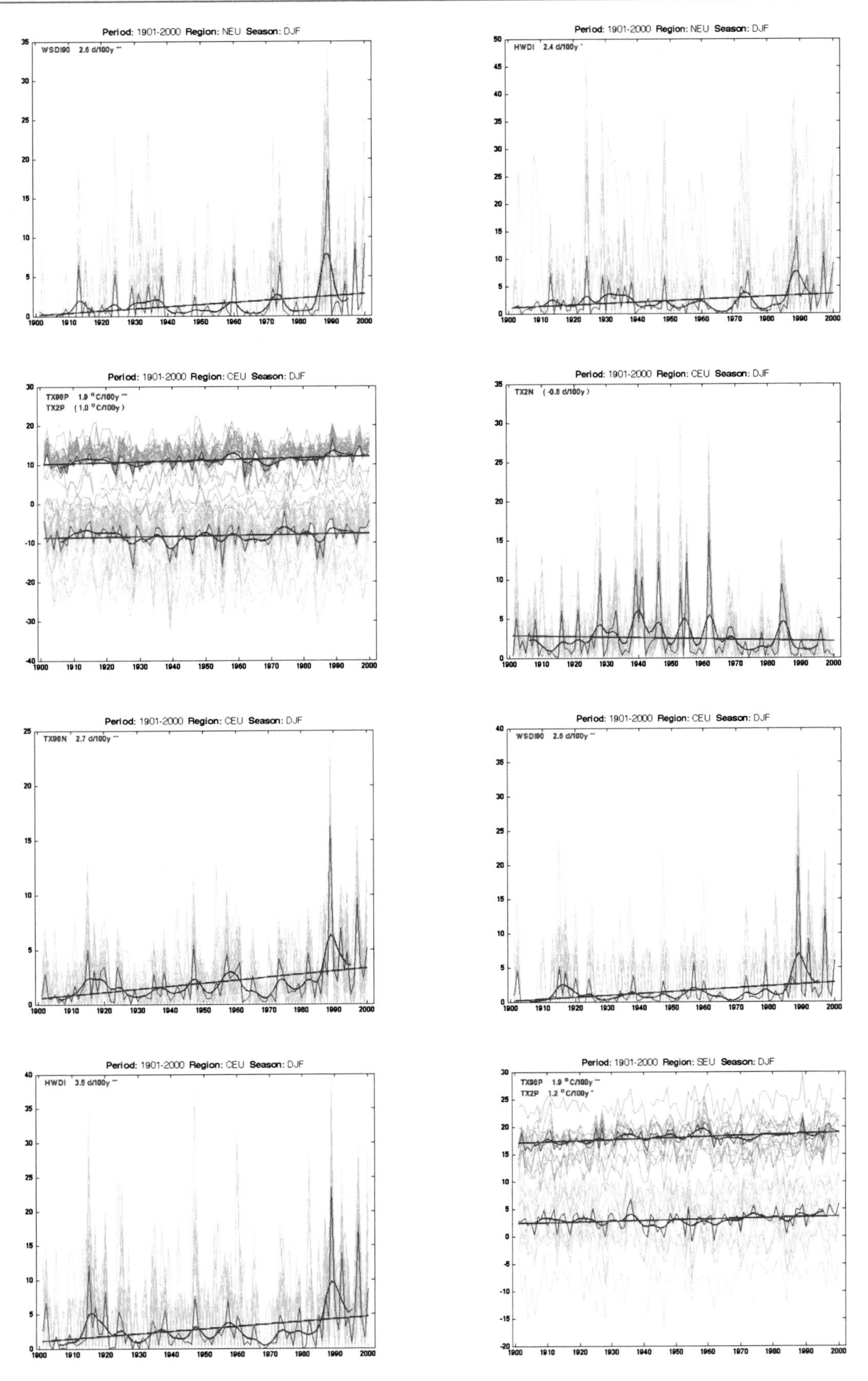

Fig. 3.151 1901–2000 DJF Tmax NEU

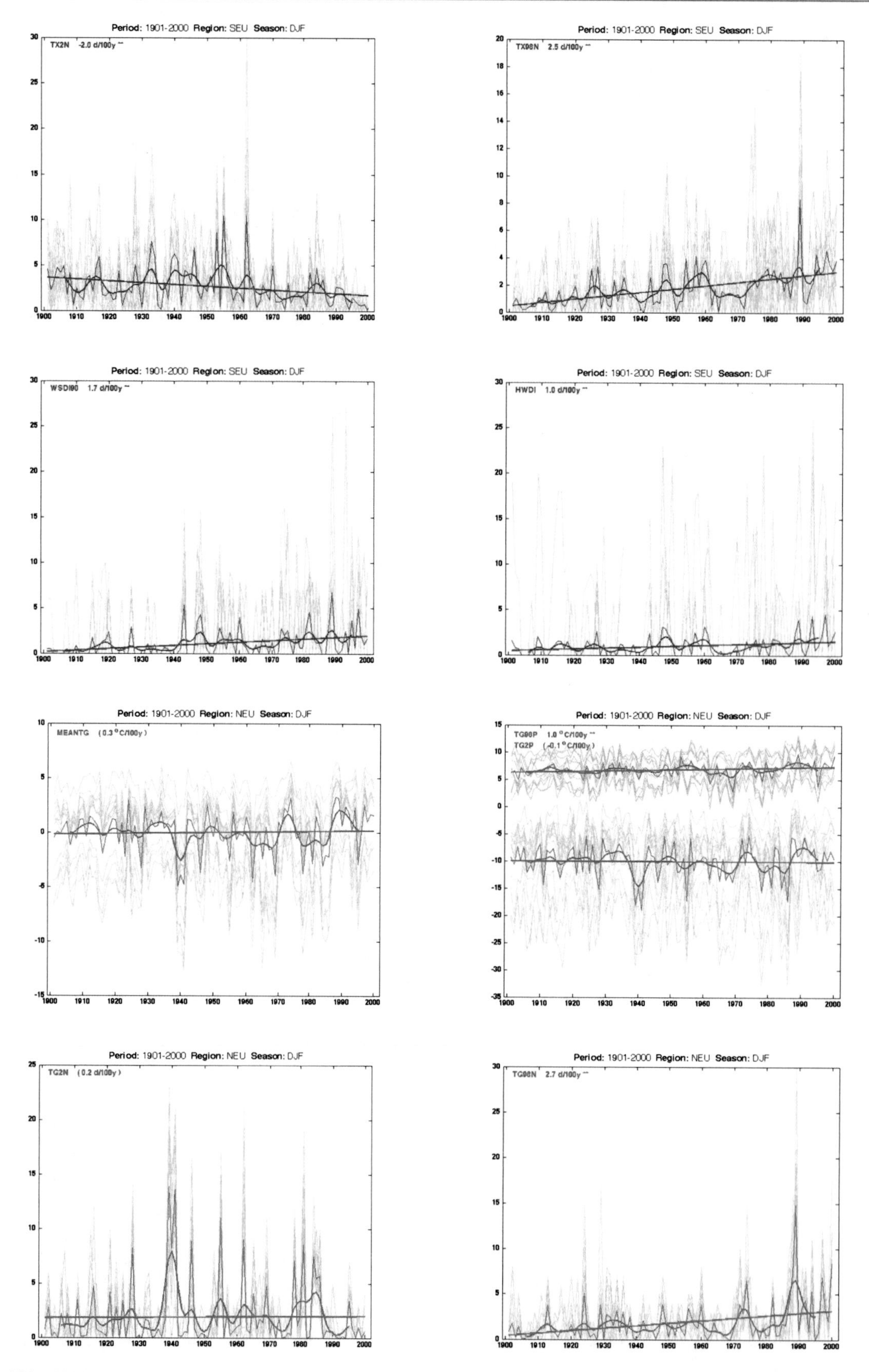

Fig. 3.152 1901–2000 DJF Tmax SEU

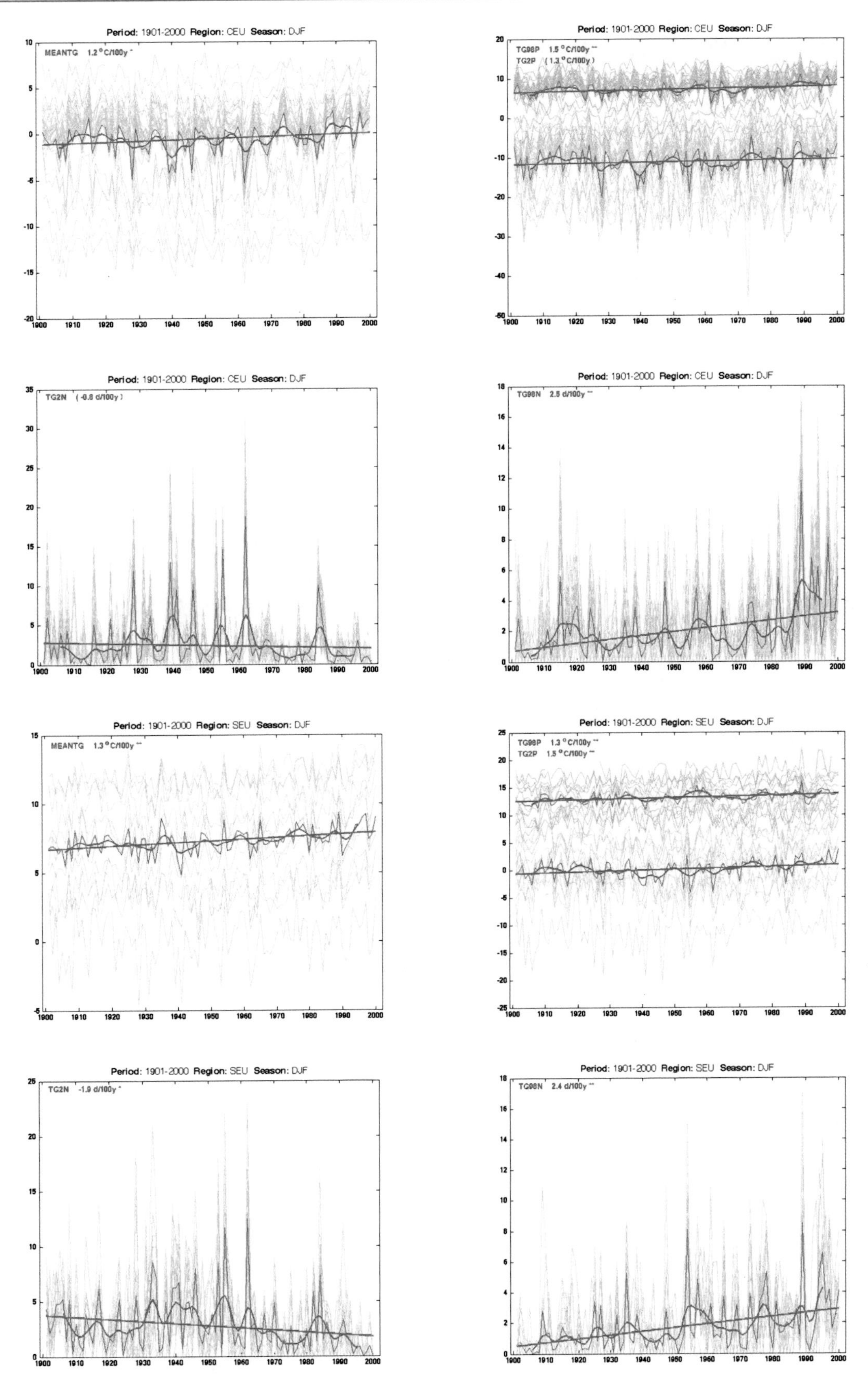

Fig. 3.153 1901–2000 DJF Tmean CEU

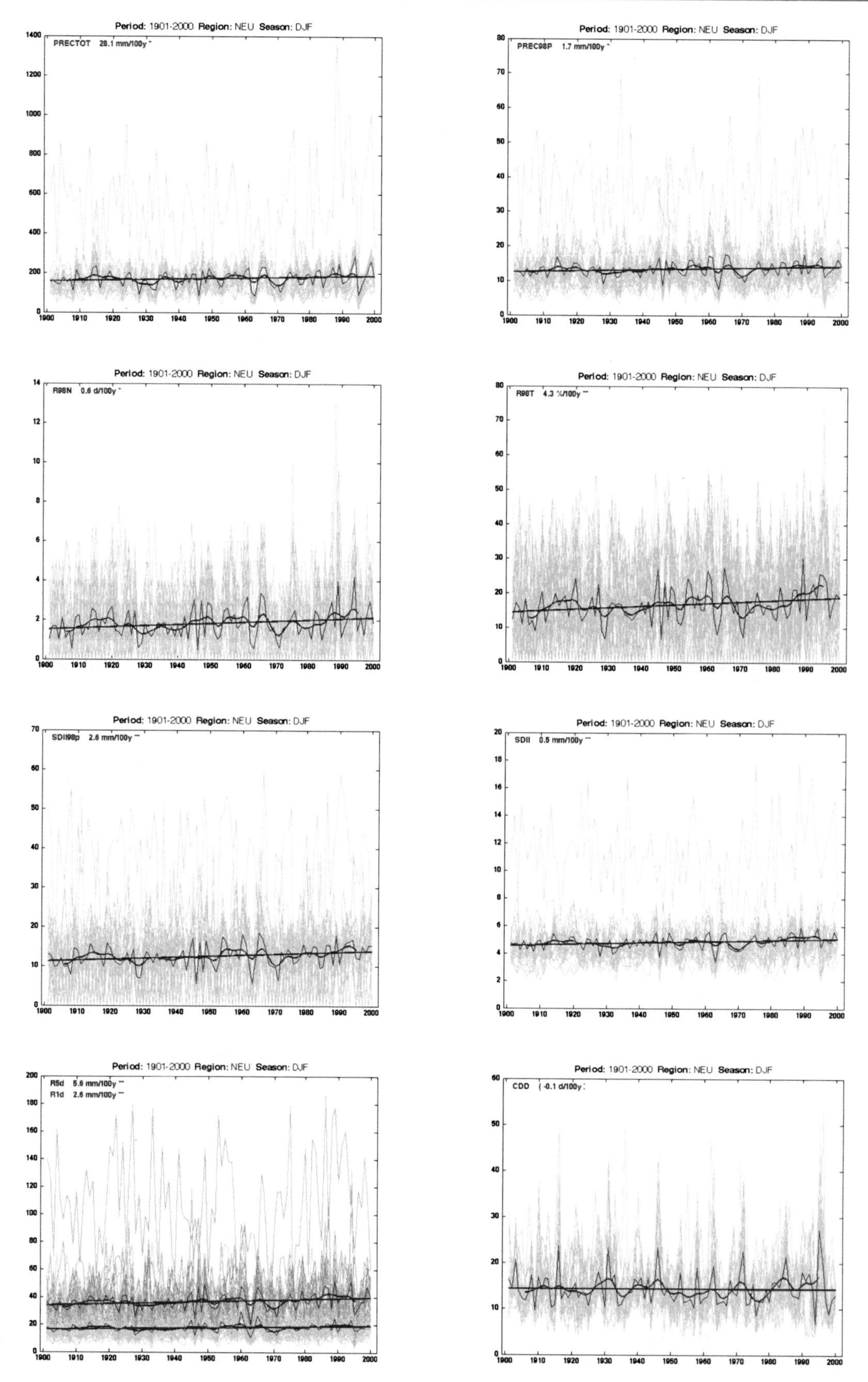

Fig. 3.154 1901–2000 DJF Prec NEU

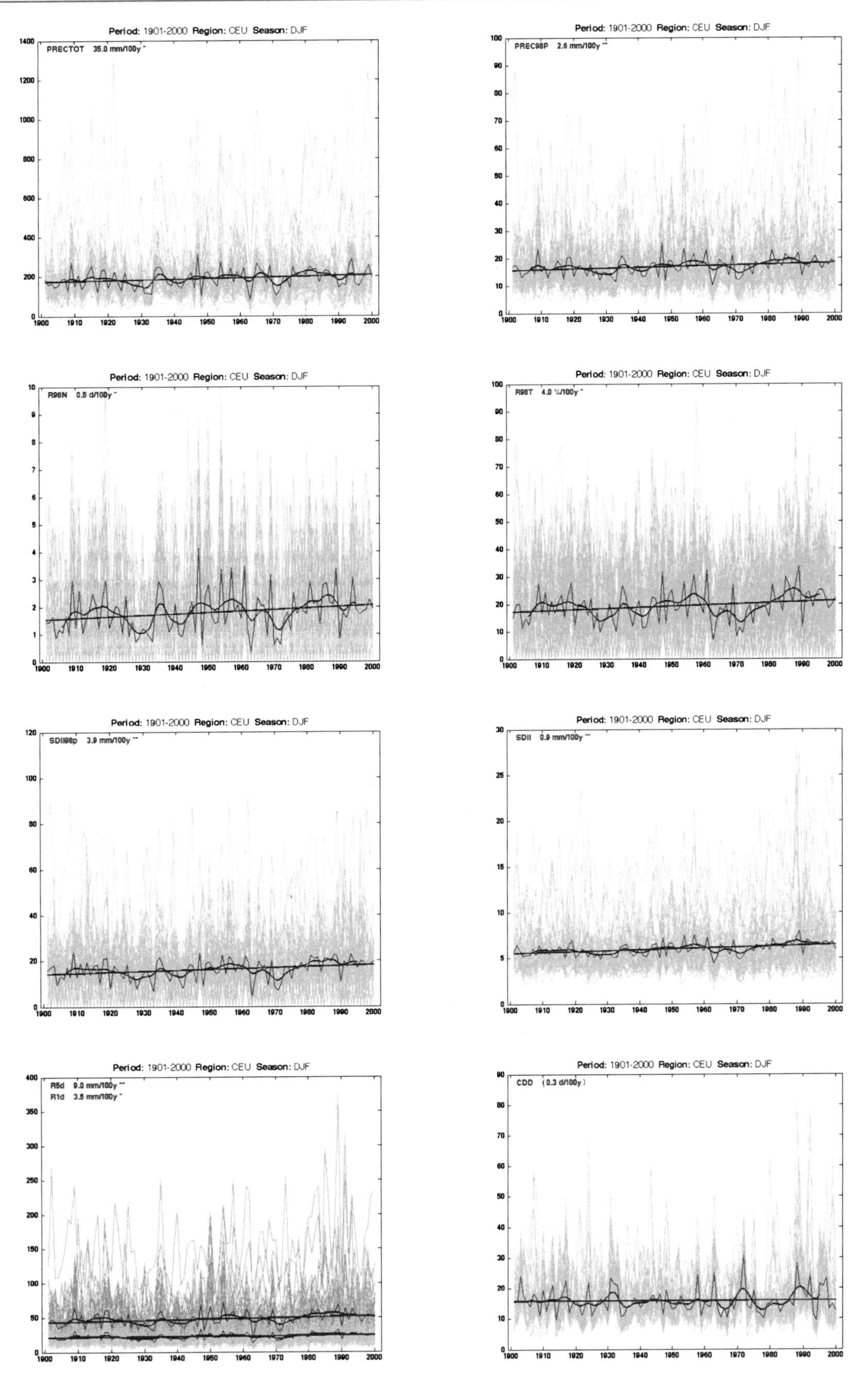

Fig. 3.155 1901–2000 DJF Prec CEU

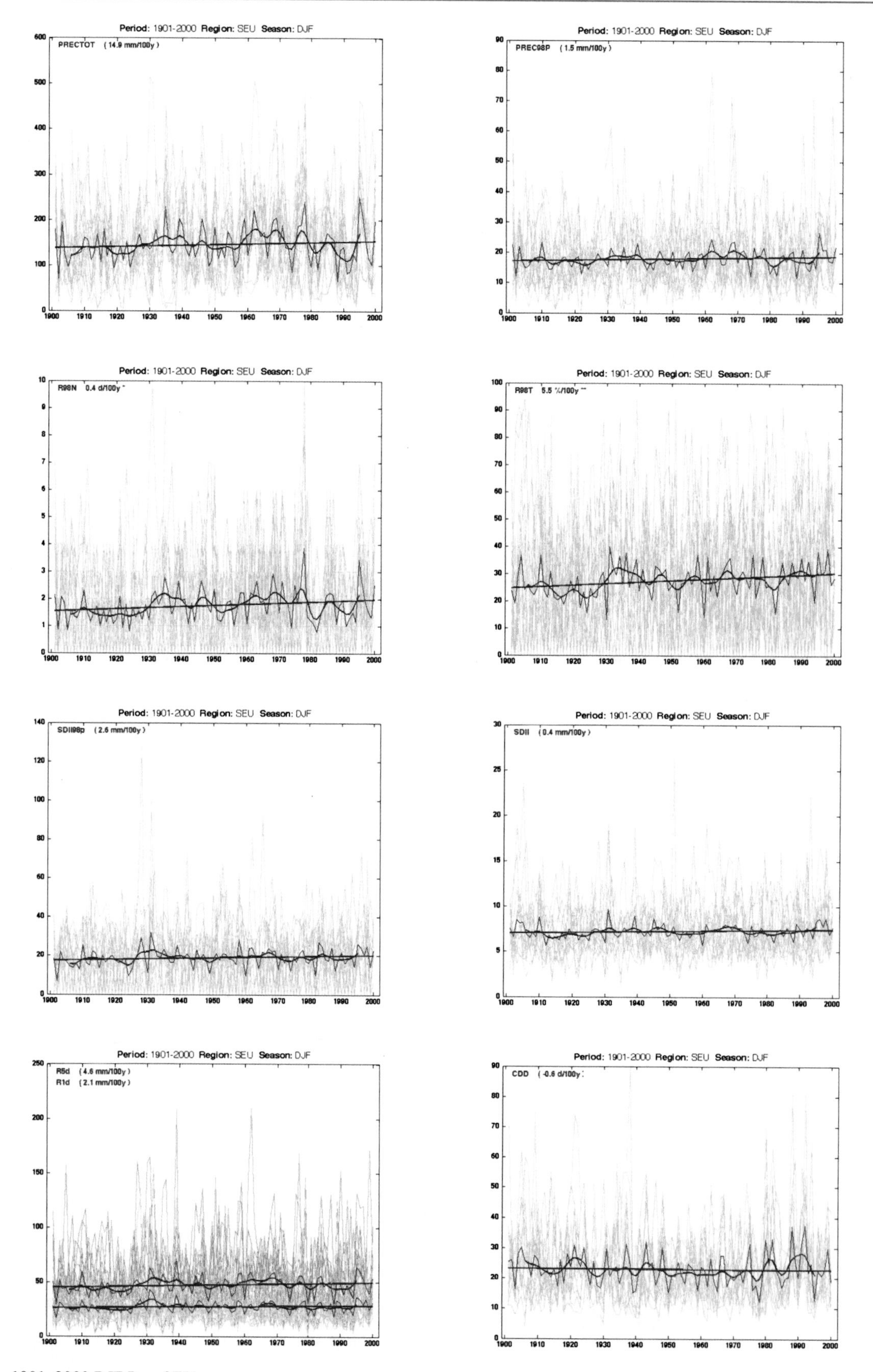

Fig. 3.156 1901–2000 DJF Prec SEU

4 Summary of the Estimated Trends

While the previous chapter can be used to find trends for all individual indices at all stations/regions, this chapter attempts to provide an overview of the trends for the three periods. Since there are many maps and figures in the previous chapter, summary statistics of the results are necessary to obtain an overview. This chapter gives statistics of the trends in terms of the trend estimates, fraction of positive and negative trends and their significance (see chapter 3 for definitions) in the form of tables.

4.1 Temporal Trends of Indices at the Stations

4.1.1 1801–2000

For indices based on Tmin/Tmax there are only three stations, while for indices based on Tmean we have seven stations. There is no precipitation record for this period. Table 4.1 summarizes the portions of the negative and positive trends. Statistically significant trends at two significance levels are indicated separately.

In summary, an overall tendency towards higher temperature means and extreme levels (percentiles), an increase in the frequency and duration of warm extremes, and a decrease in the frequency of cold ones, which is consistent with global warming can be seen. This is especially true for winter, but summer is always an exception. In fact, one third of the stations show a weakly decreasing Tmin and Tmax and the negative trends for Tmean appeared at more stations and become significant. As a result, the heat wave index (HWDI) in summer shows a slight negative trend (not significant) at one third of the stations. However, all the stations experience an increasing trend of warm spell duration (WSDI90). Thus, extremely high temperatures in summer for most stations still show an upward trend.

4.1.2 1851–2000

Nine stations with Tmin/Tmax observations and 13 stations with Tmean observations were analyzed for this period. The trends and the indication of significance are summarized in Table 4.2. Regarding the temperature indices we find similar seasonal patterns for this period as for the longer period described above. However, the fraction of stations with negative trends is smaller, and the number of significant positive trends is increased. This indicates that the overall warming rate over this period is larger than that in the previous period.

Nine precipitation stations are available and the summary statistics for the stations are shown in Table 4.2. In general there is a clear tendency towards higher precipitation totals, increased extreme levels, more frequent and more intensive rainfall events. However, the fraction of significant trends (both positive and negative) for the precipitation indices is much lower than these for temperature indices.

There are more stations with increasing trend of total precipitation in cold seasons (SON and DJF) than those in warm seasons (MAM and JJA), whereas the fraction of increasing trend of heavy precipitation (PREC98P) is the largest in summer and lowest in spring. Almost half of the stations have negative trends of PREC98P in spring, although the negative trends are not as significant as the positive trends. Since the heavy precipitation events in summer are often associated with convection, this indicates that convective precipitation has most likely increased at most of the stations. In terms of number of days with precipitation rate larger than the 98th percentiles over the reference period, most stations experience an increasing trend in autumn and winter, while spring and summer both have a larger fraction of the stations with negative trends. In terms of the measures for daily and 5day extremely heavy precipitation (SDII98P, SDII, R5d, R1d), the number of stations having positive trends are larger than that with negative trends for all seasons. Only a small fraction of stations show significant negative trends in summer and winter. These results show that climate in Europe has become wetter in general and there is an overall increase of heavy precipitation both in terms of frequency and intensity across the seasons. This tendency is confirmed by the overall negative trends in the drought index CDD which shows the max number of consecutive dry days. This feature is most obvious in winter.

D. Chen et al., *European Trend Atlas of Extreme Temperature and Precipitation Records*, DOI 10.1007/978-94-017-9312-4_4

Table 4.1 Fraction (%) of positive and negative trends and their significance levels (*=95% level, **=99% level) under the period 1801–2000

1801-2000		MAM						JJA						SON						DJF					
		pos	pos*	pos**	neg	neg*	neg**	pos	pos*	pos**	neg	neg*	neg**	pos	pos*	pos**	neg	neg*	neg**	pos	pos*	pos**	neg	neg*	neg**
Tmin/Tmax	MEANTN	100	66.7	33.3	0.0	0.0	0.0	66.7	66.7	66.7	33.3	33.3	0.0	100	66.7	66.7	0.0	0.0	0.0	100	100	100	0.0	0.0	0.0
	MEANTX	100	100	66.7	0.0	0.0	0.0	66.7	66.7	66.7	33.3	0.0	0.0	100	100	66.7	0.0	0.0	0.0	100	100	66.7	0.0	0.0	0.0
	TN2P	100	66.7	0.0	0.0	0.0	0.0	0.0	0.0	0.0	100	66.7	33.3	100	33.3	33.3	0.0	0.0	0.0	100	66.7	33.3	0.0	0.0	0.0
	TN98P	66.7	33.3	33.3	33.3	0.0	0.0	100	66.7	66.7	0.0	0.0	0.0	100	66.7	66.7	0.0	0.0	0.0	100	66.7	66.7	0.0	0.0	0.0
	TX2P	100	33.3	33.3	0.0	0.0	0.0	33.3	0.0	0.0	66.7	66.7	66.7	66.7	33.3	33.3	33.3	0.0	0.0	100	66.7	66.7	0.0	0.0	0.0
	TX98P	100	100	66.7	0.0	0.0	0.0	100	66.7	66.7	0.0	0.0	0.0	100	100	33.3	0.0	0.0	0.0	100	100	100	0.0	0.0	0.0
	TN2N	0.0	0.0	0.0	100	33.3	33.3	66.7	33.3	0.0	33.3	0.0	0.0	0.0	0.0	0.0	100	33.3	33.3	0.0	0.0	0.0	100	66.7	33.3
	TN98N	100	33.3	33.3	0.0	0.0	0.0	66.7	66.7	66.7	33.3	0.0	0.0	100	33.3	33.3	0.0	0.0	0.0	100	66.7	66.7	0.0	0.0	0.0
	TX2N	33.3	0.0	0.0	66.7	33.3	33.3	66.7	33.3	0.0	33.3	0.0	0.0	66.7	0.0	0.0	33.3	0.0	0.0	0.0	0.0	0.0	100	66.7	0.0
	TX98N	100	100	100	0.0	0.0	0.0	100	66.7	66.7	0.0	0.0	0.0	100	100	100	0.0	0.0	0.0	100	100	100	0.0	0.0	0.0
	HWDI	100	100	0.0	0.0	0.0	0.0	66.7	66.7	66.7	33.3	0.0	0.0	100	0.0	0.0	0.0	0.0	0.0	100	33.3	33.3	0.0	0.0	0.0
	WSDI90	100	66.7	33.3	0.0	0.0	0.0	100	66.7	33.3	0.0	0.0	0.0	100	33.3	0.0	0.0	0.0	0.0	100	66.7	33.3	0.0	0.0	0.0
	CSDI10	0.0	0.0	0.0	100	33.3	33.3	33.3	0.0	0.0	66.7	33.3	33.3	0.0	0.0	0.0	100	66.7	0.0	0.0	0.0	0.0	100	33.3	33.3
Tmean	MEANTG	85.7	71.4	57.1	14.3	14.3	14.3	57.1	28.6	14.3	42.9	14.3	14.3	85.7	71.4	71.4	14.3	14.3	0.0	85.7	85.7	71.4	14.3	0.0	0.0
	TG2P	100	57.1	42.9	0.0	0.0	0.0	42.9	14.3	0.0	57.1	28.6	14.3	100	57.1	42.9	0.0	0.0	0.0	100	71.4	71.4	0.0	0.0	0.0
	TG98P	85.7	14.3	14.3	14.3	14.3	14.3	57.1	57.1	42.9	42.9	14.3	14.3	57.1	42.9	14.3	42.9	14.3	14.3	85.7	85.7	71.4	14.3	0.0	0.0
	TG2N	14.3	0.0	0.0	85.7	57.1	57.1	42.9	0.0	0.0	57.1	28.6	28.6	14.3	0.0	0.0	85.7	42.9	42.9	14.3	0.0	0.0	85.7	57.1	42.9
	TG98N	85.7	42.9	42.9	14.3	14.3	14.3	57.1	57.1	28.6	42.9	28.6	14.3	85.7	57.1	57.1	14.3	14.3	0.0	85.7	85.7	71.4	14.3	0.0	0.0

Table 4.2 Fraction (%) of positive and negative trends and their significance levels (*=95% level, **=99% level) under the period 1851–2000

1851-2000		MAM						JJA						SON						DJF					
		pos	pos*	pos**	neg	neg*	neg**	pos	pos*	pos**	neg	neg*	neg**	pos	pos*	pos**	neg	neg*	neg**	pos	pos*	pos**	neg	neg*	neg**
Tmin/Tmax	MEANTN	100	88.9	88.9	0.0	0.0	0.0	100.0	88.9	77.8	0.0	0.0	0.0	100	88.9	88.9	0.0	0.0	0.0	100	88.9	77.8	0.0	0.0	0.0
	MEANTX	100	88.9	77.8	0.0	0.0	0.0	66.7	44.4	44.4	33.3	0.0	0.0	100	100.0	88.9	0.0	0.0	0.0	100	77.8	66.7	0.0	0.0	0.0
	TN2P	88.9	88.9	55.6	11.1	0.0	0.0	66.7	44.4	44.4	33.3	0.0	0.0	88.9	66.7	66.7	11.1	0.0	0.0	100	77.8	66.7	0.0	0.0	0.0
	TN98P	88.9	66.7	44.4	11.1	0.0	0.0	100	77.8	66.7	0.0	0.0	0.0	88.9	55.6	55.6	11.1	0.0	0.0	100	88.9	88.9	0.0	0.0	0.0
	TX2P	100	66.7	55.6	0.0	0.0	0.0	55.6	33.3	22.2	44.4	22.2	11.1	100	66.7	55.6	0.0	0.0	0.0	100	44.4	33.3	0.0	0.0	0.0
	TX98P	100	55.6	33.3	0.0	0.0	0.0	77.8	55.6	44.4	22.2	0.0	0.0	88.9	22.2	22.2	11.1	0.0	0.0	100	77.8	77.8	0.0	0.0	0.0
	TN2N	0.0	0.0	0.0	100	66.7	66.7	33.3	0.0	0.0	66.7	44.4	44.4	11.1	0.0	0.0	88.9	77.8	66.7	11.1	0.0	0.0	88.9	66.7	33.3
	TN98N	100	100.0	88.9	0.0	0.0	0.0	100	77.8	55.6	0.0	0.0	0.0	88.9	77.8	66.7	11.1	0.0	0.0	100	88.9	66.7	0.0	0.0	0.0
	TX2N	11.1	0.0	0.0	88.9	66.7	55.6	33.3	22.2	0.0	66.7	44.4	22.2	0.0	0.0	0.0	100	66.7	44.4	0.0	0.0	0.0	100	44.4	22.2
	TX98N	100	66.7	55.6	0.0	0.0	0.0	77.8	55.6	55.6	22.2	0.0	0.0	100	66.7	44.4	0.0	0.0	0.0	100	88.9	88.9	0.0	0.0	0.0
	HWDI	88.9	33.3	11.1	11.1	0.0	0.0	88.9	22.2	0.0	11.1	0.0	0.0	77.8	0.0	0.0	11.1	0.0	0.0	88.9	44.4	33.3	0.0	0.0	0.0
	WSDI90	100	44.4	33.3	0.0	0.0	0.0	66.7	11.1	11.1	33.3	0.0	0.0	100	22.2	11.1	0.0	0.0	0.0	100	77.8	44.4	0.0	0.0	0.0
	CSDI10	0.0	0.0	0.0	100.0	77.8	55.6	22.2	0.0	0.0	77.8	22.2	22.2	0.0	0.0	0.0	100.0	44.4	33.3	0.0	0.0	0.0	100	33.3	33.3
Tmean	MEANTG	92.3	92.3	84.6	7.7	0.0	0.0	84.6	61.5	46.2	15.4	7.7	7.7	100	92.3	76.9	0.0	0.0	0.0	100	69.2	69.2	0.0	0.0	0.0
	TG2P	100	84.6	69.2	0.0	0.0	0.0	69.2	46.2	38.5	30.8	0.0	0.0	100	84.6	69.2	0.0	0.0	0.0	100	84.6	30.8	0.0	0.0	0.0
	TG98P	84.6	38.5	38.5	15.4	7.7	7.7	84.6	38.5	38.5	15.4	7.7	0.0	84.6	38.5	7.7	15.4	0.0	0.0	100	84.6	76.9	0.0	0.0	0.0
	TG2N	0.0	0.0	0.0	100	92.3	76.9	30.8	0.0	0.0	69.2	61.5	46.2	0.0	0.0	0.0	100	69.2	69.2	0.0	0.0	0.0	100	38.5	23.1
	TG98N	92.3	76.9	53.8	7.7	0.0	0.0	84.6	46.2	46.2	15.4	7.7	7.7	92.3	69.2	30.8	7.7	0.0	0.0	100	84.6	76.9	0.0	0.0	0.0
precipitation	PRECTOT	66.7	33.3	22.2	33.3	0.0	0.0	66.7	0.0	0.0	33.3	11.1	0.0	88.9	33.3	22.2	11.1	0.0	0.0	88.9	55.6	44.4	11.1	0.0	0.0
	PREC98P	55.6	22.2	22.2	44.4	0.0	0.0	77.8	0.0	0.0	22.2	11.1	0.0	66.7	33.3	33.3	33.3	0.0	0.0	66.7	33.3	22.2	33.3	0.0	0.0
	R98N	66.7	44.4	22.2	33.3	0.0	0.0	55.6	0.0	0.0	44.4	0.0	0.0	88.9	33.3	33.3	11.1	0.0	0.0	88.9	33.3	0.0	11.1	0.0	0.0
	R98T	77.8	22.2	22.2	22.2	0.0	0.0	55.6	0.0	0.0	44.4	0.0	0.0	88.9	33.3	0.0	11.1	0.0	0.0	77.8	0.0	0.0	22.2	0.0	0.0
	SDII98p	77.8	22.2	0.0	22.2	0.0	0.0	55.6	0.0	0.0	44.4	0.0	0.0	66.7	22.2	0.0	33.3	0.0	0.0	100	33.3	0.0	0.0	0.0	0.0
	SDII	66.7	22.2	22.2	33.3	0.0	0.0	77.8	11.1	0.0	22.2	11.1	11.1	55.6	33.3	22.2	44.4	0.0	0.0	66.7	44.4	22.2	33.3	11.1	11.1
	R5d	77.8	22.2	22.2	22.2	0.0	0.0	77.8	11.1	0.0	22.2	0.0	0.0	66.7	22.2	22.2	33.3	0.0	0.0	88.9	33.3	22.2	11.1	0.0	0.0
	R1d	55.6	22.2	22.2	44.4	0.0	0.0	55.6	11.1	0.0	44.4	0.0	0.0	77.8	22.2	22.2	22.2	0.0	0.0	77.8	33.3	0.0	22.2	0.0	0.0
	CDD	33.3	0.0	0.0	66.7	0.0	0.0	33.3	0.0	0.0	66.7	0.0	0.0	33.3	0.0	0.0	66.7	0.0	0.0	22.2	0.0	0.0	77.8	0.0	0.0

4.1.3 1901–2000

For this period we have 58 stations with Tmin and Tmax observations, 54 stations with Tmean record, and 86 precipitation stations. Table 4.3 shows the percentage of positive and negative trends and their significance. For the 100-year period we see the same tendency towards higher temperature means and extreme levels, more frequent warm extremes, and less frequent cold extremes. Compared with the two longer periods the tendency has become even clearer.

For precipitation indices, positive trends are in the majority, where generally about two thirds of the trends are positive and one third is negative. For all indices and seasons we find a roughly similar partitioning between positive and negative trends. The fraction of positive significant trends ranges from 6–24 % whereas the fraction of negative significant trends is at most 6 % for all indices. In short, as for the 150-year period, we find even here signals that point to a wetter climate with higher extreme levels and more frequent and more intensive rainfall events. However, the seasonal distributions of the trends do not fully follow the pattern shown in the 150 year period. For example, the highest fraction of stations with positive trends of the 98th percentile precipitation (PREC98P) occurs now in spring, instead of in summer. Also, slightly more than half of the stations show positive trends in CDD, which displays a different pattern as compared to that over the 150 year period.

4.2 Temporal trends of indices for the regions

In this section we provide trend values and their significance level, as calculated over the period 1901–2000 for all indices defined for the three regional averages. As has been described in the data section, the station density of the regions

Table 4.3 Fraction (%) of positive and negative trends and their significance levels (*=95 % level, **=99% level) under the period 1901–2000

1901-2000		MAM						JJA						SON						DJF					
		pos	pos*	pos**	neg	neg*	neg**	pos	pos*	pos**	neg	neg*	neg**	pos	pos*	pos**	neg	neg*	neg**	pos	pos*	pos**	neg	neg*	neg**
Tmin/Tmax	MEANTN	84.2	61.4	45.6	15.8	1.8	0.0	91.2	68.4	56.1	8.8	3.5	3.5	93.0	66.7	47.4	7.0	0.0	0.0	87.7	35.1	24.6	12.3	0.0	0.0
	MEANTX	91.2	50.9	36.8	8.8	3.5	1.8	87.7	56.1	33.3	12.3	3.5	3.5	93.0	66.7	45.6	7.0	3.5	0.0	96.5	49.1	24.6	3.5	0.0	0.0
	TN2P	78.9	22.8	10.5	21.1	0.0	0.0	68.4	31.6	22.8	31.6	7.0	5.3	77.2	24.6	14.0	22.8	0.0	0.0	75.4	26.3	14.0	24.6	1.8	0.0
	TN98P	63.2	14.0	3.5	36.8	12.3	7.0	87.7	54.4	43.9	12.3	0.0	0.0	80.7	36.8	17.5	19.3	1.8	0.0	96.5	54.4	33.3	3.5	0.0	0.0
	TX2P	93.0	26.3	15.8	7.0	0.0	0.0	70.2	24.6	19.3	29.8	7.0	3.5	98.2	22.8	7.0	1.8	0.0	0.0	78.9	12.3	5.3	21.1	0.0	0.0
	TX98P	54.4	15.8	8.8	45.6	7.0	3.5	84.2	40.4	29.8	15.8	5.3	3.5	68.4	24.6	17.5	31.6	7.0	3.5	94.7	80.7	52.6	5.3	0.0	0.0
	TN2N	17.5	3.5	1.8	82.5	38.6	26.3	26.3	7.0	5.3	73.7	40.4	33.3	21.1	1.8	0.0	78.9	42.1	35.1	31.6	0.0	0.0	68.4	8.8	7.0
	TN98N	87.7	54.4	35.1	12.3	1.8	0.0	94.7	54.4	36.8	5.3	0.0	0.0	86.0	52.6	31.6	14.0	0.0	0.0	94.7	68.4	47.4	5.3	0.0	0.0
	TX2N	8.8	0.0	0.0	91.2	38.6	26.3	14.0	3.5	1.8	86.0	38.6	31.6	3.5	0.0	0.0	96.5	49.1	36.8	24.6	0.0	0.0	75.4	10.5	5.3
	TX98N	86.0	43.9	31.6	14.0	3.5	1.8	84.2	40.4	28.1	15.8	3.5	3.5	82.5	35.1	21.1	17.5	7.0	5.3	96.5	82.5	66.7	3.5	0.0	0.0
	HWDI	84.2	22.8	8.8	15.8	0.0	0.0	73.7	21.1	10.5	26.3	3.5	1.8	64.9	5.3	1.8	28.1	3.5	1.8	89.5	36.8	12.3	5.3	0.0	0.0
	WSDI90	82.5	24.6	8.8	17.5	0.0	0.0	75.4	24.6	10.5	24.6	3.5	3.5	73.7	14.0	1.8	26.3	3.5	1.8	96.5	49.1	17.5	3.5	0.0	0.0
	CSDI10	19.3	0.0	0.0	80.7	22.8	7.0	28.1	0.0	0.0	71.9	19.3	14.0	17.5	0.0	0.0	82.5	35.1	17.5	29.8	0.0	0.0	70.2	7.0	1.8
Tmean	MEANTG	98.1	61.1	44.4	1.9	0.0	0.0	98.1	70.4	55.6	1.9	0.0	0.0	98.1	75.9	63.0	1.9	0.0	0.0	92.6	38.9	33.3	5.6	0.0	0.0
	TG2P	92.6	35.2	11.1	7.4	0.0	0.0	85.2	29.6	20.4	14.8	1.9	1.9	96.3	20.4	5.6	3.7	0.0	0.0	77.8	24.1	9.3	20.4	0.0	0.0
	TG98P	48.1	7.4	7.4	51.9	7.4	1.9	88.9	48.1	33.3	11.1	1.9	0.0	72.2	22.2	11.1	27.8	0.0	0.0	96.3	66.7	42.6	1.9	0.0	0.0
	TG2N	3.7	0.0	0.0	96.3	51.9	38.9	9.3	1.9	0.0	90.7	57.4	38.9	3.7	0.0	0.0	96.3	68.5	46.3	29.6	0.0	0.0	68.5	5.6	3.7
	TG98N	90.7	44.4	35.2	9.3	0.0	0.0	92.6	59.3	35.2	7.4	0.0	0.0	92.6	40.7	24.1	7.4	0.0	0.0	98.1	83.3	64.8	0.0	0.0	0.0
precipitation	PRECTOT	68.0	14.0	4.0	32.0	3.0	2.0	48.0	6.0	4.0	52.0	3.0	2.0	67.0	21.0	12.0	33.0	3.0	0.0	78.0	27.0	10.0	22.0	2.0	1.0
	PREC98P	80.0	7.0	1.0	20.0	2.0	0.0	60.0	9.0	5.0	40.0	4.0	1.0	66.0	19.0	9.0	34.0	3.0	1.0	72.0	20.0	8.0	28.0	2.0	1.0
	R98N	72.0	11.0	3.0	28.0	2.0	2.0	59.0	13.0	6.0	41.0	1.0	1.0	65.0	22.0	10.0	35.0	2.0	0.0	77.0	18.0	8.0	23.0	1.0	1.0
	R98T	75.0	5.0	2.0	25.0	1.0	0.0	67.0	14.0	5.0	33.0	1.0	0.0	68.0	18.0	10.0	32.0	1.0	0.0	71.0	17.0	5.0	29.0	4.0	0.0
	SDII98p	56.0	8.0	3.0	44.0	0.0	0.0	58.0	11.0	3.0	42.0	1.0	0.0	66.0	12.0	5.0	34.0	1.0	0.0	66.0	19.0	6.0	34.0	0.0	0.0
	SDII	77.0	19.0	8.0	23.0	3.0	0.0	63.0	14.0	9.0	37.0	4.0	2.0	66.0	24.0	15.0	34.0	5.0	1.0	77.0	38.0	22.0	23.0	2.0	1.0
	R5d	70.0	10.0	3.0	30.0	3.0	0.0	67.0	11.0	4.0	33.0	1.0	1.0	67.0	20.0	8.0	33.0	2.0	1.0	79.0	20.0	10.0	21.0	2.0	1.0
	R1d	65.0	11.0	3.0	35.0	0.0	0.0	59.0	8.0	5.0	41.0	0.0	0.0	67.0	13.0	6.0	33.0	1.0	0.0	73.0	21.0	9.0	27.0	3.0	0.0
	CDD	58.0	5.0	0.0	42.0	1.0	0.0	71.0	1.0	1.0	29.0	1.0	1.0	57.0	1.0	0.0	43.0	3.0	0.0	48.0	0.0	0.0	52.0	2.0	0.0

Table 4.4 Regional (Northern, Central and Southern Europe) averaged trends for all the indices under period 1901–2000. (trend significance *=95% level, **=99% level). Red/blue/brown/green colours indicate changes towards warmer/colder/drier/wetter conditions

	unit	MAM			JJA			SON			DJF		
	100 yr^{-1}	NEU	CEU	SEU	NEU	CEU	SEU	NEU	CEU	SEU	NEU	CEU	SEU
MEANTN	[°C]	1.09 **	0.85 **	0.46 *	0.70 **	0.92 **	0.75 **	1.11 **	1.04 **	0.60 **	0.48	1.19 *	0.99 **
MEANTX	[°C]	0.94 **	0.65	1.63 **	0.85 *	0.59	1.63 **	0.85 **	1.00 **	1.66 **	0.48	1.11 *	1.41 **
TN2P	[°C]	1.50	1.26	0.83	0.60	0.24	0.76 *	1.12	0.92	0.68	0.12	1.53	1.36 *
TN98P	[°C]	0.23	0.02	0.19	0.75 *	1.17 **	0.62 **	0.84 **	0.71 *	0.17	0.98 **	1.46 **	0.75 *
TX2P	[°C]	1.61 *	0.89	1.66 **	0.93 **	-0.08	1.65 **	1.13	0.90	1.45 **	0.34	1.03	1.22 *
TX98P	[°C]	0.40	-0.33	0.86	0.96	0.98 *	1.62 **	0.22	0.11	1.57 **	1.20 **	1.94 **	1.87 **
TN2N	[d]	-2.01 **	-2.05 **	-1.82 **	-1.50 **	-2.09 **	-2.52 **	-2.11 **	-2.70 **	-1.48 **	-0.07	-1.28	-1.64 *
TN98N	[d]	1.83 **	1.57 **	1.30 *	1.71 **	2.44 **	1.81 **	1.76 **	1.12 **	1.09 *	2.46 **	2.03 **	1.48 **
TX2N	[d]	-2.08 **	-0.96 *	-2.89 **	-2.48 **	-0.52	-2.87 **	-2.96 **	-2.03 *	-2.39 **	-0.21	-0.77	-1.97 **
TX98N	[d]	1.13	0.75	2.44 **	1.54 *	0.64	2.93 **	0.92	0.29	1.90 **	2.58 **	2.69 **	2.50 **
HWDI	[d]	0.94	2.21 *	2.94 *	1.32	0.31	1.51 **	0.12	-0.14	1.17 *	2.40 *	3.49 **	1.02 **
WSDI90	[d]	1.34	1.33	3.13 **	1.14	0.24	2.72 **	0.58	0.00	1.80 **	2.59 **	2.64 **	1.75 **
CSDI10	[d]	-1.29	-2.05 **	-0.75	-1.87 **	-1.20 **	-1.58 **	-2.10 **	-2.68 **	-1.05 *	-0.11	-1.97	-1.71
MEANTG	[°C]	0.85 *	0.82 *	1.04 **	0.73 *	0.96 **	1.19 **	0.97 **	1.15 **	1.11 **	0.33	1.16 *	1.25 **
TG2P	[°C]	1.52	1.21	1.36 **	0.89 **	0.40	1.25 **	1.29	0.97	1.13 *	-0.13	1.27	1.55 **
TG98P	[°C]	0.04	-0.27	0.51	0.81	1.15 **	1.17 **	0.51	0.44	0.85 *	1.05 **	1.52 **	1.29 **
TG2N	[d]	-2.18 *	-1.85 **	-2.50 **	-2.47 **	-1.59 **	-2.23 **	-3.38 **	-3.24 **	-1.71 **	0.17	-0.82	-1.94 *
TG98N	[d]	1.66 *	1.51 *	2.35 **	1.91 *	2.15 **	3.14 **	1.35 *	1.02 *	1.97 **	2.73 **	2.48 **	2.39 **
PRECTOT	[mm]	15.15	23.99	-5.92	-2.23	2.14	1.89	47.21 **	16.61	-14.98	28.14 *	34.99 *	14.90
PREC98P	[mm]	1.07	2.22 *	0.09	0.06	1.49	0.11	2.49 **	2.08 *	-2.32	1.69 *	2.64 **	1.50
R98N	[d]	0.40	0.40	0.02	0.07	0.19	0.23	0.77 **	0.37	-0.07	0.60 *	0.53 *	0.42 *
R98T	[%]	2.80	3.27 *	2.66	0.79	3.69 **	5.67	4.28 **	3.67 **	0.20	4.34 **	3.99 *	5.47 **
SDII98p	[mm]	1.20	2.28	0.20	-0.31	4.34 **	0.28	3.13 **	3.81 **	-1.08	2.60 **	3.88 **	2.64
SDII	[mm]	0.32 **	0.69 **	0.01	0.09	0.52 **	-0.22	0.57 **	0.63 **	0.07	0.49 **	0.94 **	0.39
R5d	[mm]	3.68 *	5.35 *	-0.37	0.91	6.80 *	-0.15	7.44 **	4.03	1.52	5.55 **	9.04 **	4.57
R1d	[mm]	1.77 *	3.34 **	0.72	0.26	4.02 **	-0.64	2.72 *	3.26 *	0.50	2.60 **	3.52 *	2.07
CDD	[d]	0.55	0.25	1.65	1.05	0.89	-1.19	-0.75	0.97	2.43	-0.09	0.28	-0.63

varies, which makes the representativeness of the regional means different. This should be kept in mind when evaluating the results. The trends of the temperature and precipitation indices are listed in Table 4.4.

For temperature-based indices, there is a clear and significant signal of warming in all seasons and regions with only three exceptions for CEU in spring (TX98P), summer (TX2P) and autumn (HWDI). However, these cooling trends are very small and not statistically significant. As a result, generally more significant positive trends are found in southern Europe than in the other two regions. This is especially true for summer. With respect to seasonal differences, spring shows the negligible regional differences.

For precipitation indices of all the regions and seasons, there is an overall increasing trend for all heavy rainfall events indicated by the green colour, although there are a few exceptions. The exceptions are concentrated in SEU; however, none of these negative trends are statistically significant. Interestingly, the only index for dry condition (CDD) shows more drying events than wetting ones, indicating a likely shift of rainfall towards heavy ones. Most significant positive trends appear in cold seasons for NEU and CEU. In summer only CEU experiences significant changes. In general, SEU experiences only a few significant changes.

5 Summary and Conclusions

The selected 27 indices are based on daily temperature and precipitation data at some European climate stations (ranging from 3 to 86 depending on period and variable), which have among the most extensive climate records in the world. The stations used here are required to have daily records starting before 1901. Seasonal linear trends of the indices during three periods (1801–2000, 1851–2000, and 1901–2000) were estimated by simple regression. The significance of the trends was determined by a t-test (see earlier) of the estimated trend. For the most data-rich period 1901–2000, the stations are grouped into three regions (northern, central and southern Europe) and regional means are calculated as an arithmetic mean of all the stations in each region. The long term trends of the temperatures and precipitation at these stations are shown in figures and tables which provide valuable information for past extreme climatic condition based on reliable instrumental records over Europe.

In summary, the estimated trends for the temperature indices indicate a shift in the frequency distribution of temperature. Higher frequency and greater amplitude of warm and hot extremes were detected for all the three periods. At the same time, cold extremes have become rarer. A large number of these trends are found to be statistically significant at the 5% level. On the other hand, the pattern of the trends is much more heterogeneous and less significant for the precipitation indices than for the temperature ones. Nevertheless, a tendency towards increased precipitation intensity, not necessarily combined with increased precipitation totals, was established. There is strong evidence that climate in Europe has changed during the three periods analyzed, such that the occurrence and intensity of warm temperature extremes have increased. Precipitation extremes have also changed with a likely shift of the rainfall moving towards higher precipitation rate.

Based on the summary statistics of the estimated trends, the following conclusions can be highlighted:

- The majority of the trends estimated for temperature indices over all the three periods is positive and a large part of these positive trends are statistically significant. In terms of regional difference, SEU stands out as a region which experiences higher and more significant warming trends, particularly in summer.
- The increased/decreased frequency and intensity of high/low temperature extremes are associated with increased mean temperatures. Extremely cold days and nights have become fewer whereas extremely warm and hot days and nights occurred more often.
- The majority of the trends for precipitation indices suggest increased rainfall amount, increased extreme level and frequency, although there are large regional differences. There are also some differences in the trends of the indices among the three time periods. Over the recent 100 years, NEU has most significant increases, especially in autumn, while there is practically no significant change in SEU. In terms of seasonal distribution, cold seasons (SON and DJF) show more significant changes than those of warm seasons (MAM and JJA).
- Generally, similar patterns of trends with regard to season for all indices over different periods of time are established. This is particularly true for the temperature indices. The trends for the last 100 years are often higher and more significant than the two longer time periods, indicating higher speed of change over the most recent 100 years.

D. Chen et al., *European Trend Atlas of Extreme Temperature and Precipitation Records*, DOI 10.1007/978-94-017-9312-4_5

Appendix

Tables A1, A2, A3, A4, and A5 display trends of the selected indices estimated for all stations, all three periods and all seasons. Red/blue/brown/green colors indicate changes towards warmer/colder/drier/wetter conditions.

Table A1 Period 1801–2000. Trends for Tmin and Tmax indices for all seasons and stations

	MAM			JJA			SON			DJF		
	Brussles-Uccle (B)	Prague (CZ)	Milano (I)	Brussles-Uccle (B)	Prague (CZ)	Milano (I)	Brussles-Uccle (B)	Prague (CZ)	Milano (I)	Brussles-Uccle (B)	Prague (CZ)	Milano (I)
MEANTN	0.54 **	0.34 *	0.13	0.85 **	-0.21 *	0.26 **	0.87 **	0.09	0.45 **	0.69 **	0.92 **	0.78 **
MEANTX	0.49 **	0.95 **	0.37 *	0.63 **	0.74 **	-0.02	0.35 *	0.57 **	0.36 **	0.46 *	1.16 **	0.86 **
TN2P	0.31	0.95 *	0.50 *	-0.32	-0.54 **	-0.42 *	0.39	0.65	0.66 **	0.88	1.49 *	2.02 **
TN98P	1.19 **	-0.08	0.03	1.73 **	0.14	0.66 **	1.43 **	0.03	0.45 **	1.34 **	0.79 **	0.27
TX2P	0.07	1.23 **	0.01	-0.52 **	0.35	-0.94 **	-0.12	0.96 **	0.08	0.31	2.00 **	0.82 **
TX98P	1.47 **	0.98 **	0.60 *	1.63 **	1.44 **	0.21	1.33 **	0.67 *	0.56 *	1.17 **	1.47 **	0.99 **
TN2N	-0.14	-1.23 **	-0.44	0.10	0.50 *	-0.41	-0.60	-0.24	-1.09 **	-0.28	-1.43 *	-3.51 **
TN98N	1.32 **	0.47	0.68	1.37 **	-0.26	1.34 **	1.02 **	0.15	0.62	1.13 **	1.11 **	0.75
TX2N	-0.23	-1.92 **	0.09	0.18	-0.51	0.42 *	0.42	-0.80	0.30	-0.29	-1.43 *	-1.22 *
TX98N	1.45 **	1.16 **	1.31 **	1.48 **	1.32 **	0.23	1.20 **	0.86 **	1.10 **	1.38 **	1.66 **	1.03 **
HWDI	1.12 *	1.25 *	0.96 *	1.22 **	1.08 **	-0.07	0.31	0.35	0.16	0.52	2.58 **	0.15
WSDI90	0.89 *	0.56	1.27 **	1.19 **	0.61 *	0.46	0.63 *	0.52	0.79	0.25	1.39 **	0.68 *
CSDI10	-0.58	-1.67 **	-0.76	-2.06 **	0.11	-0.24	-1.40 *	-0.91	-1.46 *	-0.52	-1.38	-2.78 **

D. Chen et al., *European Trend Atlas of Extreme Temperature and Precipitation Records,* DOI 10.1007/978-94-017-9312-4

Table A2 Period 1801–2000. Trends for Tmean indices for all seasons and stations

	MAM							JJA						
	Brussels-Uccle (B)	Prague (CZ)	Hohen-peissenberg (D)	Milano (I)	Stockholm (S)	Uppsala (S)	CET (UK)	Brussels-Uccle (B)	Prague (CZ)	Hohen-peissenberg (D)	Milano (I)	Stockholm (S)	Uppsala (S)	CET (UK)
MEANTG	0.52 **	0.57 **	-0.95 **	0.20	0.68 **	0.67 **	0.25 *	0.75 **	0.12	-1.13 **	0.10	-0.19	-0.20	0.22 *
TG2P	0.33	1.03 *	0.36	0.11	2.14 **	2.29 **	0.68 **	-0.03	-0.20	-0.48 *	-0.82 **	0.34	0.47 *	0.16
TG98P	1.23 **	0.35	-1.75 **	0.28	0.14	0.00	0.21	1.75 **	0.69 **	-1.50 **	0.44 **	-0.23	-0.31	0.41 *
TG2N	-0.70	-1.61 **	0.26	-0.51	-2.15 **	-2.56 **	-1.81 **	-1.56 **	0.16	0.38	0.27	-0.50	-0.73	-1.45 **
TG98N	1.51 **	0.95 **	-3.01 **	1.18 **	0.38	0.32	0.59	1.67 **	0.68 *	-4.00 **	1.00 *	-0.67	-1.03 *	1.04 **

	SON							DJF						
	Brussels-Uccle (B)	Prague (CZ)	Hohen-peissenberg (D)	Milano (I)	Stockholm (S)	Uppsala (S)	CET (UK)	Brussels-Uccle (B)	Prague (CZ)	Hohen-peissenberg (D)	Milano (I)	Stockholm (S)	Uppsala (S)	CET (UK)
MEANTG	0.61 **	0.23	-0.41 *	0.39 **	0.45 **	0.43 **	0.53 **	0.58 **	0.98 **	-0.08	0.79 **	0.73 **	0.72 *	0.51 **
TG2P	0.18	0.76 *	0.26	0.37	1.59 **	1.55 **	0.78 **	0.66	1.87 **	0.28	1.35 **	2.10 **	2.23 **	1.02 **
TG98P	1.26 **	0.10	-0.74 **	0.51 *	-0.19	-0.41	0.41 *	0.94 **	1.07 **	-0.17	0.53 *	0.51 **	0.61 **	0.41 **
TG2N	-0.44	-0.45	0.23	-0.27	-2.51 **	-1.83 **	-2.55 **	-0.24	-1.31 *	0.17	-2.34 **	-1.57 **	-0.98 **	-0.53
TG98N	1.37 **	0.61 **	-0.47 *	0.95 **	0.32	0.16	0.77 **	1.18 **	1.53 **	-0.06	1.05 **	0.79 **	0.99 **	0.64 *

Table A3 Period 1851–2000. Trends for Tmin and Tmax indices for all seasons and stations

	MAM								
	Brussels-Uccle (B)	Prague (CZ)	Helsinki (FIN)	Jena (D)	Bologna (I)	Milano (I)	Cadiz (SP)	Uppsala (S)	Armagh (UK)
MEANTN	1.33 **	1.65 **	2.33 **	1.19 **	1.12 **	0.39 **	0.56 **	2.15 **	0.62 **
MEANTX	1.59 **	2.78 **	1.87 **	1.30 **	0.43	0.93 **	0.70 **	1.10 **	0.38 *
TN2P	1.59 *	3.11 **	4.57 **	3.23 **	1.40 **	1.24 *	0.77 *	4.49 **	-0.05
TN98P	1.10 *	0.03 **	1.78 **	0.21	0.75 *	-0.08 *	0.97 **	1.20 **	0.79 **
TX2P	1.11 *	3.69 **	3.12 **	2.10 **	0.34	1.02 *	1.06 **	2.76 **	0.63 *
TX98P	3.32 **	1.48 *	2.55 **	0.91	0.13	0.00 *	1.53 **	0.82	0.86 *
TN2N	-3.39 **	-3.88 **	-6.86 **	-2.95 **	-3.43 **	-1.02 *	-0.06	-6.24 **	-0.05
TN98N	2.38 **	2.14 **	1.86 **	1.05 **	1.70 **	2.10 *	2.30 **	1.63 **	1.66 **
TX2N	-1.73 *	-6.22 **	-4.37 **	-2.63 **	0.05	-1.51 **	-1.33	-1.68 **	-1.01
TX98N	3.06 **	2.29 **	1.51 **	0.97 **	0.87	1.96 *	2.07 **	0.44	0.58
HWDI	2.87 *	4.07 **	0.67	0.85	0.16	1.51 *	0.61 *	0.63	-0.03
WSDI90	2.26 *	2.26 **	1.34 **	0.56	0.31	2.03 *	2.39 **	0.67	0.90
CSDI10	-3.28 *	-5.14 **	-6.30 **	-2.69 **	-4.33 **	-0.79 *	-0.33	-5.40 **	-1.03 *

	JJA								
	Brussels-Uccle (B)	Prague (CZ)	Helsinki (FIN)	Jena (D)	Bologna (I)	Milano (I)	Cadiz (SP)	Uppsala (S)	Armagh (UK)
MEANTN	0.79 **	0.43 *	0.92 **	0.59 **	0.71 **	0.12 **	0.72 **	1.51 **	0.52 **
MEANTX	1.09 **	1.76 **	0.75 **	0.95 **	-0.05	-0.12 **	0.09	-0.28	0.28
TN2P	0.01 **	-0.27 **	1.43 **	0.86 **	0.27	-0.49 **	0.91 **	2.23 **	-0.43
TN98P	1.82 **	0.77 **	0.56	0.70 **	0.60 *	0.34 **	1.36 **	1.16 **	0.96 **
TX2P	-1.01 *	1.38 **	1.78 **	0.11	-1.85 **	-0.42 **	0.25	0.92 *	-0.06
TX98P	3.13 **	2.04 **	0.36	1.66 **	0.18	-0.54 **	1.71 **	-0.46	0.78 *
TN2N	0.29 **	-0.04 **	-4.24 **	-2.49 **	-1.64 **	-0.29 **	0.21	-6.01 **	0.19
TN98N	2.39 **	1.76 *	0.54	1.62 **	2.21 *	1.80 **	1.73 **	1.92 **	1.60 **
TX2N	-0.60 **	-2.52 **	-2.58 **	-0.76 *	0.57 *	0.32 **	1.12 *	-0.78 *	-0.34
TX98N	2.82 **	2.43 **	0.18	1.65 **	0.49	-0.44 **	1.71 **	-0.75	1.48 **
HWDI	2.53 *	1.69 **	0.04	1.21	0.89	-0.85 **	0.65	0.15	1.03 *
WSDI90	2.48 **	0.83 **	0.90	0.87	-0.36	-0.07 **	0.69	-0.09	1.02
CSDI10	-0.60 **	0.45 **	-3.36 **	-0.16	-0.17	0.28 **	-0.02	-3.33 **	-0.27

	SON								
	Brussels-Uccle (B)	Prague (CZ)	Helsinki (FIN)	Jena (D)	Bologna (I)	Milano (I)	Cadiz (SP)	Uppsala (S)	Armagh (UK)
MEANTN	1.06 **	1.43 **	1.06 **	1.20 **	0.85 **	0.46 **	0.74 **	1.53 **	0.88 **
MEANTX	1.93 **	1.85 **	0.87 **	1.17 **	0.58 *	0.74 **	0.96 **	0.57 **	0.48 **
TN2P	1.98 **	3.31 **	3.56 **	3.27 **	0.81	1.76 **	0.08	3.80 **	-0.22
TN98P	0.57 **	0.50 **	0.92 **	0.81 **	0.25	-0.12 **	1.16 **	0.88 **	1.02 **
TX2P	1.74 *	3.62 **	2.89 **	2.35 **	0.34	1.74 **	0.58	1.92 **	0.39
TX98P	2.94 **	1.03 **	0.42	1.07	0.28	0.09 **	1.64 **	-0.78	0.48
TN2N	-2.04 *	-3.98 **	-3.27 **	-2.34 **	-2.85 **	-2.86 **	0.26	-3.48 **	-0.32
TN98N	1.33 **	1.19 **	1.10 *	1.31 **	1.17	-0.66 **	1.76 **	1.46 **	1.73 **
TX2N	-0.89 **	-4.82 **	-3.88 **	-1.99 **	-0.21	-0.93 **	-1.23 *	-2.03 **	-1.07 *
TX98N	2.08 **	1.56 **	0.93 *	0.86 *	0.86	1.28 **	1.70 **	0.32	1.04 **
HWDI	0.27 **	0.30 **	0.02	0.53	-0.09	0.09 **	0.17	0.34	
WSDI90	0.93 **	0.09 **	0.85	0.86 *	1.16	0.98 **	1.03 **	0.26	0.14
CSDI10	-1.94 **	-4.50 **	-3.36 **	-1.49	-2.41 *	-2.60 **	-0.05	-2.90 **	-0.93

	DJF								
	Brussels-Uccle (B)	Prague (CZ)	Helsinki (FIN)	Jena (D)	Bologna (I)	Milano (I)	Cadiz (SP)	Uppsala (S)	Armagh (UK)
MEANTN	1.03 *	2.23 **	1.87 **	1.48 **	1.42 **	1.45 **	0.86 **	1.46 **	0.27
MEANTX	1.74 **	2.60 **	1.37 **	0.98 *	1.13 **	2.03 **	0.99 **	0.69	0.14
TN2P	3.01 **	3.09 *	2.62 **	2.90 **	1.61 **	3.54 **	0.05	3.72 **	0.61
TN98P	1.25 **	1.87 **	0.89 **	1.05 **	1.29 **	0.25 **	1.60 **	0.99 **	1.03 **
TX2P	1.56 **	3.95 **	2.42 *	0.94	0.32	2.11 **	1.18 **	0.91	0.12
TX98P	2.85 **	2.59 **	1.22 **	1.73 **	1.70 **	1.26 **	1.67 **	1.20 **	0.22
TN2N	-1.72 **	-3.32 *	-1.87 **	-2.19 *	-3.57 **	-6.91 **	0.23	-1.46 *	-0.75
TN98N	1.36 *	2.59 **	2.29 **	1.64 **	1.29 *	0.41 **	1.95 **	2.01 **	1.72 **
TX2N	-2.20 *	-3.08 *	-1.44 *	-0.20	-0.62	-3.50 **	-2.28 **	0.00	-0.62
TX98N	3.19 **	3.40 **	2.01 **	2.12 **	1.83 **	1.73 **	1.95 **	1.61 **	0.38
HWDI	1.67 *	6.15 **	3.76 **	3.71 **	0.66	0.51 **		2.15	0.20
WSDI90	1.73 **	3.40 **	3.02 **	2.08 **	0.79	1.59 *	1.82 **	1.83 *	0.01
CSDI10	-1.97 **	-3.31 **	-1.67	-1.32	-6.29 **	-6.11 **	-0.68	-1.04	-2.07 **

Table A4 Period 1851–2000. Trends for Tmean indices for all seasons and stations

	MAM												
	Brussels-Uccle (B)	Prague (CZ)	Helsinki (FIN)	Hohen-peissenberg (D)	Jena (D)	Stykkisholmur (IS)	Bologna (I)	Milano (I)	St Petersburg (RU)	Cadiz (SP)	Stockholm (S)	Uppsala (S)	CET (UK)
MEANTG	1.46 **	2.15 **	2.45 **	-0.32 **	1.49 **	1.09 **	0.77 **	0.54 *	1.97 **	0.63 **	1.74 **	1.76 **	0.81 **
TG2P	1.48 *	3.20 **	4.29 **	1.71 *	2.72 **	3.27 **	0.76	0.74 *	3.34 **	1.10 **	4.86 **	4.60 **	1.79 **
TG98P	2.19 **	0.96 **	3.43 **	-1.78 **	1.19 **	0.51	0.50	-0.09 *	1.69 **	1.13 **	0.87 **	0.80 **	0.43 **
TG2N	-3.67 **	-4.67 **	-8.21 **	-1.05 **	-3.86 **	-5.97 **	-0.96 *	-1.66 *	-3.76 **	-1.59 **	-3.67 **	-5.02 **	-4.29 **
TG98N	3.29 **	2.20 **	2.00 **	-1.53 **	1.62 **	0.44	2.17 **	1.73 *	1.42 **	2.34 **	1.59 *	1.48 *	1.33 **

	JJA												
	Brussels-Uccle (B)	Prague (CZ)	Helsinki (FIN)	Hohen-peissenberg (D)	Jena (D)	Stykkisholmur (IS)	Bologna (I)	Milano (I)	St Petersburg (RU)	Cadiz (SP)	Stockholm (S)	Uppsala (S)	CET (UK)
MEANTG	0.94 **	1.07 **	1.39 **	-1.27 **	1.14 **	0.32 *	0.33	-0.05 **	0.60 **	0.40 **	0.17 **	0.18 **	0.55 *
TG2P	-0.11 **	0.49 **	1.92 **	-0.65 **	0.61 *	0.89 **	-0.70	-0.89 **	1.44 **	0.86 **	0.83 **	1.42 **	0.59 *
TG98P	2.72 **	1.72 **	1.49 **	-1.36 *	1.85 **	0.21	0.36	0.00 **	0.12	1.55 **	0.26 **	-0.29 **	0.79 *
TG2N	-2.66 **	-1.34 *	-5.20 **	0.49 *	-2.25 **	-2.37 **	0.22	0.03 **	-1.09 *	0.39	-0.76 **	-2.43 **	-2.81 **
TG98N	3.18 **	2.22 **	1.58 **	-5.87 **	2.25 **	0.50	2.26	1.43 **	0.44	1.93 **	0.81 **	-0.66 **	2.29 **

	SON												
	Brussels-Uccle (B)	Prague (CZ)	Helsinki (FIN)	Hohen-peissenberg (D)	Jena (D)	Stykkisholmur (IS)	Bologna (I)	Milano (I)	St Petersburg (RU)	Cadiz (SP)	Stockholm (S)	Uppsala (S)	CET (UK)
MEANTG	1.49 **	1.44 **	1.11 **	0.42 **	1.23 **	0.43 *	0.72 **	0.56 *	0.77 **	0.85 **	1.10 **	1.26 **	1.30 **
TG2P	1.88 **	3.30 **	3.32 **	1.22 **	2.81 **	1.40 **	0.95 *	1.45 *	2.92 **	0.58	2.90 **	3.72 **	1.49 **
TG98P	1.39 *	0.70 **	0.85 *	-0.61 **	0.81 *	0.05	0.21	0.04 *	0.88 *	1.29 **	0.38 **	-0.24 **	0.62 **
TG2N	-3.13 **	-4.03 **	-4.94 **	-1.41 **	-3.72 **	-1.95 **	-1.02	-1.68 *	-2.79 **	-1.11	-5.55 **	-4.39 **	-5.50 **
TG98N	2.09 **	1.48 **	1.03 *	-0.04 **	1.00 **	0.65	1.41 *	0.70 *	0.86 *	1.73 **	0.71 **	1.35 *	1.25 *

	DJF												
	Brussels-Uccle (B)	Prague (CZ)	Helsinki (FIN)	Hohen-peissenberg (D)	Jena (D)	Stykkisholmur (IS)	Bologna (I)	Milano (I)	St Petersburg (RU)	Cadiz (SP)	Stockholm (S)	Uppsala (S)	CET (UK)
MEANTG	1.39 **	2.18 **	1.66 **	0.77 **	1.27 **	1.19 **	1.25 **	1.67 **	1.36 **	0.93 **	0.82 *	0.96 **	0.52 **
TG2P	2.32 *	3.41 **	2.62 **	1.09 **	2.01 *	2.96 **	1.10 *	2.40 **	2.41 *	0.75 *	2.85 **	2.96 *	1.13 **
TG98P	1.66 **	2.03 **	1.22 **	1.84 **	1.26 **	0.74 **	1.34 **	0.84 **	0.81 **	1.15 **	0.77 **	1.36 **	0.58 *
TG2N	-1.50 **	-2.67 **	-1.55 *	-0.45 **	-1.48	-4.23 **	-2.30 **	-5.03 **	-1.65 *	-1.27	-1.21 **	-0.53 **	-1.14 *
TG98N	2.41 **	3.47 **	2.49 **	1.75 **	1.98 **	1.23 **	1.37 **	2.04 *	1.99 **	1.86 **	0.90 **	1.84 **	0.46 *

Table A5 Period 1851–2000. Trends for Precipitation indices for all seasons and stations

	MAM									JJA								
	Prague (CZ)	Helsinki (FIN)	Jena (D)	Bologna (I)	De Kooy (NL)	Eelde (NL)	Groningen (NL)	Uppsala (S)	Armagh (UK)	Prague (CZ)	Helsinki (FIN)	Jena (D)	Bologna (I)	De Kooy (NL)	Eelde (NL)	Groningen (NL)	Uppsala (S)	Armagh (UK)
PRECTOT	3.09	-5.37	14.78	-8.45	14.36 *	37.18 **	37.18 **	-4.40	13.21	19.06	6.38	-2.86	15.53	-3.78	4.67	4.67	4.35	-24.87 *
PREC98P	0.44	-0.70	1.31	-1.24	0.63	3.30 **	3.30 **	-0.13	-0.05	2.47	0.65	0.16	2.59	-0.40	0.08	0.08	1.12	-2.15 *
R98N	-0.01	-0.04	0.60 *	0.13	0.54 *	0.88 **	0.88 **	-0.13	0.02	0.50	0.11	-0.13	0.37	-0.37	0.33	0.33	-0.14	-0.59
R98T	1.84	-0.79	4.65	0.83	3.03	6.90 **	6.90 **	-0.44	0.01	5.49	2.84	-1.83	3.35	-2.85	1.95	1.95	-0.80	-3.02
SDII98p	3.04	-0.70	1.28	-2.87	0.47	3.97 *	3.97 *	0.47	0.61	2.31	2.79	-2.74	0.41	-0.20	-0.74	-0.74	1.23	1.03
SDII	0.15	-0.05	0.23	-0.26	0.21	0.64 **	0.64 **	-0.42	0.01	0.62 *	0.10	0.29	0.33	-0.23	0.01	0.01	0.01	-0.60 **
R5d	3.75	0.11	2.58	-3.91	0.00	10.17 **	10.17 **	-0.55	1.85	8.75 *	5.54	3.24	7.62	-1.63	3.43	3.43	3.29	-4.73
R1d	1.36	-0.32	1.31	-3.62	-0.69	4.67 **	4.67 **	-0.68	0.36	5.75 *	1.45	-0.68	1.52	0.41	-0.06	-0.06	1.35	-1.42
CDD	-1.51	0.52	-0.50	0.25	0.05	-1.26	-1.26	-0.05	-0.28	1.80	1.38	0.37	-3.08	-1.76	-0.13	-0.13	-0.49	-0.67

	SON									DJF								
	Prague (CZ)	Helsinki (FIN)	Jena (D)	Bologna (I)	De Kooy (NL)	Eelde (NL)	Groningen (NL)	Uppsala (S)	Armagh (UK)	Prague (CZ)	Helsinki (FIN)	Jena (D)	Bologna (I)	De Kooy (NL)	Eelde (NL)	Groningen (NL)	Uppsala (S)	Armagh (UK)
PRECTOT	0.92	20.49	0.59	-22.10	27.62 *	36.20 **	36.20 **	9.11	14.34	-1.88	17.93 *	19.36 **	15.38	25.15 **	54.96 **	54.96 **	8.16	19.57
PREC98P	-0.04	1.02	-1.03	-0.17	3.05 **	3.78 **	3.78 **	0.65	0.78	-0.33	0.80	1.38 *	-0.43	1.13	3.64 **	3.64 **	0.00	1.36
R98N	0.22	0.46	0.26	0.18	0.75 **	0.59 **	0.59 **	0.04	-0.02	-0.24	0.29	0.43	0.36	0.11	0.61 *	0.61 *	0.02	0.62 *
R98T	0.27	2.24	1.91	1.66	4.99 *	5.18 *	5.18 *	-0.22	0.01	-2.33	0.29	1.97	2.75	1.14	4.27	4.27	-2.10	4.24
SDII98p	-1.75	1.87	0.22	-4.80	-0.19	4.41 *	4.41 *	1.14	1.03	0.37	1.18	1.35	1.24	3.03 *	3.89 *	3.89 *	0.82	2.51
SDII	-0.24	0.22	0.05	-0.15	0.52 *	0.69 **	0.69 **	-0.36	-0.13	-0.13	0.30 *	0.23	-0.18	0.36 *	0.81 **	0.81 **	-0.80 **	0.21
R5d	-0.54	3.71	-0.01	-2.98	4.48	9.69 **	9.69 **	4.40	3.86	-0.67	1.95	3.42	2.73	4.92 *	9.36 **	9.36 **	0.59	4.27
R1d	-2.12	1.74	0.61	-4.16	0.58	4.46 **	4.46 **	1.84	1.39	-0.55	1.74	1.48	-0.19	2.39 *	3.24 *	3.24 *	0.03	1.93
CDD	-1.42	0.68	0.37	0.27	-1.35	-1.12	-1.12	-0.74	-0.63	-1.23	0.11	-1.37	-2.40	0.03	-1.49	-1.49	-0.62	-1.54

References

Achberger, C., & Chen, D. (2006). *Trend of extreme precipitation in Sweden and Norway during 1961–2004*. Research report C72 (p. 58). Gothenburg: Department of Earth Sciences.

Arndt, D. S., Baringer, M. O., Johnson, M. R., Alexander, L. V., Diamond, H. J., Fogt, R. L., Levy, J. M., Richter-Menge, J., Thorne, P. W., Vincent, L. A., Watkins, A. B., & Willett, K. M. (2010). State of the climate in 2009. *Bulletin of the American Meteorological Society, 91*(7), S1–S218.

Auer, I., Böhm, R., Jurkovic, A., Lipa, W., Orlik, A., Potzmann, R., Schöner, W., Ungersböck, M., Matulla, C., Briffa, K., Jones, P., Efthymiadis, D., Brunetti, M., Nanni, T., Maugeri, M., Mercalli, L., Mestre, O., Moisselin, J. M., Begert, M., Müller-Westermeier, G., Kveton, V., Bochnicek, O., Stastny, P., Lapin, M., Szalai, S., Szentimrey, T., Cegnar, T., Dolinar, M., Gajic-Capka, M., Zaninovic, K., Majstorovic Z., & Nieplova E. (2007). HISTALP—Historical instrumental climatological surface time series of the Greater Alpine Region. *International Journal of Climatology, 27*(1), 17–46.

Beniston, M. (2004). The 2003 heat wave in Europe. A shape of things to come? *Geophysical Research Letters, 31,* 2022–2026.

Beniston, M., & Stephenson, D. B. (2004). Extreme climatic events and their evolution under changing climatic conditions. *Global and Planetary Change, 44,* 1–9.

Beniston, M., Stephenson, D. B., Christensen, O. B., Ferro, C. A. T., Frei, C., Goyette, S., Halsnaes, K., Holt, T., Jylha, K., Koffi, B., Palutikof, J., Scholl, R., Semmler, T., & Woth, K. (2007). Future extreme events in European climate: An exploration of regional climate model projections. *Climatic Change, 81,* 71–95.

Brown, B. G., & Katz, R. W. (1995). Regional analysis of temperature extremes: Spatial analog for climate change? *Journal of Climate, 8,* 108–119.

Brunet, M., & Jones, P. (2011). Data rescue initiatives: Bringing historical climate data into the 21st century. *Climate Research, 47*(1–2), 29–40.

Brunsden, D. (1999). Some geomorphological considerations for the future development of landslide models. *Geomorphology, 30,* 13–24.

Chen, D., Walther, A., Moberg, A., Jones, P. D., Jacobeit J., & Lister, D. (2006). *Trend atlas of the EMULATE indices*. Research Report C73 (p. 798). Gothenburg: Department of Earth Sciences.

Christidis, N., Stott, P. A., Brown, S., Hegerl, G. C., & Caesar, J. (2005). Detection of changes in temperature extremes during the second half of the 20th century. *Geophysical Research Letters, 32*(20), 1–4.

Donat, M. G., Alexander, L. V., Yang, H., Durre, I., Vose, R., Dunn, R. J. H., Willett, K. M., Aguilar, E., Brunet, M., Caesar, J., Hewitson, B., Jack, C., Klein Tank, A. M. G., Kruger, A. C., Marengo, J., Peterson, T. C., Renom, M., Oria Rojas, C., Rusticucci, M., Salinger, J., Elrayah, A. S., Sekele, S. S., Srivastava, A. K., Trewin, B., Villarroel, C., Vincent, L. A., Zhai, P., Zhang, X., & Kitching S. (2013). Updated analyses of temperature and precipitation extreme indices since the beginning of the twentieth century: The HadEX2 dataset. *Journal of Geophysical Research D: Atmospheres, 118*(5), 2098–2118.

Easterling, D. R., Meehl, G. A., Parmesan, C., Changnon, S. A., Karl, T. R., & Mearns, L. O. (2000). Climate extremes: Observations, modeling, and impacts. *Science, 289,* 2068–2074.

Field, C. B., Barros, V., Stocker, T. F., Qin, D., Dokken, D. J., Ebi, K. L., Mastrandrea, M. D., Mach, K. J., Plattner, G.-K., Allen, S. K., Tignor, M., & Midgley, P. M. (2012). *Managing the risks of extreme events and disasters to advance climate change adaptation*. A special report of working groups I and II of the intergovernmental panel on climate change. Cambridge: Cambridge University Press.

Fowler, H. J., & Kilsby, C. G. (2003). A regional frequency analysis of United Kingdom extreme rainfall from 1961–2000. *International Journal of Climatology, 23,* 1313–1334.

Frich, P., Alexander, L. V., Della-Marta, P., Gleason, B., Haylock, M., Klein-Tank, A., & Peterson, T. (2002). Observed coherent changes in climatic extremes during the second half of the 20th Century. *Climate Research, 19,* 193–212.

Groisman, P. Y., Karl, T. R., Easterling, D. R., Knight, R. W., Jameson, P. F., Hennessy, K. J., Suppiah, R., Page, C., Wibig, J., Fortuniak, K., Razuaev, V., Douglas, A., FÃ¸rland, E., & Zhai, P. (1999). Changes in the probability of heavy precipitation: Important indicators of climatic change. *Climatic Change, 42,* 243–283.

Groisman, P. Y., Knight, R. W., Easterling, D. R., Karl, T. R., Hegerl, G. C., & Razuvaev, V. N. (2005). Trends in intense precipitation in the climate record. *Journal of Climate, 18,* 1326–1350.

Hansen, J., Sato, M., & Ruedy, R. (2012). Perception of climate change. *Proceedings of the National Academy of Sciences of the United States of America, 109*(37), E2415–E2423.

Haylock, M. R., & Goodess, C. M. (2004). Interannual variability of European extreme winter rainfall and links with mean large-scale circulation. *International Journal of Climatology, 24,* 759–776.

Houghton, J. T., Ding, Y., Griggs, D. J., Noguer, M., van der Linden, P. J., Dai, X., IPCC. (2002). Workshop report of the IPCC workshop on changes in extreme weather and climate events. 11–13 June, Beijing.

Hundecha, Y., & Bárdossy, A. (2005). Trends in daily precipitation and temperature extremes across western Germany in the second half of the 20th century. *International Journal of Climatology, 25,* 1189–1202.

IPCC. (2001). *Climate change 2001: The scientific basis. Contribution of working group I to the third assessment*. Report of the intergovernmental panel on climate change (IPCC). UK: Cambridge University Press.

IPCC. (2007). *Climate change 2007: The physical science basis. Contribution of working group I to the fourth assessment*. Report of the intergovernmental panel on climate change. Cambridge: Cambridge University Press.

D. Chen et al., *European Trend Atlas of Extreme Temperature and Precipitation Records,* DOI 10.1007/978-94-017-9312-4

Jacobeit, J., Rathmann, J., Philipp, A., & Jones, P. D. (2009). Central European precipitation and temperature extremes in relation to large-scale atmospheric circulation types. *Meteorologische Zeitschrift, 18*(4), 397–410.

Jones, P. D., Horton, E. B., Folland, C. K., Hulme, M., Parker, D. E., & Basnett, T. A. (1999). The use of indices to identify changes in climatic extremes. *Climatic Change, 42,* 131–149.

Karl, T. R., & Easterling, D. R. (1999). Climate extremes: Selected review and future research directions. *Climatic Change, 42,* 309–325.

Katz, R. W., & Brown, B. G. (1992). Extreme events in a changing climate: Variability is more important than averages. *Climatic Change, 21,* 289–302.

Klein-Tank, A. M. G., & Können, G. P. (2003). Trends in indices of daily temperature and precipitation extremes in Europe. *Journal of Climate, 16,* 3665–3680.

Klein-Tank, A. M. G., Wijngaard, J., & van Engelen, A. (2002a). *Climate of Europe: Assessment of observed daily temperature and precipitation extremes*. De Bilt: KNMI.

Klein-Tank, A. M. G., Wijngaard, J. B., Können, G. P., Böhm, R., Demarée, G., Gocheva, A., Mileta, M., Pashiardis, S., Hejkrlik, L., Kern-Hansen, C., Heino, R., Bessemoulin, P., Müller-Westermeier, G., Tzanakou, M., Szalai, S., Palsdottir, T., Fitzgerald, D., Rubin, S., Capaldo, M., Maugeri, M., Leitass, A., Bukantis, A., Aberfeld, R., van Engelen, A. F. V., Forland, E., Mietus, M., Coelho, F., Mares, C., Razuvaev, V., Nieplova, E., Cegnar, T., López, J. A., Dahlström, B., Moberg, A., Kirchhofer, W., Ceylan, A., Pachaliuk, O., Alexander, L. V., & Petrovic, P. (2002b). Daily dataset of 20th-century surface air temperature and precipitation series for the European climate assessment. *International Journal of Climatology, 22,* 1441–1453.

Mearns, L. O., Katz, R. W., & Schneider, S. H. (1984). Extreme high temperature events: Changes in their probabilities and changes in mean temperature. *Journal of Climate and Applied Meteorology, 23,* 1601–1613.

Moberg, A., & Bergström, H. (1997). Homogenization of Swedish temperature data. Part III: The long temperature records from Uppsala and Stockholm. *International Journal of Climatology, 17*(7), 667–699.

Moberg, A., & Jones, P. D. (2005). Trends in indices for extremes in daily temperature and precipitation in central and Western Europe 1901–1999. *International Journal of Climatology, 25,* 1149–1172.

Moberg, A., Jones, P. D., Lister, D., Walther, A., Brunet, M., Jacobeit, J., Alexander, L. V., Della-Marta, P. M., Luterbacher, J., Yiou, P., Chen, D. L., Tank, A. M. G. K., Saladie, O., Sigro, J., Aguilar, E., Alexandersson, H., Almarza, C., Auer, I., Barriendos, M., Begert, M., Bergstrom, H., Bohm, R., Butler, C. J., Caesar, J., Drebs, A., Founda, D., Gerstengarbe, F. W., Micela, G., Maugeri, M., Osterle, H., Pandzic, K., Petrakis, M., Srnec, L., Tolasz, R., Tuomenvirta, H., Werner, P. C., Linderholm, H., Philipp, A., Wanner, H., & Xoplaki, E. (2006). Indices for daily temperature and precipitation extremes in Europe analyzed for the period 1901–2000. *Journal of Geophysical Research-Atmospheres, 111*(D22106). doi:10.1029/2006JD007103.

Morak, S., Hegerl, G. C. & Kenyon, J. (2011). Detectable regional changes in the number of warm nights. *Geophysical Research Letters, 38*(17). doi:10.1029/2011GL048531.

Pellikka, P. & Järvenpää, E. (2003). Forest stand characteristics and snow and wind induced forest damage in boreal forests. Proceedings of the international conference on wind effects on trees (pp. 269–276), 16–18 September. Karlsruhe: B. Ruck.

Peterson, T. C., Folland, C., Gruza, G., Hogg, W., Mokssit, A., & Plummer, N. (2001). Report on the activities of the working group on climate change detection and related rapporteurs 1998–2001.

Peterson, T. C., P., Stott, A., & Herring, S. (2012). Explaining extreme events of 2011 from a climate perspective. *Bulletin of the American Meteorological Society, 93*(7), 1041–1067.

Schmidli, J., & Frei, C. (2005). Trends in heavy precipitation and wet and dry spells in Switzerland during the 20th century. *International Journal of Climatology, 25,* 753–771.

Schuster, R. L. (1996). Socioeconomic significance of landslides. In A. K. Turner & R. L. Schuster (Eds.), *Landslides: Investigation and mitigation* (pp. 12–35). Washington, DC: National Academy Press.

Solantie, R. (1994). Effect of weather and climatological background on snow damage of forests in southern Finland in November 1991. *Silva Fennica, 28*(3), 203–211.

Trenberth, K. E. (1999). Conceptual framework for changes of extremes of the hydrological cycle with climate change. *Climatic Change, 42*(1), 327–339.

Vautard, R., & Yiou, P. (2009). Control of recent European surface climate change by atmospheric flow. *Geophysical Research Letters, 36*(22). L22702, doi:10.1029/2009GL040480.

Venema, V. K. C., Mestre, O., Aguilar, E., Auer, I., Guijarro, J. A., Domonkos, P., Vertacnik, G., Szentimrey, T., Stepanek, P., Zahradnicek, P., Viarre, J., Müller-Westermeier, G., Lakatos, M., Williams, C. N., Menne, M. J., Lindau, R., Rasol, D., Rustemeier, E., Kolokythas, K., Marinova, T., Andresen, L., Acquaotta, F., Fratianni, S., Cheval, S., Klancar, M., Brunetti, M., Gruber, C., Prohom Duran, M., Likso, T., Esteban, P., & Brandsma, T. (2012). Benchmarking homogenization algorithms for monthly data. *Climate of the Past, 8*(1), 89–115.

Wigley, T. M. L. (1985). Impact of extreme events. *Nature, 316,* 106–107.

WISE. (1999). Workshop report of the workshop on economic and social impacts of climate extremes: Risks and benefits, 14–16 October.

Zhang, X., Wang, J., Zwiers, F. W., & Groisman, P. Y. (2010). The influence of large-scale climate variability on winter maximum daily precipitation over north America. *Journal of Climate, 23*(11), 2902–2915.

Zwiers, F. W., Zhang, X., & Feng, Y. (2011). Anthropogenic influence on long return period daily temperature extremes at regional scales. *Journal of Climate, 24*(3), 881–892.